T0189627

CHEMICAL PROCESS IN LIQUID AND SOLID PHASES

Properties, Performance and Applications

CHEMICAL PROCESS IN LIQUID AND SOLID PHASES

Properties, Performance and Applications

Edited by
Alfonso Jiménez, PhD, and Ali Pourhashemi, PhD

Gennady E. Zaikov, DSc, and A. K. Haghi, PhD
Reviewers and Advisory Board Members

Apple Academic Press

TORONTO   NEW JERSEY

Apple Academic Press Inc. | Apple Academic Press Inc.
3333 Mistwell Crescent | 9 Spinnaker Way
Oakville, ON L6L 0A2 | Waretown, NJ 08758
Canada | USA

©2014 by Apple Academic Press, Inc.

First issued in paperback 2021

Exclusive worldwide distribution by CRC Press, a member of Taylor & Francis Group

No claim to original U.S. Government works

Library of Congress Control Number: 2013943023

Library and Archives Canada Cataloguing in Publication

Chemical process in liquid and solid phases: properties, performance and applications/ edited by Alfonso Jiménez, PhD and Ali Pourhashemi, PhD; reviewers and advisory board members, A.K. Haghi, PhD and Gennady E. Zaikov, DSc.

Includes bibliographical references and index.
ISBN 978-1-926895-51-2
1. Chemical processes. 2. Biochemistry. 3. Polymers. I. Jiménez, Alfonso, 1965-, editor of compilation II. Pourhashemi, Ali, editor of compilation

TP155.7.C44 2013   660'.2   C2013-904224-5

Apple Academic Press also publishes its books in a variety of electronic formats. Some content that appears in print may not be available in electronic format. For information about Apple Academic Press products, visit our website at **www.appleacademicpress.com** and the CRC Press website at **www.crcpress.com**

ISBN 13: 978-1-77463-280-2 (pbk)
ISBN 13: 978-1-926895-51-2 (hbk)

ABOUT THE EDITORS

Alfonso Jiménez, PhD

Professor of Analytical Chemistry and Materials Science, University of Alicante, Spain

Alfonso Jiménez, PhD, is currently a professor of analytical chemistry and food science and technology at the University of Alicante in Spain since 2001. He is the Head of the Polymer and Nanomaterials Analysis Group. He has edited 13 books on polymer degradation, stabilization biodegradable, and sustainable composites, and has written 67 research papers published in journals in the area of analytical chemistry and polymer science. He has chaired organizing committees for several international conferences on biopolymers and polymer chemistry, in particular BIOPOL-2007 and BIOPOL-2009 held in Alicante, Spain. His main research areas include chemical characterization of polymers and biopolymers, environmentally friendly additives in polymers, characterization of biodegradable polymers and sustainable composites, modification of PLA for flexible films manufacturing, natural antioxidants for active packaging, and TPUs obtained from vegetable oils.

Ali Pourhashemi, PhD

Professor, Department of Chemical and Biochemical Engineering, Christian Brothers University, Memphis, Tennessee, USA

Ali Pourhashemi, PhD, is currently a professor of chemical and biochemical engineering at Christian Brothers University (CBU) in Memphis, Tennessee. He was formerly the department chair at CBU and also taught at Howard University in Washington, DC. He taught various courses in chemical engineering, and his main area has been teaching the capstone process design as well as supervising industrial internship projects. He is a member of several professional organizations, including the American Institute of Chemical Engineers. He is on the international editorial review board of the *International Journal of Chemoinformatics and*

Chemical Engineering and is an editorial member of the *International of Journal of Advanced Packaging Technology.* He has published many articles and presented at many professional conferences.

REVIEWERS AND ADVISORY BOARD MEMBERS

Gennady E. Zaikov, DSc

Gennady E. Zaikov, DSc, is Head of the Polymer Division at the N. M. Emanuel Institute of Biochemical Physics, Russian Academy of Sciences, Moscow, Russia, and Professor at Moscow State Academy of Fine Chemical Technology, Russia, as well as Professor at Kazan National Research Technological University, Kazan, Russia. He is also a prolific author, researcher, and lecturer. He has received several awards for his work, including the the Russian Federation Scholarship for Outstanding Scientists. He has been a member of many professional organizations and on the editorial boards of many international science journals.

A. K. Haghi, PhD

A. K. Haghi, PhD, holds a BSc in urban and environmental engineering from University of North Carolina (USA); a MSc in mechanical engineering from North Carolina A&T State University (USA); a DEA in applied mechanics, acoustics and materials from Université de Technologie de Compiègne (France); and a PhD in engineering sciences from Université de Franche-Comté (France). He is the author and editor of 65 books as well as 1000 published papers in various journals and conference proceedings. Dr. Haghi has received several grants, consulted for a number of major corporations, and is a frequent speaker to national and international audiences. Since 1983, he served as a professor at several universities. He is currently Editor-in-Chief of the *International Journal of Chemoinformatics and Chemical Engineering* and *Polymers Research Journal* and on the editorial boards of many international journals. He is also a faculty member of University of Guilan (Iran) and a member of the Canadian Research and Development Center of Sciences and Cultures (CRDCSC), Montreal, Quebec, Canada.

CONTENTS

List of Contributors .... *xiii*

List of Abbreviations .... *xix*

Preface .... *xxi*

1. **Thermodynamics as a Basic of Inanimate and Animate Nature** 1
   G. P. Gladyshev and M. D. Goldfein

2. **Nanotherapeutics: A Novel Approach of Target Based Drug Delivery** 17
   Anamika Singh and Rajeev Singh

3. **Phosphorus-Containing Polypeptides** 27
   Victor Kablov, Dmitry Kondrutsky, Maria Sudnitsina, Lidiya Zimina, and Marina Arţsis

4. **Free Radical Initiation in Polymers Under the Action of Nitrogen Oxides** 41
   E. Ya. Davydov, I. S. Gaponova, T. V. Pokholok, G. B. Pariyskii, L.A. Zimina, M. I. Artsis, and G. E. Zaikov

5. **The Structure and Properties of Blends of Poly(3-hydroxybutyrate) With an Ethylene-Propylene Copolymer** 65
   A. A. Ol'khov, Yu. V. Tertyshnaya, L. S. Shibryaeva, A. L. Iordanskii, and G. E. Zaikov

6. **Analysis Methods of Some Nanocomposites Structure** 75
   Kozlov Georgii Vladimirovich, Yanovskii Yurii Grigor'evich, Zaikov Gennady Efremovich, Eli M. Pearce, and Stefan Kubica

7. **Synthesis, Properties and Applications Ozone and Its Compounds** 109
   S. K. Rakovsky, S. Kubica, and G. E. Zaikov

8. **Functional Models of Fe(Ni) Dioxygenases: Supramolecular Nanostructures Based on Catalytic Active Nickel and Iron Heteroligand Complexes** 185
   Ludmila I. Matienko, Vladimir I. Binyukov, Larisa A. Mosolova, Elena M. Mil, and Gennady E. Zaikov

9. **Association between Calcium Chloride and Caffeine as Seen by
    Transport Techniques and Theoretical Calculations**.......................... 205

    Marisa C. F. Barros, Abilio J. F. N. Sobral, Luis M. P. Verissimo,
    Victor M. M. Lobo, Artur J. M. Valente, Miguel A. Esteso, and Ana C. F. Ribeiro

10. **Synthesis of Biologically Awake Antioxidants in Reactions of
    Esterification 2-(N-acetylamid)-3-(3',5'-di-tert.butyl-4'-
    hydroxyphenyl)-Propionic Acid** 221

    G. E. Zaikov, A. A. Volodkin, L. N. Kurkovskaja, N. M. Evteeva, S. M. Lomakin,
    L. J. Gendel, and E. J. Parshina

11. **Attempting to Consider Mechanism of Originating of Damages of
    Chromosome in the Different Phases of Mitotic Cycle** 231

    Larissa I. Weisfeld

12. **Biology of Development of Phytopathogenic Fungi of Fusarium
    Link and Resistance of Cereals to it in Climatic Conditions of
    Tyumen Region** 251

    N. A. Bome, A. Ja. Bome, and N. N. Kolokolova

13. **Realization of Potential Possibilities of a Genotype at Level
    of Phenotype** 259

    N. A. Bome, S. A. Bekuzarova, L. I. Weisfeld, and I. A. Cherkashina

14. **COMMENTARY: Cooperation of High Schools and Scientific
    Institutions on the Way of Education of Scientific Shots to
    Modern Conditions** 273

    N. I. Kozlowa, A. E. Rybalko, K. P. Skipina, and L. G. Kharuta

15. **Fluid Flow and Control of Bending Instability During
    Electrospinning** 281

    A. Pourhashemi and A. K. Haghi

16. **Relaxation Parameters of Polymers** 313

    Ninel N. Komova, Gennady E. Zaikov, and Alfonso Jimenez

17. **The Morphological Features of Poly(3-hydroxybutyrate) with an
    Ethylene-Propylene Copolymer Blends** 329

    A. A. Ol'khov, L. S. Shibryaeva, Yu. V. Tertyshnaya, A. L. Iordanskii,
    and G. E. Zaikov

18. **Perspectives of Application Multi-Angle Laser Light Scattering
    Method for Quality Control of Medicines** 339

    A. V. Karyakin, E. D. Skotselyas, and P. A. Flegontov

19. **Interaction and Structure Formation of Gelatin Type a with Thermo Aggregates of Bovine Serum Albumin** 353

Y. A. Antonov and I. L. Zhuravleva

20. **Wild Orchids of Colchis Forests and Save Them as Objects of Ecoeducation, and Producers of Medicinal Substances** 373

E. A. Averjanova, L. G. Kharuta, A. E. Rybalko, and K. P. Skipina

21. **Express Assessment of Cell Viability in Biological Preparations** 387

L. P. Blinkova, Y. D.Pakhomov, O. V. Nikiforova, V. F. Evlashkina, and M. L. Altshuler

22. **Activity of Liposomal Antimicrobic Preparations Concerning *Staphylococcus Aureus*** 393

N. N. Ivanova, G. I. Mavrov, S. A. Derkach, and E. V. Kotsar

23. **Polyelectrolyte Microsensors as a New Tool for Metabolites' Detection** 399

L. I. Kazakova, G. B. Sukhorukov, and L. I. Shabarchina

24. **Selection of Medical Preparations For Treating Lower Parts of the Urinary System** 419

Z. G. Kozlova

25. **Improvement of the Functional Properties of Lysozyme by Interaction with 5-Methylresorcinol** 427

E. I. Martirosova and I. G. Plashchina

26. **Introductions in Culture *In Vitro* Rare Bulbous Plants of the Sochi Black Sea Coast (*Scilla, Muscari, Galanthus*)** 439

A. O. Matskiv, A. A. Rybalko, and A. Rybalko

27. **Change of Some Physico-Chemical Properties of Ascorbic Acid and Paracetamol High-Diluted Solutions at Their Joint Presence** 451

F. F. Niyazy, N. V. Kuvardin, E. A. Fatianova, and S. Kubica

28. **The Methods of the Study the Processes of the Issue to Optical Information Biological Object** 461

U. A. Pleshkova and A. M. Likhter

29. **Research Update on Conjugated Polymers** 473

A. Jimenez amd G. E. Zaikov

30. Experimental and Theoretical Study of the Effectiveness of Centrifugal Separator 503

R. R. Usmanova and G. E. Zaikov

Index 515

LIST OF CONTRIBUTORS

M. L. Altshuler
FGBU I. Mechnikov Research Institute for Vaccines and Sera of RAMS. Moscow. 105064 Maliy Kazyonniy side st,. 5a. E-mail: labpitsred@yandex.ru

Y. A. Antonov
N. M. Emanuel Institute of Biochemical Physics, Russian Academy of Sciences, Kosygin Stir. 4. 119334 Moscow, Russia. E-mail: chehonter@yandex.ru

M. I. Artsis
N. M. Emanuel Institute of Biochemical Physics, Russian Academy of Sciences. 4 Kosygin Street. 119334 Moscow, Russia

E. A. Averjanova
Sochi Branch of the Russian Geographical Society, 354024, Sochi, Kurortny pr., 113, E-mail: geo@opensochi.org

M. C. F. Barros
Department of Chemistry, University of Coimbra, 3004 - 535 Coimbra, Portugal

S. A. Bekuzarova
Doctor of Agricultural Sciences, Honored the Inventor of the Russian Federation, Professor, Federal State Institution of Highest Professional Education Gorsky State Agrarian University, Kirov St., 37, 362040,Vladikavkaz, Republic of North Ossetia Alania, Russia. E-mail: ggau@globalalania.ru

V. I. Binyukov
N. M. Emanuel Institute of Biochemical Physics, Russian Academy of Sciences, ul. Kosygina 4, 119334 Moscow, Russian Federation

L. P. Blinkova
FGBU I. Mechnikov Research Institute for Vaccines and Sera of RAMS. Moscow. 105064 Maliy Kazyonniy side st,. 5a, E-mail: labpitsred@yandex.ru

A. Ja. Bome
Tyumen Basing Point of the All-Russia Research Institute of Plant Growing of N.I. Vavilov, Bol'shaya Morskaya St., 190000, St. Petersburg, Russia. E-mail: office@vir.nv.ru

N. A. Bome
Department of Botany and Plant Growing, Professor, Federal State Institution of Highest Professional Education Tyumen State University, Semakova St., 1, 625003, Tyumen, Russia, E-mail: president@utmn.ru

I. A. Cherkashina
Science Federal State Budgetary Institution, The Tobolsk Complex, Scientific Station of the "Ural Branch of the Russian Academy of Sciences," Academician Yuri Osipov St., 15b, 626150, Tobolsk, Russia, E-mail: tbsras@rambler.ru

Ya. Davydov
N. M. Emanuel Institute of Biochemical Physics, Russian Academy of Sciences. 4 Kosygin Street. 119334 Moscow, Russia

S. A. Derkach
Mechnicov Institute of Microbiology and Immunology of Academy of Medical Sciences of Ukraine

M. A. Esteso
Departamento de Química Física, Facultad de Farmacia, Universidad de Alcalá, 28871. Alcalá de Henares, Madrid, Spain

V. F. Evlashkina
FGBU I. Mechnikov Research Institute for Vaccines and Sera of RAMS. Moscow. 105064 Maliy Kazyonniy side st,. 5a, E-mail: labpitsred@yandex.ru

N. M. Evteeva
Institute of Biochemical Physics of N.M. Emanuelja of the Russian Academy, Sciences, The Russian Federation, 119991 Moscow, Kosygina, 4

E. A. Fatianova
Department of General and Inorganic Chemistry, South-West State University, Kursk, Russia

P. A. Flegontov
Scientific Center of Hematology, Ministry of Health and Social Development of the Russian Federation, 125167, Russia, Moscow, Novyi Zykovskyi 4, E-mail: il@blood.ru, flegant@yandex.ru

I. S. Gaponova
N.M. Emanuel Institute of Biochemical Physics, Russian Academy of Sciences. 4 Kosygin Street. 119334 Moscow, Russia

L. J. Gendel
Institute of Biochemical Physics of N.M. Emanuelja of the Russian Academy, Sciences, The Russian Federation, 119991 Moscow, Kosygina, 4

G. P. Gladyshev
N. N. Semenov Institute of Chemical Physics, Russian Academy of Science, 4 Kosygin St., Moscow, 119334, Russia

M. D. Goldfein
Saratov State University, 83 Astrakhanskaya St., Saratov, 410012, Russia

Y. Y. Grigor'evich
Institute of Applied Mechanics of Russian Academy of Sciences, Leninskii pr., 32 a, Moscow 119991, Russian Federation, E-mail: iam@ipsun.ras.ru

A. K. Haghi
University of Guilan, Rasht, Iran, E-mail: Haghi@Guilan.ac.ir

A. L. Iordanskii
Semenovl Institute of Chemical Physics, Russian Academy of Sciences, ul. Kosygina 4, Moscow, 119991 Russia

N. N. Ivanova
Institute of Dermatology and Venerology of Academy of Medical Sciences of Ukraine St. Chernishevskaya 7/9, 61057 Kharkov, Ukraine, E-mail: jet-74@mail.ru

A. Jiménez
Alicante University, Chemical Faculty, Alicante, Spain, E-mail: AlfJimenez@ua.es

V. Kablov
Volzhsky Polytechnic Institute (Branch of) Volgograd State Technical University, 42-a, Engels St., Volzhsky 404120, Russia, E-mail: Kablov@volpi.ru

A. V. Karyakin
Scientific Center of Hematology, Ministry of Health and Social Development of the Russian Federation, 125167, Russia, Moscow, Novyi Zykovskyi 4, E-mail: il@blood.ru, flegant@yandex.ru

L. I. Kazakova
Institute of Theoretical and Experimental Biophysics, Russian Academy of Science, 142290 Pushchino, Moscow Region 142290, Russia, E-mail: shabarchina@rambler.ru

L. G. Kharuta
Sochi Institute of Russian People's Friendship University, 354340, Sochi, Kuibyshev St., 32, E-mail: sfrudn@rambler.ru

N. N. Kolokolova
Associate Professor of the Chair for Botany, Biotechnology and Landscape Architecture of Tyumen State University, Tyumen, Russia

N. N. Komova
Moscow State Academy of Fine Chemical Technology, 124 Vernadsky avenue, Moscow 119571, Russia, E-mail: komova_@mail.ru

D. Kondrutsky
Volzhsky Polytechnic Institute (Branch of) Volgograd State Technical University, 42-a, Engels srt., Volzhsky 404120, Russia

E. V. Kotsar
Institute of Dermatology and Venerology of Academy of Medical Sciences of Ukraine, St. Chernishevskaya ,7/9, 61057 Kharkov, Ukraine, E-mail: jet-74@mail.ru

Z. G. Kozlova
Emanuel Institute of Biochemical Physics of Russian Academy of Sciences, 119334, Moscow, 4 Kosygin St. E-mail: yevgeniya-s@inbox.ru

N. I. Kozlowa
Sochi Institute of Russian People's Friendship University, Kuibyshev St. 32, E-mail: sfrudn@rambler.ru

S. Kubica
Institute for Engineering of Polymer Materials and Dyes, 55 M. Sklodowskiej-Curie St., 87-100 Torun, Poland, E-mail: S.Kubica@impib.pl

L. N. Kurkovskaja
Institute of Biochemical Physics of N.M. Emanuelja of the Russian Academy, Sciences, The Russian Federation, 119991 Moscow, Kosygina, 4

N. V. Kuvardin
Department of General and Inorganic Chemistry, South-West State University, Kursk, Russia

A. M. Likhter
Astrakhan State University Bld. 20a, Tatischeva St., Astrakhan, 414056 Russian Federation, E-mail: pjulia@pisem.net

V. M. M. Lobo
Department of Chemistry, University of Coimbra, 3004 - 535 Coimbra, Portugal

S. M. Lomakin
Institute of Biochemical Physics of N.M. Emanuelja of the Russian Academy, Sciences, The Russian Federation, 119991 Moscow, Kosygina, 4

E. I. Martirosova
Emanuel Institute of Biochemical Physics RAS, Kosygin Street 4, Moscow 119334, Russia, E-mail: ms_martins@mail.ru

L. I. Matienko
N. M. Emanuel Institute of Biochemical Physics, Russian Academy of Sciences, ul. Kosygina 4, 119334, Moscow, Russian Federation. Fax (7-495) 137 41 01, Tel. (7-495) 939 71 40, E-mail: matienko@sky.chph.ras.ru

A.O Matskiv
Sochi Institute of Russian People's Friendship University, Kuibyshev St. 32, E-mail: sfrudn@rambler.ru

G. I. Mavrov
Institute of Dermatology and Venerology of Academy of Medical Sciences of Ukraine St. Chernishevskaya 7/9, 61057 Kharkov, Ukraine, E-mail: jet-74@mail.ru

E. M. Mil
N. M. Emanuel Institute of Biochemical Physics, Russian Academy of Sciences, ul. Kosygina 4, 119334 Moscow, Russian Federation

L. A. Mosolova
N. M. Emanuel Institute of Biochemical Physics, Russian Academy of Sciences, ul. Kosygina 4, 119334 Moscow, Russian Federation

O. V. Nikiforova
FGBU I. Mechnikov Research Institute for Vaccines and Sera of RAMS. Moscow. 105064 Maliy Kazyonniy Side St. 5a, E-mail: labpitsred@yandex.ru

F. F. Niyazy
Department General and Inorganic chemistry, South-West State University, Kursk, Russia, E-mail: FarukhNiyazi@yandex.ru

A. A. Ol'khov
Moscow State Academy of Fine Chemical Technology, pr. Vernadskogo 86, Moscow, 117571 Russia

Y. D. Pakhomov
FGBU I. Mechnikov Research Institute for Vaccines and Sera of RAMS. Moscow. 105064 Maliy Kazyonniy Side st,. 5a, E-mail: labpitsred@yandex.ru

G. B. Pariyskii
N. M. Emanuel Institute of Biochemical Physics, Russian Academy of Sciences. 4 Kosygin Street. 119334 Moscow, Russia

E. J. Parshina
Institute of Biochemical Physics of N.M. Emanuelja of the Russian Academy, Sciences, The Russian Federation, 119991 Moscow, Kosygina, 4

E. M. Pearce
New York State University in Brooklyn, 333 Jay St., Six Metrotech Center, Brooklyn, NYC, USA,
E-mail: EPearceg@gmail.edu

I. G. Plashchina
Emanuel Institute of Biochemical Physics RAS, Kosygin Street 4, Moscow 119334, Russia, E-mail:
ms_martins@mail.ru

U. Pleshkova
Astrakhan State University Bld. 20a, Tatischeva St., Astrakhan, 414056 Russian Federation, E-mail:
pjulia@pisem.net,

V. Pokholok
N. M. Emanuel Institute of Biochemical Physics, Russian Academy of Sciences. 4 Kosygin Street.
119334 Moscow, Russia

A. Pourhashemi
Christian Brothers University, Memphis, TN, USA

S. K. Rakovsky
Institute of Catalysis, Bulgarian Academy of Sciences, 7 Nezabravka St., Sofia 1113, Bulgaria,
E-mail: Rakovsky@ic.bas.bg

A. C. F. Ribeiro
Department of Chemistry, University of Coimbra, 3004 - 535 Coimbra, Portugal, E-mail: anacfrib@
ci.uc.pt

A. E. Rybalko
Sochi Institute of Russian people's friendship university, Kuibyshev St. 32, E-mail: sfrudn@rambler.
ru

Anamika Singh
Division of Reproductive and Child Health, Indian Council of Medical Research, New Delhi, India

Rajeev Singh
Division of Reproductive and Child Health, Indian Council of Medical Research, New Delhi, E-mail:
10rsingh@gmail.com

L. I. Shabarchina
Institute of Theoretical and Experimental Biophysics Russian Academy of Science, 142290
Pushchino, Moscow region 142290, Russia, E-mail: shabarchina@rambler.ru

L. S. Shibryaeva
Emanuel Institute of Biochemical Physics, Russian Academy of Sciences, ul. Kosygina 4, Moscow,
119991 Russia

K. P. Skipina
Sochi Institute of Russian People's Friendship University, 354340, Sochi, Kuibyshev St., 32, Russia,
E-mail: sfrudn@rambler.ru

D. Skotselyas
Scientific Center of Hematology, Ministry of Health and Social Development of the Russian Federation, 125167, Russia, Moscow, Novyi Zykovskyi 4, E-mail: il@blood.ru, flegant@yandex.ru

A. J. F. N. Sobral
Department of Chemistry, University of Coimbra, 3004 - 535 Coimbra, Portugal

M. Sudnitsina
Volzhsky Polytechnic Institute (branch of) Volgograd State Technical University, 42-a, Engels srt., Volzhsky 404120, Russia

G. B. Sukhorukov
School of Engineering and Materials Science, Queen Mary University of London, London, UK

Yu. V. Tertyshnaya
Emanuel Institute of Biochemical Physics, Russian Academy of Sciences, ul. Kosygina 4, Moscow, 119991 Russia

R. R. Usmanova
Ufa State Technical University of Aviation; 12 Karl Marks St., Ufa 450000, Bashkortostan, Russia, E-mail: Usmanovarr@mail.ru

A. J. M. Valente
Department of Chemistry, University of Coimbra, 3004 - 535 Coimbra, Portugal

L. M. P. Verissimo
Department of Chemistry, University of Coimbra, 3004 - 535 Coimbra, Portugal

K. G. Vladimirovich
Kh. M. Berbekov Kabardino-Balkarian State University, Department of Chemistry, 360004, Nalchik, Chernyshevskogo, 173, Russia, E-mail: I_dolbin@mail.ru

A. A. Volodkin
Institute of Biochemical Physics of N.M. Emanuelja of the Russian Academy, Sciences, The Russian Federation, 119991 Moscow, Kosygina, 4

L. I. Weisfeld
N. M. Emanuel Institute of Biochemical Physics RAS, Moscow, Russia, E-mail: liv11@yandex.ru

G. E. Zaikov
N. M. Emanuel Institute of Biochemical Physics, Russian Academy of Sciences, ul. Kosygina 4, 119334 Moscow, Russian Federation. Tel. (7-495) 939 71 40; Fax (7-495) 137 41 01, E-mail: chembio@sky.chph.ras.ru

I. L. Zhuravleva
N. M. Emanuel Institute of Biochemical Physics, Russian Academy of Sciences, Kosygin Stir. 4. 119334 Moscow, Russia

L. A. Zimina
N. M. Emanuel Institute of Biochemical Physics, Russian Academy of Sciences, 4 Kosygin Street, 119334 Moscow, Russia

LIST OF ABBREVIATIONS

AFM	atomic force microscopy
ANN	artificial neural networks
AO	antioxidants
BAS	bioanalytical system
BCS	biocybernetical system
CAN	ceric ammonium nitrate
CIP	complex immunoglobulin product
CR	complement receptor
CvME	caveolae-mediated endocytosis
DMSO	dimethylsulfoide
ECDs	electrochromic display devices
EP	European pharmacopoeia
FcR	Fc receptor
GC	glassy carbon
HES	hydroxylethyl starch
LbL	layer-by-layer
MALLS	multi-angle laser light scattering detector
ME	molecular electronics
MIC	minimum inhibitory concentrations
NC	nanocrystal
PABA	para-aminobenzoic acid
PAN	polyacrylonitrile
PTFE	polytetrafluoroethylene
PU	polyurethane
RES	reticuloendothelial system
RI	refractive index
RME	receptor-mediated endocytosis
SEC	size-exclusion chromatography
SEM	scanning electron microscopy
SLN	solid lipid nanoparticles

SSD	supersmall doses effect
TEM	transmission electron microscopy
THF	tetrahydrofurane
UV	ultraviolet

PREFACE

This new book offers research and updates on the chemical process in liquid and solid phases. The collection of topics in this book reflects the diversity of recent advances in chemical processes with a broad perspective that will be useful to scientists as well as for graduate students and engineers. The book will help to fill the gap between theory and practice in industry. This book:

- covers important issues related to the chemical process in liquid and solid phases
- introduces the reader to the basic issues in polymer processes, in particular, processing, properties, and applications
- provides a broad survey of various topics related to polymer research from the R&D stage to industrial developments and applications
- gives a unified perspective to varied topics intended for a broad spectrum of potential readers, from students in chemistry and materials engineering to researchers in the field of polymers and composites
- presents interesting and updated results from leading research groups in polymer science and technology
- discusses problems related to polymer environmental issues and particularly waste treatment

In the first chapter, thermodynamics as a basic of inanimate and animate natures is discussed. In this chapter the concept which allows the investigation of the behavior of open (biological) systems by using the laws of unbalanced (classical) thermodinamics is examined. Nanotherapeutics, as a novel approach of target-based drug delivery, is presented in Chapter 2.

In Chapter 3 results of phosphorus-containing polypeptides on the base of some substrates (peptides, gelatin, and fish collagen) synthesis research are presented. The chemical structure of a product, sorption properties, and thermal-oxidative degradation are also investigated. Free radical initiation in polymers under the action of nitrogen oxides is investigated in Chapter 4. In this connection, the research of the mechanism of nitrogen oxide reactions with various organic compounds including polymers are important for definition of stability of these materials in polluted atmosphere. Chapter 5 is concerned with the study of the structural features of PHB-EPC blends and their thermal degradation. New methods of some nanocomposites structure are analyzed in Chapter 6. The ozone application is based on its powerful oxidative action. The ozonolysis of chemical and biochemical compounds, as a rule, takes place with high rates at low temperatures and activation energies. This provides a basis for the development of novel and improved technologies, which find wide application in ecology, chemical, pharmaceutical and perfume industries, cosmetics, cellulose, paper and sugar industries, flotation, microelectronics, veterinary and human medicine, agriculture, foodstuff industry, and many others. Therefore, synthesis, properties and applications of ozone and its compounds are surveyed in Chapter 7 in detail. We have offered a new approach to research of mechanism of catalysis with heteroligand complexes of nickel (iron) which can be considered as model $Ni^{II}(Fe^{II})$ ARD in Chapter 8.

In Chapter 9 we have shown the association between calcium chloride and caffeine as seen by transport techniques and theoretical calculations. In Chapter 10 we have shown how it is possible to assume that properties esters 2-(N-acetylamid)-3- (3', 5 '-di-tert.butyl-4 '-hydroxyphenyl)- propionic acid are bound to geometry of communications between kernels of atoms in a molecule and possibility of electron transition with an antioxidant on the lowest vacant orbital of molecule peroxy radical. Scientists have drawn attention to the chemical compounds that cause heritable changes. There remains an unsolved question of mechanism of their action on the chromosome. The action of many compounds is similar to ionizing radiation; it cause mutations of genes, disruptions of cell division, and rearrangement of chromosomes. In Chapter 11 we attempt to consider the mechanism of originating of damages of chromosome in the different phases of mitotic cycle. In Chapter 12 we studied the biology of the development of phytopathogenic fungi of the *Fusarium* link and resistance of cereals to it in climatic conditions of the Tyumen region. A significant role of para-aminobenzoic acid (PABA) treatment is shown in Chapter 13 for the development of stable plant growing in Russia. PABA has a positive influence on the sowing qualities of seeds, morphological traits of plants, and the chemical composition of spring wheat's leaves. The increase of flaglist square and pigment content in it is especially important for spring crops due to the short spring season in West Siberia, in particular in the Tyumen region. Ways of use and concentration of growth regulator are determined, which ensure fuller realization of ecological and biological potential of spring wheat cultivars in sub-taiga zone of the Tyumen region. Taking into account that the experiment has been held on soils with low content of micro- and macro-elements, humus and organic acids, one can assume that protective properties of the substance increase in extreme environmental conditions.

A commentary on educational systems is presented in Chapter 14. We have presented why electrospinning is a very simple and versatile method for creating polymer-based high-functional and high-performance nanofibers that can revolutionize the world of structural materials in Chapter 15. The process is versatile in that there is a wide range of polymer and biopolymer solutions or melts that can spin. The electrospinning process is a fluid dynamics related problem. In order to control the property, geometry, and mass production of the nanofibers, it is necessary to understand quantitatively how the electrospinning process transforms the fluid solution through a millimeter diameter capillary tube into solid fibers, which are four to five orders smaller in diameter. When the applied electrostatic forces overcome the fluid surface tension, the electrified fluid forms a jet out of the capillary tip toward a grounded collecting screen. Although electrospinning gives continuous nanofibers, mass production and the ability to control nanofibers properties are not obtained yet. The combination of both theoretical and experimental approaches seems to be a promising step for a better description of the electrospinning process. Applying simple models of the process would be useful in atoning for the lack of systematic, fully characterized experimental observations and the theoretical aspects in predicting and controlling effective parameters. The analysis and comparison of the model with experiments identify the critical role of the spinning fluid's parameters. The theoretical and quantitative tools developed in different models provide semi-empirical methods for predicting ideal electrospinning process or electrospun fiber properties. In each model, the researcher tried to improve the existing models or changed the tools in electrospinning by using another view. Therefore, it was attempted to have a whole view on important models after investigation about basic objects. A real mathematical model, or, more accurately, a real physical model, might initiate a revolution in the understanding of dynamic and quantum-like phenomena in the electrospinning process. A new theory is much needed to bridge the gap between Newton's world and the quantum world.

In Chapter 16 we present the relaxation parameters of polymers for readers. Chapter 17 explains the morphological features of poly(3-hydroxybutyrate) with an ethylene-propylene copolymer blends. Chapter 18 focuses on application of a multi-angle laser light scattering method for quality control of medicines. In Chapter 19 we investigated the interaction and structure formation of gelatin type A with thermo aggregates of bovine serum albumin. Presentation of the wild orchids of the Colchic forests and to save them as objects of ecoeducation and producers of medicinal substances is the main aim of Chapter 20.

Chapter 21 reviews the assessment of cell viability in biological preparations while Chapter 22 considers activity of liposomal antimicrobic preparations concerning *Staphylococcus aureus*. The aim of Chapter 23 is to demonstrate a particular example of a sensor system, which combines catalytic activity for urea and at the same time enabling monitoring enzymatic reaction by optical recording. The proposed sensor system is based on multilayer polyelectrolyte microcapsules containing urease and a pH-sensitive fluorescent dye, which translates the enzymatic reaction into a fluorescently registered signal.

In Chapter 24 we focused on the selection of medical preparations for treating lower parts of the urinary system. Improvement of the functional properties of lysozyme by interaction with 5-methylresorcinol is investigated in detail in Chapter 25. In Chapter 26, we have developed regimes for the micropropagation in the culture of cultivars that have been developed for lily bulbs and rare plants of the North Caucasus (*Muscari, Scilla, Galanthus*) on various types of explants scales of bulbs collected in the wild Sochi suburban forests, and micropropagation of plants grown in culture *in vitro*. Various plant growth regulators in MS basal medium were used for initiation of organogenesis and the formation of bulblets tested, including cytokinins-benzyladenine (BA) and auxin-naphthaleneacetic acid. Within 8 weeks of the multiplication factor was a lily – 1:10, *Muscari*

– 1:21 and *Scilla* – 3:5. We discussed the changes of some physico-chemical properties of ascorbic acid and paracetamol high-diluted solutions at their joint presence in Chapter 27. In Chapter 28 the methods of the study of the processes of the issue of optical information biological object is presented. In Chapter 29 we have presented research update on conjugated polymers in detail. Experimental and theoretical study of the effectiveness of centrifugal separators are presented in the last chapter.

Topics of this new book provides physical principles in explaining and rationalizing chemical and biological phenomena and covers the physical chemical process in liquid and solid phases from a modern point of view. This book also focuses on topics with more advanced methods.

— Alfonso Jiménez, PhD, and Ali Pourhashemi, PhD

CHAPTER 1

THERMODYNAMICS AS A BASIC OF INANIMATE AND ANIMATE NATURE

G. P. GLADYSHEV and M. D. GOLDFEIN

CONTENTS

1.1   Introduction 2
1.2   General Methodolocical Aspects of Modern Thermodynamics 2
1.3   Thermodynamics of the Development of Biological Systems 7
Keywords 15
References 15

1.1   INTRODUCTION

In this chapter the concept which allows to investigate the behavior of open (biological) systems by using the laws of unbalanced (classical) ther-modinamics is examined. The second law of thermodynamics is applicable to open systems as well, provided that only in case of considering of the whole system (continuum) "open system – environment" the total entropy will always to increase. The law of temporary hierarchies allows to resolve quasi-closed thermodinamic systems in open biosystems and to investigate their development (ontogenesis) and evolution (philogenesis) by studying changes in the value of Gibbs' specific fuctions of formation of any submolecular structure (including a population).

1.2   GENERAL METHODOLOCICAL ASPECTS OF MODERN THERMODYNAMICS

The concepts of classical and quantum physics allow one, either exactly or with a certain probability, to predict the state of macrobodies or microparticles. In particular, this concerns mechanical movement regularities, which can be described by using spatial-temporal, coordinates, the vales of mass, velocity, pulse, wave characteristics, the knowledge of the fundamental type of interaction. However, there exist some processes whose features can be explained by neither classical physics nor quantum representations. E.g. the existence of bodies to them in different aggregation states, the appearance of elastic forces at deformations of systems, possible transformation of some compounds into others, etc. As a rule, these and similar processes are accompanied by transition of systems from one state to another one with changes in thermal energy. Just such processes and most general thermal properties of macroscopic bodies are studied by the section of physics and chemistry called *thermodynamics* [1, 2, 9–11].

Thermodynamics studies a system, i.e., a body or group of bodies capable of changing under the influence of physical or chemical processes. Everything surrounding the system forms the external medium (environment). By a "body" they usually understand any substance having a certain volume and characterized by some physical properties. Thermodynamic systems can be homogeneous and heterogeneous. Every system is related with such a notion as the condition of a system, which represents a set of the quantitative values of all its thermodynamic properties. At change of at least one of these properties the system passes from one state to another one. Usually, thermodynamic processes proceed at constancy of certain parameters of the system.

The bases of classical thermodynamics have been formulated in the works by Clausius, van't Hoff, Arrhenius, Gibbs, Helmholtz, le Chatelier, et al., which describes processes depending on changes of the properties of systems in an equilibrium state. In other words, the thermodynamic description of systems and phenomena is based on the idea of an equilibrium state. Thermodynamics answers the question: where a process is directed before equilibrium is reached. Therefore it does not deal, in an explicit form, with time as a physical parameter and considers no mechanisms of processes. If a system infinitely slowly passes from one state to another one through a continuous series of equilibrium states and if the maximum work is thus done, such a process is called *thermodynamically equilibrium process*. Between two next states of equilibrium, the values of any functions of state of the system differ by an infinitesimal value, and it is always possible to return the system to its initial state by infinitesimal changes of the functions of state. The system having made an equilibrium process can return to its initial state, having passed the same equilibrium states in the opposite direction, as in the direct one. Therefore, as a result of a *thermodynamically reversible process* the system and environment return to their original state, i.e., no changes remain in the system and environment. On the other hand, a process after which the system passes through a series of non-equilibrium states is called *thermodynamically non-equilibrium process*. If the system and environment cannot return to their original state, i.e., any changes remain in them, such a process is considered *thermodynamically irreversible*.

Thermodynamics only considers the initial and final states of a system, which are characterized by special thermodynamic parameters. Usually, temperature, pressure, volume, and such *characteristic functions* as enthalpy, entropy, internal energy, Helmholtz' free energy, Gibbs' thermodynamic potential act as such parameters of state. Just these functions and their values characterize the thermodynamic state of a system and cause the probability of proceeding of this or that process accompanied by changes of thermal energy. The functions of state allow establishing the direction of spontaneous processes and determining the degree of their completeness in real thermodynamic systems. For example, Gibbs' thermodynamic potential G can be used for exploring equilibrium (and quasi-equilibrium, as will be shown further) processes and closed (quasi-closed) systems, in which transformations proceed at constant temperature and pressure. Helmholtz' free energy F is applicable to studying similar processes and systems at constancy of temperature and volume.

Thermodynamics also features being a phenomenological science. This means that all physical quantities, functions and laws of thermodynamics are based on experience only, i.e., there is no strict theory concerning them. The sense of then empirical provisions of thermodynamics can be characterized as follows. (1) Unlike mechanical movement, all spontaneous thermal processes are irreversible (and, first of all, this applies to heat propagation in the environment). (2) According to various forms of movement and various types of energy there are also various forms of energy exchange. In classical thermodynamics, only two forms of energy transfer are resolved, namely, work and heat. Any kind of energy is an unequivocal function of the state of the system, i.e., the change of the energy does not depend on the way of the system's transition from one state to another one. At the same time, heat and work are unequal forms of energy transfer as work can be directly used for replenishment of the stock of any kind of energy, whilst heat – on replenishment of the stock of the system's internal energy only. (3) Any physicochemical system left to itself (i.e., in absence of external forces and fields) always aspires to pass to the state of thermodynamic equilibrium; this state is characterized by the uniformity of distribution of temperature, pressure, density, and the concentration of components. (4) The characteristic functions of the state of system are interrelated by certain ratios, for example:

$$\Delta F = \Delta U - T\Delta S;\ \Delta G = \Delta H - T\Delta S,$$

where ΔU, ΔH, and ΔS are the changes of the internal energy, enthalpy, and entropy, respectively; T is absolute temperature, which points to their interconditionality and universality relative to the description of both initial and final state of the system.

The physical sense of the concepts of classical (equilibrium) thermo-dynamics is directly related with its three basic laws, which are general laws of nature.

The second law of thermodynamics means that the processes of energy transformation can occur spontaneously only provided that energy passes from its concentrated (ordered) form to a diffused (disordered) one. Such energy redistribution in the system is characterized by a quantity which has been named as *entropy*, which, as a function of state of the thermody-namic system (the more energy irreversible dissipates as heat, the higher entropy is. Whence it follows that any system whose properties change in time aspires to an equilibrium state at which the entropy of the system takes its maximum value. In this connection, the second law of thermo-dynamics is often called the law of increasing entropy, and entropy (as a physical quantity or as a physical notion) is considered as a measure of disorder of a physicochemical system.

In connection with the notion of entropy, the notion of time gets a re-newed sense in thermodynamics. As was already told, the direction of time is ignored in classical mechanics, and it is possible to determine the state of a mechanical system both in the past and in the future. In thermodynam-ics, time acts as an irreversible process of entropy increasing in the system. i.e., the higher the entropy, the longer time interval has been passed by the system in its development.

It is also necessary to mean that there are four classes of thermody-namic systems in nature:

(a) isolated systems (upon transition of such systems from one state in another one no transfer of energy, substance, and information through the system's borders proceeds);
(b) adiabatic systems (no heat exchange with the environment only);
(c) closed systems (no substance transfer only);
(d) open systems (which exchange substance, energy, and information with the environment).

Such a classification of thermodynamic systems, at first sight, leads to a seeming contradiction between the general principle of relativity in nature and the validity of the second law of thermodynamics for whole nature. The matter is that, ignoring exchange processes, the second law of thermodynamics is directly applicable to systems (a) and (b) only, for which entropy either is constant, or (which is much more often) increases: $\Delta S \geq 0$. This means that in such systems, the changes are characterized by increasing disorder (chaos) in them until the system will not reach thermodynamic equilibrium (or its entropy will not reach its maximum value). In open systems entropy may decrease. The latter fact, first of all, concerns biological systems, i.e., living organisms that are open non-equilibrium systems. Such systems are characterized by gradients of the concentration of chemicals, temperatures, pressure, and other physical and chemical quantities. In this regard, a seeming contradiction appears between the second law of thermodynamics and biological ordering.

The recovery from such a contradiction can be found in the use of the concepts of modern (i.e., non-equilibrium) thermodynamics, which describe the behavior of open systems always exchanging energy, substance, and information with the environment. Schematically, the mechanism of interaction of an open system with the environment can be represented as follows. First, such a system at its functioning takes substances, energy, and information from the environment, and then gives it the used ones (naturally, in other types and forms). Second, all these processes are also accompanied by entropy production (i.e., energy losses and the formation of various wastes). Third (and this is most important), the formed entropy (characterizing a certain degree of disorder), without collecting in the system, is released into the environment, and the system takes from it new substances, energy, and information necessary for its optimum performance. All this allows drawing a conclusion that the open system, as a result of exchange processes, can be in a non-equilibrium state as well, for whose description new laws (in particular, non-equilibrium thermodynamics) are necessary.

The concepts developing non-equilibrium thermodynamics in its application to biological systems were developed by de Donder, de Groot, Mazur, Glansdorff, and Prigogine.

Another recovery from the specified contradiction is connected with the development of the concept of thermodynamics based on the possibility to apply classical (i.e., equilibrium) thermodynamics to real (open) systems. The matter is that many facts are now known to allow, in a certain degree, to neglect the irreversibility of some physical processes and chemical transformations (i.e., to study them in a certain approximation). In these cases it is merely necessary to establish such ranges of changes of the properties of a system, within which it is possible to consider them equilibrium ones. Usually, such states are called as quasi-equilibrium, when a process studied by changes of some parameter is steady and has such a time, in comparison with which the times of other processes in the system are either very long or negligible. Such a concept underlies *the law of temporary hierarchies* [1, 2], which allows to resolve quasi-closed thermodynamic systems (subsystems) in open biosystems and to investigate their development (ontogenesis) and evolution (philogenesis) by studying changes in the value of Gibbs' specific (per unit of volume or mass) functions of formation of a given highest monohierarchical structure from the monohierarchical structures of a lowest level. Proceeding from these considerations, it is possible to make two important conclusions:

1. In every specific case, depending on the chosen time scale, (or process) this state of the system can be considered (with various degrees of the accuracy of approximation) equilibrium or non-equilibrium, i.e., to apply various ways of description depending on the character of the task to be solved;

2. The second law of thermodynamics is applicable to real (open) systems as well, provided that the whole system (continuum) "open system–environment" is considered; for such a set, the total entropy will always increase, as for the first component $\Delta S_1 < 0$, for the second component $\Delta S_2 > 0$, and at $|\Delta S_1| \ll |\Delta S_2|$ we have:

$$\Delta S = (\Delta S_1 + \Delta S_2) > 0.$$

1.3 THERMODYNAMICS OF THE DEVELOPMENT OF BIOLOGICAL SYSTEMS

The total number of live organisms in every biocenosis, the rate of their development and reproduction depend on the quantity of energy arriving into the ecological system, and the intensity of circulation of chemical elements and their compounds in it, which can be reproduced and used many times. Substance circulations and energy transformation in nature obey the fundamental laws of thermodynamics. At the same time, live organisms are open non-equilibrium macrosystems with gradients of the concentration of chemical substances, temperature, pressure, and electric potential. The possibility of spontaneous emergence of ordered dissipative structures owing to cooperative movement of large groups of molecules is shown in the works by Prigogine, Glansdorff et al., who have created the bases of nonlinear non-equilibrium thermodynamics. Such self-organizing processes underlie a new interdisciplinary scientific lead called synergetics. Whence it follows that the emergence of a certain order from disordered systems leads to a seeming contradiction between the second law of thermodynamics and the ordering of biological systems (entropy reduction). However, as was already told, if open systems are expanded by adding the environment to them, the second law of thermodynamics becomes applicable to such a continuum.

So, to prove that life in the environment does not contradict the second law of thermodynamics, it is necessary to proceed from that: (i) entropy has a thermodynamic and (or) statistical sense only:

$$S_2 - S_1 = \int (dQ/T)_{rev}; \; S = k \ln W,$$

where the subscript *rev* indicates the reversible nature of the process; dQ is the change of the heat quantity; T temperature; k Boltzmann's constant; W the number of the microstates realizing the current macrostate of the system; and (ii) according to the second law of thermodynamics, any evolving system aspires to an equilibrium state, in which all its physical parameters accept constant values; the state of equilibrium is reached when the entropy of the "system–environment" set becomes maximum.

In recent years, achievements of the physics of biopolymers have promoted the development of the latest representations of the possibility to study biological systems and their evolution by means of classical thermodynamic; techniques (Danielle, Denbigh, Mitchell, Gladyshev, etc.). The thermodynamic theory of life system evolution [1–8, 11] gives answers to the questions: why evolution proceeds? Where it is directed to? What is its motive power?

By now, a rather large amount of experimental data is accumulated to specify that phylogenesis, ontogenesis, and biological evolution as a whole have a strongly pronounced thermodynamic orientation. In a qualitative relation, the sense of evolution of live beings can be understood from the viewpoint of thermodynamics within the *structure formation model* (structures of highest hierarchies emerge from those of lowest hierarchies). According to this model, every natural hierarchical structure has the energy of its formation, average lifetime (relaxation), biomass volume, etc. This allows studying live systems as quasi-closed ones in some approximation. The common material at building of hierarchical structural levels in the course of biological evolution is chemical substance. Therefore, the physical theory of evolution of live beings is based on analysis and generalization of experimental data connected with transformations of separate chemical substances in the processes of transition from a simple structure to a more complex one. The most noteworthy facts consist in the following.

1. During their development, such biosystems as an organella, a cell, an organism's tissue, the biomass of a population, etc., are enriched with power-rich chemical substances to force out water. Therefore, the kinetic features of an organism's aging correlate with the changes of water and fat in it. The same concerns the water content in the brain of a live being (depending on the extent of its relative development). Despite the growth of the power consumption of the biosystem, its chemical stability decreases, and the thermodynamic stability of the supramolecular structures of the body tissues increases. It is also necessary to note that the quantity of water, organic and inorganic substances in biotissues are directly related with the values of the Gibbs specific functions G_{im} and G_{ch} of formation of supramolecular and chemical structures, respectively.

2. Data on the changes of the temperature of melting (denaturation) of biotissues in the processes of ontogenesis and phylogenesis are known.

3. Weber—Fechner's law relates the reactions of biological systems to external influences of the environment (light, noise, ionizing radiation, mechanical tension, psychostresses, etc.).

4. A correlation between the characteristic sizes of structures of various hierarchies and the average duration of their life in a biomass has been experimentally established: lower-hierarchy structures live much shorter than higher-hierarchy ones of in biosystems

(Fig. 1). As was shown above, atoms, common molecules, macro-molecules, microbodies, cells, tissues … an organism, population etc. are meant under hierarchical structures. This, in essence, means that the evolution of lower-hierarchy structures proceeds much quicker than that of the organism itself, whose biomass serves the environment for its atoms, molecules, cells, etc. Just in this connection, for every biosystem it is possible to isolate its habitation medium, i.e., a peculiar thermostat with certain parameters. Such quasi-closed subsystems possessing their own thermostats, uniting, form a thermodynamic system of a higher hierarchical level.

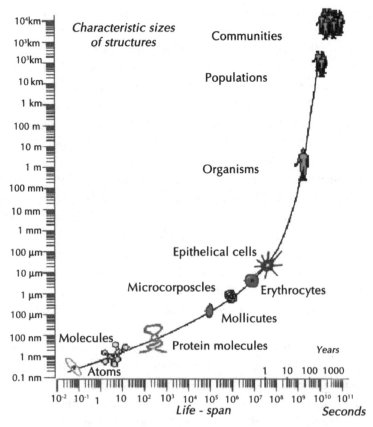

FIGURE 1 Correlation between the characteristic sizes of the structures of various hierarchies and their lifetime in the biomass. The dependence is expressed in logarithmic scales.

The main idea of the thermodynamic model of evolution of live beings, uniting all the listed facts, consists in that the Gibbs specific function of formation of a structure (ΔG) serves a quantitative measure of all changes (including the thermodynamic stability of the chemical, supramolecular and other structures). From the offered model, the existence of a correlation follows between the Gibbs specific function of formation of chemical substances from atoms and molecules and the specific Gibbs function of formation of supramolecular structures from these substances. Therefore, the changes in G at formation of any supramolecular structure, including a population, serve as a measure of the extent of its evolutionary transformation.

Therefore, the life expectancy of biostructures (τ), arising on aggregation of chemical substances, supramolecular and other formations, is determined by the change of the Gibbs function G_{im} (P = const, T = const) at formation of these structures:

$$\tau \sim \exp(\Delta G_{im} / RT),$$

where the symbol "–" means the specific character of G, the subscript im belongs to intermolecular interactions. According to the model,

$$\Delta G_{im} \to min.$$

Generalizing the foregoing, it is possible to present the model of biological evolution as given as follows. Under the influence of solar energy, thermodynamically stable substances convert into various products of photosynthesis, which, as a result of spontaneous "dark reactions" and according to the laws of chemical thermodynamics, are transformed into various substances. Further, under the laws of the so-called local thermodynamics of supramolecular processes (owing to the aspiration of the Gibbs specific function of biostructures to its minimum), more stable suprastructures are selected, which are accumulated in microvolumes and macrovolumes of systems. First of all, nucleic acids are selected, whose structure and composition (because of the action of thermodynamic factors) slowly adapt to the environment, including the nature of proteins, whose structure is determined by DNA itself. This explains the feedback existence (having a thermodynamic basis as well) between the structures of proteins and DNA. Concurrently to synthesis processes, the disintegration processes of chemical compounds proceed. However, live systems resist to it and aspire to preserve their state, which also has the thermodynamic nature, i.e., biosystems reproduce perishing supramolecular structures, and thermodynamics promotes selecting the most stable of them. So, it is thermodynamically favorable for macromolecular chains to couple with similar chains and to surround itself with the updated "young" substance of live organisms. Evolution selects those thermodynamically preferable ways of processes, which promote cell fission and DNA preservation. All this proceeds on the background of parameter fluctuations of the environment (thermostats), which, together with other factors, provides life maintenance.

The physical essence of the model of evolution of biological systems can be understood on an example of the interconnected processes of formation (polymerization) and destruction (depolymerization) of such a known polymer as poly(acrylonitrile). If one analyzes the whole sequence of the processes including heterogeneous formation of solid poly(acrylonitrile) (accompanied by precipitation), its depolymerization (under the influence of light), polymerization of the again formed monomer etc., then, in a thermodynamic relation, these processes (of course, in a certain approximation) model the development of an organism and its evolution.

The proposed thermodynamic theory of biological evolution has a number of very interesting and important consequences for both science development and human life. One of them concerns the predefiniteness of the functions of DNA macromolecules in the emergence and development of live beings. If one assumes the thermodynamic orientation of life evolution, three conditions are necessary for DNA function realization, namely: (i) the existence of an aqueous medium, (ii) the chemical stability of AT and CG pairs, iii) surrounding of the DNA double helix with thermodynamically stabile supramolecular structures. Therefore, the structure of the double helix in comparison with other supramolecular formations features an increased thermodynamic stability (expressed by the value of G_{im}), which causes the accuracy of replication and the high resistance to mutations. Just this "thermodynamic conservatism" of DNA determines the role of this biopolymer as the keeper of genetic information. Another major consequence of applied character consists in the possibility not only to estimate physiological age by means of Gibbs's function but also to regulate it. The average life expectancy of an organism is determined by its hereditary signs and physicochemical and social conditions of the environment. Every instant of life (age) corresponds to the state of the organism characterized by a certain G_{im} value. Therefore, any change in the conditions of the environment leads to a change in the Gibbs specific function upwards or downwards. It is clear that at reduction of the value of Gibbs' function (due to the transition of the biological system from one state to another one or, from the viewpoint of thermodynamics, from one thermostat to another one), an increase in the average life expectancy should follow.

The possibility of regulation of the physiological age of an organism is due to the correlation between the chemical and supramolecular components of the Gibbs specific function of formation of food biomass. In particular, if food possesses a lower caloric content, the value of Gibbs' specific function becomes more and more positive (or less negative). On the other hand, at an identical caloric content of food, depending on the nature of the consumed biomass, various effects of "rejuvenescence" are reached. The gerontological action of some medicines can be explained (at a quantitative level) similarly.

Thus, it is possible to make some conclusions of both fundamental and practical character on the basis of the modern concepts of the thermodynamics and macrokinetics of live matter.

1. Life in the Universe (including Earth) arises and develops in full accordance with the universal laws of nature, in particular, the preservation laws, the second law of thermodynamics, and the law of temporary hierarchies.

2. All biological systems obey *the substance stability principle*, whose sense is that any structural unit of live matter (an atom, a molecule,..., a cell,..., a population...) is potentially thermodynamically limited in its interaction with doth the structures of its "own" hierarchy and the structures of adjacent hierarchies. One of this principle's consequences is the inverse relationship between the stability of chemical bonds in molecules and the stability of the corresponding intermolecular bonds (characteristic of supramolecular structures).

3. The evolution processes of animate nature, consisting of quasi-closed systems close to equilibrium, can be studied by the methods

of phenomenological thermodynamics. This is confirmed by many experimental data which point to that live systems can participate in evolution not only under the influence of expansion work but also under the influence of such types of work (energy) as chemical, electromagnetic, gravitational, osmotic, etc.

4. Special "thermodynamically focused" diets promote reduction of the aging rate of live beings, including the man, and improvement of the quality of their life. *The anti-gerontological indicator* of any foodstuff (of either vegetative or animal origin) is caused by its chemical composition and the supramolecular structure contained therein, and also the ontogenetic and phylogenetic age of this product and the source of its origin (features of the environment). The quantitative characteristic of *the gerontological healthiness* of this or that foodstuff is, first of all, the minimum value of the specific Gibbs function of formation of its supramolecular structure.

5. The basic recommendations on dietary nutrition, directed to reduction of the risk of oncological diseases and their treatment, consist in the following:

    – a sharp decrease in the use of sugar, milk, and dairy products;
    – an increased quantity of green vegetables and some fruits of the bean family (soy, haricot, peas, etc.);
    – certain restriction of the consumption of meat with preference to be given to cold water fish species (herring, tuna, cod, trot, salmon, etc.);
    – of the highest gerontological and anti-cancer value are some food oils with a low hardening temperature whose value depends on the age and habitat medium of plants (minus 15°C – minus 30°C) (hempseed, linseed, sea-buckthorn, cedar, sunflower-seed, soybean oils);
    – a sharp (threefold, fivefold, and even tenfold) increase in the consumption of the daily dose of vitamins C, E, PP (e.g., nicotinamide), a, the B group, zinc and selenium.

KEYWORDS

- **Dark reactions**
- **Entropy**
- **Gerontological healthiness**
- **Open system environment**
- **Rejuvenescence**
- **Temporary hierarchies**
- **Thermodynamics**

REFERENCES

1. Gladyshev, G. P. In *Thermodynamics and Macrokinetics of Natural and Hierarchical Processes*. Moscow, Russia, 1988.
2. Gladyshev, G. P. In *Supramolecular Thermodynamics as a Key to Comprehension of the Phenomenon of Life. What is Life from the Physicochemist's View*. Moscow, Russia, 2003.
3. Gladyshev, G. P. *Biol. Bull.*, **2002**, *29*(1), 1–4.
4. Gladyshev, G. P. *Electron. J. Math. Phys. Sci.*, **2002**, Sem. *2*, 1–15.
5. Gladyshev, G. P. *Progress in Reaction Kinetics and Mechanism* (*Int. Rev. J. UK, USA*), **2003**, *28*(2), 157–188.
6. Gladyshev, G. P. *Adv. Gerontol., Russia*, **2004**, *13*, 70–80.
7. Gladyshev, G. P. *Int. J. Mod. Phys., B.*, **2004**, *18*(6), 801–825.
8. Gladyshev, G. P. *Adv. Gerontol., Russia*, **2005**, *16*, 21–29.
9. Goldfein, M. D.; Karnaukhova, L. I. *Principles of the Physics of Synthetic and Natural Macromolecules*. Saratov, Russia, 1998.
10. Goldfein, M. D.; Ivanov, A. V.; Malikov, A. N. *Concepts of Modern Natural Sciences*. Moscow, Russia, 2009.
11. Goldfein, M. D.; Ursul, A. D.; Ivanov, A. V.; Malikov, A. N. *Fundamentals Natural – Scientific Image of the World*. Saratov, Russia, 2011.

CHAPTER 2

NANOTHERAPEUTICS: A NOVEL APPROACH OF TARGET BASED DRUG DELIVERY

ANAMIKA SINGH and RAJEEV SINGH

CONTENTS

2.1 Introduction 18
2.2 Properties of Nanotherapeutics 18
    2.2.1 Types of Nanoparticles 19
        2.2.1.1 Metal-Based Nanoparticles 19
        2.2.1.2 Lipid-Based Nanoparticles 19
        2.2.1.3 Polymer-Based Nanoparticles 20
        2.2.1.4 Biological Nanoparticles 20
2.3 Nanocarriers in Drug Delivery 20
    2.3.1 Phagocytosis Pathway 21
    2.3.2 Non-Phagocytic Pathway 22
2.4 Clathrin-Mediated Endocytosis 22
2.5 Caveolae-Mediated Endocytosis 23
2.6 Macropinocytosis 24
2.7 Conclusion 25
Keywords 25
References 26

2.1   INTRODUCTION

Nanotechnology expands itself as one of the major area of science [1, 2]. An important and exciting aspect of nanomedicine is the use of nanoparticle for target specific drug delivery systems and it allows a new innovative therapeutic approaches. Due to their small size, these drug delivery systems are promising tools in therapeutic approaches such as selective or targeted drug delivery towards a specific tissue or organ. It also enhances drug transport across biological barriers and intracellular drug delivery. Nanotherapeutic agents work successfully against traditional drugs, which exhibits very less bioavailability and are often associated with gastrointestinal side effects. Nanotherapeutics improves the delivery of drugs that cannot normally be taken orally and it also improves the safety and efficacy of low molecular weight drugs. It also improves the stability and absorption of proteins that normally cannot be taken orally.

2.2   PROPERTIES OF NANOTHERAPEUTICS

A good therapeutic agent or nanoparticle should have ability to: (1) cross one or various biological membranes (e.g., mucosa, epithelium, endothelium) before (2) diffusing through the plasma membrane to (3) finally gain access to the appropriate organelle where the biological target is located.

Different methods of Drug delivery:

1. Oral Drug Delivery
2. Injection Based Drug Delivery
3. Transdermal Drug Delivery
4. Bone Marrow Infusion
5. Control Release Systems
6. Targeted Drug Delivery:

    a. Therapeutical Monoclonal
    b. Antibodies
    c. Liposomes
    d. Microparticles

     e. Modified Blood Cells

     f. Nanoparticles

7. Implant Drug Delivery System

2.2.1   TYPES OF NANOPARTICLES

2.2.1.1   METAL-BASED NANOPARTICLES

Metallic nanoparticles are either spherical metal or semiconductor particles with nanometer-sized diameters. Similar to nanoparticles "nanocrystal" (NC) or "nanocrystallite" term is used for nanoparticles made up of semiconductors. Generally semiconductor materials from the groups II-VI (CdS, CdSe, CdTe), III-V (InP, InAs) and IV-VI (PbS, PbSe, PbTe) are of particular interest. Metal Based nanoparticle are used for nanoscale transistors, biological sensors, and next generation photovoltaics [3].

2.2.1.2 LIPID-BASED NANOPARTICLES

Lipid based nanoparticles are widely used for drug targeting and drug delivery. Solid lipid nanoparticles (SLN) introduced in 1991 represent an alternative carrier system for drugs [4]. Nanoparticles made from solid lipids are attracting major attention as novel colloidal drug carrier for intravenous applications as they have been proposed as an alternative particulate carrier system. SLN are sub-micron colloidal carriers ranging from 50 to 1000 nm, which are composed of physiological lipid [5–7].

Lipid based system are many used because

1. Lipids enhance oral bioavailability and reduce plasma profile variability.
2. Better characterization of lipoid excipients.
3. An improved ability to address the key issues of technology transfer and manufacture scale-up.

2.2.1.3   POLYMER-BASED NANOPARTICLES

Polymeric nanoparticles are nanoparticles, which are prepared from polymers. Polymeric nanoparticles forms (1) the micronization of a material into nanoparticles and (2) the stabilization of the resultant nanoparticles [8]. As for the micronization, one can start with either small monomers or a bulk polymer. The drug is dissolved, entrapped, encapsulated or attached to a nanoparticles and one can obtain different nanoparticles, nanospheres or nanocapsules according to methods of preparation [9]. Gums, Gelatin Sodium alginate Albumin are used for polymer based drug delivery. Polymeric nanoparticles are prepared by Cellulosics, Poly(2-hydroxy ethyl methacrylate), Poly(N-vinyl pyrrolidone), Poly(vinyl alcohol), Poly(methyl methacrylate), Poly(acrylic acid), Polyacrylamide, Poly(ethylene-co-vinyl acetate) like polymeric materials. Polymer used in drug delivery must have following qualities like it should be chemically inert, non-toxic and free of leachable impurities [10].

2.2.1.4   BIOLOGICAL NANOPARTICLES

For proper interaction with molecular targets, a biological or molecular coating or layer is required which act as a bioinorganic interface, and it should be attached to the nanoparticle for proper interaction. These biological coating may be antibodies, biopolymers like collagen [11]. A layer of small molecules that make the nanoparticles biocompatible are also used as bioinorganic interface [12]. For detection and analysis these nanoparicles should be optically active and they should either fluoresces or change their color in different intensities of light.

2.3   NANOCARRIERS IN DRUG DELIVERY

Nanoparticles and Liposomes get absorbed by blood streams for its mode of action. For particular function theses nanoparticles must have specific structure and composition. These particles rapidly get cleared from blood stream by macrophages of reticuloendothelial system (RES) [13, 14].

Nanoparticles can be delivered to targets by Phagocytosis pathway and Non-Phagocytosis pathway

2.3.1 PHAGOCYTOSIS PATHWAY

Phagocytosis occurs by profession phagocytes and by nonprofessional phagocytes. A professional phagocyte includes macrophages, monocytes, neutrophils and dendritic cell, while non-professional phagocytes are fibroblast, epithelial and endothelial cell [15, 16].

Phagocytosis takes place by:

1. Recognition of the opsonized particles in the blood stream.
2. Adhesion of particles to macrophages
3. Ingestion of the particle

Opsonization of the nanoparticlesis a major step and it takes place before phagocytosis. In this process nanoparticles get tagged by a protein known as psonins. Due to this tagging nanoparticles make a visible complex for macrophages, this whole process takes place in blood streams. Opsonins are immunoglobins (Ig) G, M and complement components like C3, C4 and C5 [17, 18].

These activated particles then get attached to macrophages and this interaction is similar to receptor-ligand interaction [15, 19]. The most important receptor is Fc receptor (FcR) and its complement receptor (CR). This receptor attachment is mediated by Rho family GTPase and surface extension occurs, which leads to formation of pseudopodia around nanoparticles. Small actin fibers get assembled in it and engulfment of nanoparticles takes place [20, 21].

As actin is depolymerized from the phagosome, the newly denuded vacuole membrane becomes accessible to early endosomes [15, 22]. Through a series of fusion and fission events, the vacuolar membrane and its contents will mature, fusing with late endosomes and ultimately lysosomes to form a phagolysosome (Fig. 1). The rate of these events depends on the surface properties of the ingested particle, typically from half to several hours [15]. The phagolysosomes become acidified due to the vacuolar proton pump ATPase located in the membrane and acquire many enzymes, including esterases and cathepsins [16, 23].

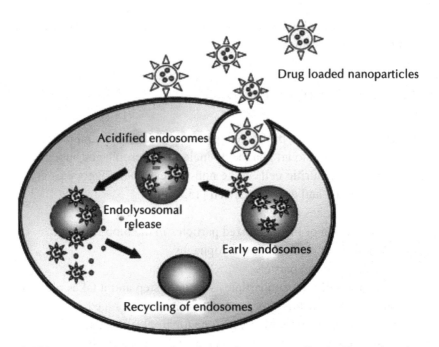

FIGURE 1   Nanoparticles enter the cell through receptor-mediated endocytosis and get localized in the endolysosomal compartment.

2.3.2   NON-PHAGOCYTIC PATHWAY

This is also known as pinocytosis or cell drinking method. It is basically taking of fluids and solutes. Very small nanoparticles get entered in well and absorbed by pinocytosis, as phagocytosis is restricted to size of particle. The process of Pinocytosis occurs in specialized cell. It may be clathrin mediated endocytosis, caveolae-mediated endocytosis, macropinocytosis and caveolae-independent endocytosis.

2.4   CLATHRIN-MEDIATED ENDOCYTOSIS

Endocytosis via clathrin-coated pits, or clathrin-mediated endocytosis (CME), occurs constitutively in all mammalian cells, and fulfills crucial physiological roles, including nutrient uptake and intracellular communication. For most cell types, CME serves as the main mechanism of internalization for macromolecules and plasma membrane constituents. CME via specific receptor-ligand interaction is the best described mechanism, to the extent that it was previously referred to as "receptor-mediated endocytosis" (RME). However, it is now clear that alternative non-specific endocytosis via clathrin-coated pits also exists (as well as receptor-mediated but clathrin-independent endocytosis). Notably, the CME, either receptor-dependent or independent, causes the endocytosed material to end up in degradative lysosomes. This has an important impact in the drug delivery field since the drug-loaded nanocarriers may be tailored in order to become metabolized into the lysosomes, thus releasing their drug content intracellularly as a consequence of lysosomal biodegradation.

2.5   CAVEOLAE-MEDIATED ENDOCYTOSIS

Although CME is the predominant endocytosis mechanism in most cells, alternative pathways have been more recently identified, caveolae-mediated endocytosis (CvME) being the major one. Caveolae are characteristic flask-shaped membrane invaginations, having a size generally reported in the lower end of the 50–100 nm range [24–27], typically 50–80 nm. They are lined by caveolin, a dimeric protein, and enriched with cholesterol and sphingolipids (Fig. 2). Caveolae are particularly abundant in endothelial cells, where they can constitute 10–20% of the cell surface [26], but also smooth muscle cells and fibroblasts. CvMEs are involved in endocytosis and trancytosis of various proteins; they also constitute a port of entry for viruses (typically the SV40 virus) [28] and receive increasing attention for drug delivery applications using nanocarriers.

Unlike CME, CvME is a highly regulated process involving complex signaling, which may be driven by the cargo itself [25, 26]. After binding to the cell surface, particles move along the plasma membrane to caveolae invaginations, where they may be maintained through receptor-ligand interactions [25]. Fission of the caveolae from the membrane, mediated by the GTPase dynamin, then generates the cytosolic caveolar vesicle, which does not contain any enzymatic cocktail. Even this pathway is employed by many pathogens to escape degradation by lysosomal enzymes. The use of nanocarriers exploiting CvME may therefore be advantageous to by-pass the lysosomal degradation pathway when the carried drug (e.g., peptides, proteins, nucleic acids, etc.) is highly sensitive to enzymes. On the whole, the uptake kinetics of CvME is known to occur at a much slower rate than that of CME. Ligands known to be internalized by CvME include folic acid, albumin and cholesterol [25].

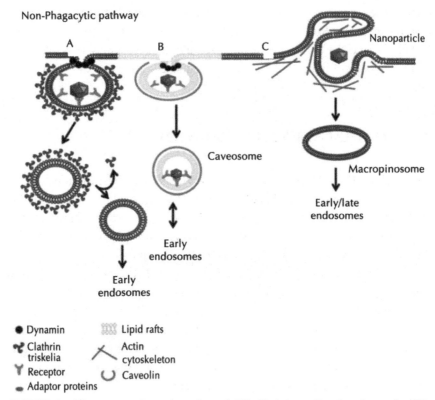

FIGURE 2   Diagrammamtic presentation of (A) Clathrin-mediated endocytosis; (B) Caveolae-mediated endocytosis; (C) Macropinocytosis.

2.6 MACROPINOCYTOSIS

Macropinocytosis is another type of clathrin-independent endocytosis pathway [29], occurring in many cells, including macrophages [24]. It occurs via formation of actin-driven membrane protusions, similarly to phagocytosis. However, in this case, the protusions do not zipper up along the ligand-coated particle; instead, they collapse onto and fuse with the plasma membrane [26] (Fig. 2). This generates large endocytic vesicles, called macropinosomes, which sample the extracellular milieu and have a size generally bigger than 1 lm [26] (and sometimes as large as 5 lm [24]). The intracellular fate of macropinosomes vary depending on the cell type, but in most cases, they acidify and shrink. They may eventually fuse with lysosomal compartments or recycle their content to the surface [24] (Fig. 2). Macropinosomes have not been reported to contain any specific coating, nor do they concentrate receptors. This endocytic pathway does not seem to display any selectivity, but is involved, among others, in the uptake of drug nanocarriers [29].

2.7 CONCLUSION

Now-a-days n-number of drugs are available in market, so nanotherapeutics is particularly very important to minimize health hazards caused by drugs. The most important aspect of nanotherapeutics is, it should be non-invasive and target oriented, it should have very less or no side effects. In-spite of using nonmaterial one can create nanodevices for better functioning. As most of the peptides and proteins have short half-lives so frequent dosing or injections are required, which is not applicable for many situations. By using nanotherapeutics one can solve these problems. Thus, Nanotherapeutics is expanding day by day in drug delivery system.

KEYWORDS

- **Drug delivery system**
- **Nanocrystallite**
- **Nanotherapeutics**
- **Receptor-mediated endocytosis**

REFERENCES

1. Ozin, G. A.; Arsenault, A. C. In Nanochemistry; RSC Publishing: Cambridge, 2005.
2. McFarland, A. D.; Haynes, C. L.; Mirkin, C. A. J. Chem. Ed. **2004**, 81, 544A.
3. Mukherjee, S.; Ray, S.; Thakur, R. S. Ind. J. Pharm. Sci., **2009**, 15, 349–358.
4. Abdul Hasan, A. S.; Priyanka, K. Sci. Revs. Chem. Commun., **2012**, 2(1), 80–102.
5. Mozafari, M. R. Nano Carrier Technol.: Frontiers Nano Ther., **2006**, 8, 41–50.
6. Li, H.; Xiaobin, Z.; Ma, Y.; Zhai, G.; Li, L. B.; Xiang, H. Lou. J. Cont. Rel., **2009**, 133, 238–244.
7. Uner, M.; Yener, G. Int. J. Nanomed., **2007**, 2(3), 289–300.
8. Anjali, K. Nanotherapeutics in Drug Delivery-A Review, Pharmatutor-ART-1291.
9. Mua, L.; Fenga, S.S. J. Controlled Release, **2003**, 86, 33–48.
10. Lu, X. Y.; Wu, D. C.; Li, Z. J.; Chen, G. Q. *Prog. Mol. Biol. Transl. Sci.*, **2011**, *104*, 299–323.
11. Sinani, V. A.; Koktysh, D. S.; Yun, B. G.; Matts, R. L.; Pappas, T. C.; Motamedi, M.; Thomas, S. N.; Kotov, N. A. *Nano Lett.*, **2003**, *3*, 1177–1182.
12. Zhang, Y.; Kohler, N.; Zhang, M. *Biomaterials*, **2002**, *23*, 1553–1561.
13. Gregoriadis, G. *Nature*, **1978**, *275*, 695–696.
14. Grislain, L.; Couvreur, P.; Lenaerts, V.; Roland, M.; Deprezdecampeneere, D.; Speiser, P. *Int. J. Pharm.*, **1983**, *15*, 335–345.
15. Aderem, A.; Underhill, D.; *Annu. Rev. Immunol.*, **1999**, *17*, 593–623.
16. Rabinovitch, M. *Trends Cell. Biol.*, **1995**, *5*, 85–87.
17. Vonarbourg, A.; Passirani, C.; Saulnier, P.; Benoit, J. *Biomaterials*, **2006**, *27*, 4356–4373.
18. Owens, D.; Peppas, N. *Int. J. Pharm.*, **2006**, *307*, 93–102.
19. Groves, E.; Dart, A.; Covarelli, V.; Caron, E. *Cell Mol. Life Sci.*, **2008**, *65*, 1957–1976.
20. Vachon, E.; Martin, R.; Plumb, J.; Kwok, V.; Vandivier, R.; Glogauer, M.; Kapus, A.; Wang, X.; Chow, C.; Grinstein, S.; Downey, G.; *Blood*, **2006**, *107*, 4149–4158.
21. Caron, E.; Hall, A. *Science*, **1998**, *282*, 1717–1721.
22. Swanson, J. A.; Baer, S. C.; *Trends Cell Biol.*, **1995**, *5*, 89–93.
23. Claus, V.; Jahraus, A.; Tjelle, T.; Berg, T.; Kirschke, H.; Faulstich, H.; Griffiths, G. *J. Biol. Chem.*, **1998**, *273*, 9842–9851.
24. Mukherjee, S.; Ghosh, R. N.; Maxfield, F. R. *Physiol. Rev.*, **1997**, *77*, 759–803.
25. Bareford, L. M.; Swaan, P. W. *Adv. Drug Deliv. Rev.*, **2007**, *59*, 748–758.
26. Conner, S. D.; Schmid, S. L. *Nature*, **2003**, *422*, 37–44.
27. Mayor, S.; Pagano, R. E. *Nat. Rev. Mol. Cell Biol.*, **2007**, *8*, 603–612.
28. Swanson, J. A.; Watts, C. *Trends Cell Biol.*, **1995**, *5*, 424–428.
29. Racoosin, E. L.; Swanson, J. A. *J Cell Sci.*, **1992**, *102*, 867–880.

CHAPTER 3

PHOSPHORUS-CONTAINING POLYPEPTIDES

VICTOR KABLOV, DMITRY KONDRUTSKY, MARIA SUDNITSINA,
LIDIYA ZIMINA, and MARINA ARTSIS

CONTENTS

3.1 Introduction 28
3.2 Experimental 28
3.3 Results and Discussion 31
    3.3.1 Synthesis of Initial Peptides 31
    3.3.2 The Phosphorus-Containing Polypeptides Reaction
           Synthesis Investigation 31
    3.3.3 Investigation of Synthesis Conditions Influence to
           Phosphorus-Containing Polypeptides Properties 32
    3.3.4 IR-Spectroscopy of Phosphorus-Containing
           Polypeptides 35
    3.3.5 Investigation of Sorption and Ion-Exchange
           Activity of Phosphorus-Containing Polypeptides 36
    3.3.6 Differential Thermal Analysis of Phosphorus-Containing
           Polypeptides 38
3.4 Conclusion 40
Keywords 40
References 40

3.1 INTRODUCTION

The raw materials amounts reduction, such as oil and gas, and the environment pollutions in view of significant volume of the non-biodegradable wastes pose an actual problem of use of the raw materials of renewed sources [1].

One of possible line of investigations is a synthesis of phosphorus-containing polymers of natural proteins. They possess sorption and ion-exchange activity, reduced combustibility, biodegradability etc.

In the present work results of phosphorus-containing polypeptides on the base of some substrates (peptides, gelatin and fish collagen) synthesis research are presented. The chemical structure of a product, sorption properties and thermal-oxidative degradation are investigated.

3.2 EXPERIMENTAL

The synthesis of peptides by enzymatic hydrolysis of proteins was carried out with gas-vortex gradientless bioreactor of type BIOK-022. As enzymes were used pepsin and trypsin. Initial concentration of the protein substrates was 20% on weight. Temperature of the synthesis was 36-40°C. Synthesis was carried out during 1 h. Received peptides molecular weight was 800–1100 g/mol of titration method with formalin [2].

The synthesis phosphorus-containing polypeptides (polyproteinyl-methenephosphinic acids) on base of peptides were carried out by the reaction phosphonomethylation. Sodium hypophosphite or phosphinic acid were added into 20% by weight water solution of peptides under agitation at 45–60°C. Then 36.7% formalin was added in the reaction solution. pH was controlled by litmus paper and one was supported in the range 3–4 by addition of hydrochloric acid. Molar ratio of peptides, formaldehyde and sodium hypophosphite (phosphinic acid) was varied in the range from 1:2:1 to 1:8:4, accordingly. Reaction was stopped by means of quick water dilution of the mixture and fast separation the insoluble products from the mixture by vacuum filter. Then the synthesized product was washed at first distilled water, then 0.1 normal solution of hydrochloric acid, again water, after that 0.1 normal solution of sodium hydroxide and again water until neutral pH. Remaining water in the product was removed by vacuum filter.

If the product was soluble that it was separated from the solution by sodium sulfate injection in the mixture with next double-ply processes soluble one in water and dissoluble in present of sodium sulfate for purpose to wash from reaction mixture leavings.

The synthesis phosphorus-containing polypeptides from fish collagen were carried out the same manner. Into 20% by dry weight water suspension of fish collagen was added the known reagents (formaldehyde and sodium hypophosphite or phosphinic acid) in the molar relations to dry collagen from 1:8:3 to 1:10:20, accordingly. pH was supported in the range 3–4. The reaction was stopped by addition 10 ml of 2 normal solution of NaOH. The product was separated and in series washed in vacuum filter by means of 0.1 normal solution of hydrochloric acid, then water, after that 0.1 normal solution of sodium hydroxide and again water until neutral pH. Remaining water in the product was removed by vacuum filter.

The template synthesis of increase capacity toward cations Cu^{2+} sorbents were carried out by addition into water suspension of fish collagen 50% cuprum sulfate water solution in mass relation of fish collage to $CuSO_4$ as 5:1. Then the synthesis was conducted in same previous manner. For more effective desorption residual cations Cu^{2+} from the polymeric die was applied double consecutive wash by 0.1 normal HCl solution, 0.1 normal NaOH solution and water until neutral pH.

A photocolorimeter FEC-56PM was used to quantitative chemical analysis to phosphorus (GOST 208512-75) [3]. A SPECORD-M82 was used to directly characterize the phosphorus-containing polypeptides. IR spectrum was performed for the product in form of thin suspension into vaseline.

Static ion-exchange capacity (SEC) toward cations Cu^{2+} versus pH was investigated for about 1 g polymer samples, which had been consecutive washed 0.1 normal solution HCl, 0.1 normal solution NaOH and next water. This treatment converts the polymer into "Na-form." The samples were dipped into 30-repeated abundance by weight 0.1 normal solution cuprum sulfate and were got necessary meaning of pH by means of addition HCl or NaOH. The samples were kept at 25°C for 3 days.

An electronic pH-meter was used to control pH of reactionary solutions. One included a glass electrode, an argentums chloride electrode and a thermo compensative electrode. They were plugged on an universal potentiometer EV-74.

The residual concentration of Cu^{2+} cation of solution was investigated after time of the sorption by the photocolorimeter FEC-56PM by use of ammonia method [4].

The SEC of polymer was calculated in agreement with expression:

$$SEC = \frac{\left(\tilde{n}_n^0\left(Cu^{2+}\right) - \tilde{n}_n^1\left(Cu^{2+}\right)\right) \cdot V_{sorb}}{m_{pol}}, mmol/g$$

where $c_n^0\left(Cu^{2+}\right)$ – initial normal concentration Cu^{2+} cation of solution before sorption, mol/l;

$c_n^1\left(Cu^{2+}\right)$ – final normal concentration Cu^{2+} cation of solution after sorption, mol/l;

V_{sorb} – volume of solution for sorption, ml;

m_{pol} – mass of phosphorus-containing polypeptides sample, g.

Water adsorption of initial phosphorus-containing polypeptides samples was calculated in agreement with expression:

$$\varphi = \frac{m_{wet} - m_{dry}}{m_{wet}} \cdot 100\%$$

where m_{wet} – mass of phosphorus-containing polypeptides sample ("Na-form") after consecutive wash by 0.1 normal solution HCl, then 0.1 normal solution NaOH and next water until neutral pH. Remaining water in the product was removed by vacuum filter;

m_{dry} – mass of phosphorus-containing polypeptides sample after exsiccation under standard condition.

Differential-thermal analysis (DTA) of product and initial peptides was carried out from 25 to 600°C using a derivatograph "MOM," Hungary. The mass of samples was 100 mg. Rate of heating was 10°C/min. Sensitivity DTA was 1/5°C.

3.3   RESULTS AND DISCUSSION

3.3.1   SYNTHESIS OF INITIAL PEPTIDES

To prepare the initial substances for the synthesis of phosphorus-containing polypeptides, the research of fermentative hydrolysis of gelatin to mixture of low-molecular oligomers (the peptides) was carried out. An action of series of hydrolytic enzymes such as pepsin and trypsin was investigated. These enzymes are high specific catalysts for selective hydrolysis of proteins. Trypsin splits bonds that have been built carboxylic group of lysine or arginine. Using trypsin conducts to sufficiently large variation of molecular mass and increases its value than in case of less specific pepsin. Pepsin requires acid condition (pH = 3–4). That pH was received by addition 0.1 normal solution HCl. Like receiving solution of peptides had acid pH and its neutralization by 0.1 normal solution of sodium hydroxide produced formation sodium chloride in reaction system that further required additional deionization. All of these facts decrease technological attraction of pepsin using.

In case of using trypsin the process was carried out in neutral or weakly basic conditions (pH=7.0–7.8). This fact presents significant advantage because the synthesis is able to carry out without using of other acid-basic regulators. Thus trypsin was selected as enzyme for hydrolysis.

The optimal concentration of gelatin for hydrolysis was 20% by dry weight. In this condition the process was more technologic because medium viscosity wasn't too high and heat exchange was better. The optimal temperature of the process was 36-40°C that well agrees with optimum of enzyme catalytic activity.

3.3.2   THE PHOSPHORUS-CONTAINING POLYPEPTIDES REACTION SYNTHESIS INVESTIGATION

The phosphorus-containing polypeptides synthesis passes via few stages with series intermediate production. Further the intermediates lead to creation cross-linked high-branched polymer what includes derivation of phosphinic acid.

At first peptide methylol-derivations synthesize by adding formaldehyde as more reactionary substance to nucleophilic centers of peptides.

In general view schema of this reaction can be introduced as

$$\text{\textasciitilde\textasciitilde}NH_2 + CH_2O \longrightarrow \text{\textasciitilde\textasciitilde}NH-CH_2-OH$$

peptid   intermediate 1

At subsequent stage methylol-derivations of peptide react to phosphinic acid that conduct to form the product of phosphonomethylation. The scheme of this reaction can be introduced as

$$\text{\textasciitilde\textasciitilde}NH-CH_2-OH + H_3PO_2 \longrightarrow \text{\textasciitilde\textasciitilde}NH-CH_2-\overset{\overset{O}{\|}}{\underset{\underset{H}{|}}{P}}-OH + H_2O$$

intermediate 1   product 1

At final stage of the synthesis a series of reactionary groups couple together and form cross-linked phosphorus-containing polypeptides. The scheme of such reaction can be introduced as

$$\text{\textasciitilde\textasciitilde}NH\text{-}CH_2OH + H-\overset{\overset{O}{\|}}{\underset{\underset{OH}{|}}{P}}-H_2C-HN\text{\textasciitilde\textasciitilde} \longrightarrow \text{\textasciitilde\textasciitilde}NH-CH_2-\overset{\overset{O}{\|}}{\underset{\underset{OH}{|}}{P}}-H_2C-HN\text{\textasciitilde\textasciitilde} + H_2O$$

cross-linked product

3.3.3 INVESTIGATION OF SYNTHESIS CONDITIONS INFLUENCE TO PHOSPHORUS-CONTAINING POLYPEPTIDES PROPERTIES

Influence of different factors such as the relation of reagents, the medium pH and the temperature to quantity of phosphorus in the phosphorus-containing polypeptides was investigated

The experiment results of phosphorus-containing polypeptides synthesis at 21°C for 24 h are presented in Table 1. We can observe that the optimal conditions of synthesis are 3-repeated abundance of the reagents and acid medium.

Table 1 shows that in neutral medium a good deal of phosphor cannot be ingrafted to peptides.

TABLE 1   Investigation of influence of the reagent relations and the medium pH to quantity of phosphorus in the product.

M_r peptides, g/mol	Molar ratio of the reagents			pH	ω(P), %	Comment
	Peptide	CH_2O	H_3PO_2			
	1	2	1		0.04	
	1	3	1,5		0.02	
1400	1	4	2	7	0.03	The form of product is insoluble gel
	1	6	3		0.03	
	1	8	4		0.03	
	1	2	1		0.50	
	1	3	1,5		0.75	
1400	1	4	2	3	0.95	The form of product is viscous liquid
	1	6	3		1.13	
	1	8	4		1.07	

In adding hydrochloric acid and decrease medium pH until 3 quantity of phosphorus in the product increased in a few orders and got at 1.1% at molar relation of peptide:formaldehyde:phosphinic acid as 1:6:3. Like this the degree of phosphonomethylation was about 50%. But such conditions weakness is soluble form of phosphorus-containing polypeptides that confuse them applications as ion-exchange materials.

Investigation of the temperature influences to quantity of phosphorus in the product showed that synthesis conducted at 50°C under similar other conditions to cause to 2.3% of phosphorus in the product. This is adequate ultimate value of the phosphonomethylation degree for peptides having similar molar weight. Using more high temperature is no purpose, because phosphinic acid is able to suffer disproportionation with elimination phosphine under heating. The experiment results are shown in Table 2.

TABLE 2   Investigation of the temperature influences to quantity of phosphorus in the product.

M_r peptides, g/mol	Molar ratio of the reagents			t, °C	pH	$\omega(P)$, %
	Peptide	CH_2O	H_3PO_2			
1400	1	6	3	21	3	1.13
				50		2.3

For receiving of product in the insoluble form, which has high phosphorus quantity, fish collagen was investigated as an initial substrate. The fish collagen has high quantity amino groups, which include to functional lateral radicals of L-lysine and L-arginine.

The fish collagen swells limited in water and produces a suspension, which stables in acid medium. The concentration of amino group according to formalin titration in the fish collagen was 1.1 mol/g. The fish collagen contains in 10% more amino group then the gelatin peptides.

To decrease hydrolysis and swelling the fish collagen during phosphonomethylation reaction sodium sulfate was added in reaction mixture as a graining reagent. The results of the experiment are presented in Table 3.

TABLE 3   Investigation of graining agent (Na_2SO_4) influence to phosphonomethylation reaction.

M_r peptides, g/mol	Molar ratio of the reagents				t^0	τ, h	pH	$\omega(P)$, %
	Na_2SO_4	Peptide	CH_2O	H_3PO_2				
900	1	1	6	3	50	1	3	1.2
	4							0.49
	6							0.34

The experimental data analysis is showed that the product produces in a good spherical granule form under increasing molar ratio of sodium sulfate. But quantity of phosphorus in the product decreased in several times in comparison with ordinary synthesis. In that case application of abundance sodium hypophosphite is more perspective way to decrease the product losses all along of acid hydrolysis during phosphonomethylation. This way lets to obtain insoluble product having high phosphorus quantity.

3.3.4 IR-SPECTROSCOPY OF PHOSPHORUS-CONTAINING POLYPEPTIDES

Infrared spectroscopy was performed for confirmation scheme of the product synthesis.

The results of IR-spectroscopy are presented in Table 4. It was founded the characteristic vibration of groups and bounds, which build by atoms of phosphorus and carbon (the bond P–C) or atoms of phosphorus and oxygen (the bond P=O) in structure of the product.

TABLE 4   The characteristic vibrations of IR-radiation absorption of reagents and product.

Functional group	Peptide	NaH$_2$PO$_2$	Phosphorus-containing polypeptides
		v, cm^{-1}	
–NH$_3^+$	3130-3030	–	3130–3030
–COO⁻	1454	–	1454
=C=O («Amide 1»)	1651	–	1651
=NH и ≡C–N= («Amide 2»)	1538	–	1538
≡C–NH-	1082	–	1082
P–C	–	–	920–930
P=O	–	1182	1153–1168
P–O–C	–	–	–
P–H	–	2320–2350	2320–2350
–OH (polyassociates)	–	–	3600–3200

At the same time signals of absorption for bound P–H disappeared and band for P–O–C was not detected. These facts show that the reaction passes via forming bounds P–C with taking bound P-H as it was shown in the scheme of phosphonomethylation reaction.

3.3.5 INVESTIGATION OF SORPTION AND ION-EXCHANGE ACTIVITY OF PHOSPHORUS-CONTAINING POLYPEPTIDES

The main purposes of this paper were the investigation sorption activity of phosphorus-containing polypeptides, the determination of full (FEC) and static exchange capacity (SEC), the exploration of influence different factors to SEC, first of all the quantity of phosphorus and the water adsorption.

The value of FEC can be calculated as

$$FEC = \frac{k \cdot v(P)}{m_{pol}}, \quad \text{mmol/g}$$

where k – number of dentate of complexing groups, and "k" shows number of ions to one reactionary center

v(P) – amount of substance ion-exchange group, mol

$$v(P) = \frac{m_{pol} \cdot \omega(P)}{M(P) \cdot 100\%}$$

m_{pol} – mass of phosphorus-containing polypeptides, g
$\omega(P)$ – phosphorus abundance,%

If we substitute v(P) into expression for FEC that we take a new expression as

$$FEC = \frac{k \cdot \omega(P) \cdot 1000}{M(P) \cdot 100\%}$$

Thus, for phosphorus-containing polypeptides theoretical determining FEC value was 0.74 mmol/g at maximal possibly quantity of phosphorus about 2.3% and k = 1.

The results of experiments for investigation SEC depending on quantity of phosphorus and value of water adsorption are presented in Table 5.

TABLE 5 Investigation of influence quantity of phosphorus and water adsorption to SEC for phosphorus-containing polypeptides on base of the fish collagen.

Mr peptides, g/mol	ω (P),%	FEC, mmol/g	φ,%	SEC, mmol/g (Cu)	k
900	0.49	0.16	84	1.87	11.7
	0.16	0.05	45	0.54	10.8

The experimental data confirm about increasing SEC under increase of quantity of phosphorus and water adsorption.

The experimental SEC value at 10–11 times more then calculated FEC at adequate ω (P) value. This fact indicates that series additional groups of residual peptides (carboxylic group, amide units and etc.) are involved in sorption process.

For increasing SEC value was investigated processes complexation and chelation of metal ions in the polymeric die to produce centers of complementary for these ions. In this case polyions and molecular sorption are able to pass with ultramicroscopic phase of salt formation into a polymeric die.

By means of the template synthesis at cation copper (II) was made an attempt to creation in the material die of the ligand sites complementary to this ion.

The using of template synthesis permitted to increase SEC at 2–3 times, but led to diminish quantity of phosphorus in the product in few times. It can be explained by inhibition of part of the reactionary centers (amino groups) of the fish collagen because the complexes of amino groups with cations Cu^{2+} form and reduction Cu^{2+} to Cu^{+} passes during the synthesis.

These secondary processes determine different color changes of reactionary mixture and the product.

At pH change ion-contained polymeric die has to transform its conformation that ion-exchanges activity changes too. For phosphorus-containing polypeptides (Na-form) template synthesized with quantity of phosphorus 0.1% decrease pH medium from 4.9 to 2.9 reduces SEC at 50% from 2.9 to 1.9 mmol/g. The experiment results about investigation of influence pH to SEC are presented in Table 6.

TABLE 6 Investigation of pH medium influence to SEC (ω(P)=0.1%) from $CuSO_4$ 0.1 normal solution.

pH	SEC, mmol/g
2.94	1.88
4.41	2.84
4.88	2.90

The registered regularity presents important advantage for phosphorus-containing polypeptides in that it allows reversible to adsorb-desorb desirable ions from aqueous solution at pH medium regulation.

The research alternative ways of the product application or utilization the material waste present actual problems.

The phosphorus-containing polypeptides includes considerable number of reactionary centers (carboxylic group, phosphorus unit, amino group, amide group) which present in well-known self-extinguish material so it is very perspective to research thermal-oxidative material destruction.

3.3.6 DIFFERENTIAL THERMAL ANALYSIS OF PHOSPHORUS-CONTAINING POLYPEPTIDES

The results of differential thermal analysis of phosphorus-containing polypeptides are presented in Figs. 1 and 2.

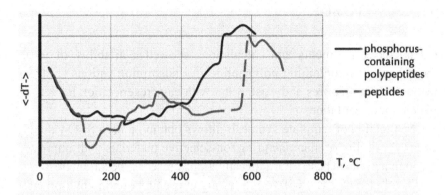

FIGURE 1   Curves of differential thermal analysis of phosphorus-containing polypeptides and initial peptides.

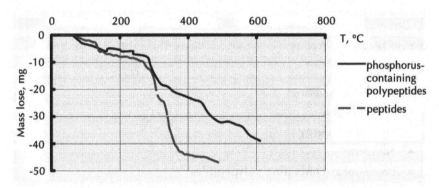

FIGURE 2   Thermo gravimetric curves of phosphorus-containing polypeptides and initial peptides.

The phosphorus-containing polypeptides stable until 270°C. Heat effect of structuring is only in that temperature range. Intensively thermal-oxidative destruction starts at 300°C and passes until 600°C. The phosphorus-containing polypeptides are more stable to thermal influence as experimental data show. That at 500°C mass loss for peptides is 49%, but for the phosphorus-containing polypeptides is only 31%. In total phosphorus-containing polypeptides have reasonably stable to thermal-oxidative influences that high value of coking part mass confirms (about 69%). This fact performs significant advantage and expands areas of application of these materials.

3.4 CONCLUSION

Completed experiments and calculations showed availability of using the phosphorus-containing polypeptides for adsorption cations Cu from technological mixtures and wastewater with easy regeneration by way of pH medium regulation.

Application of template synthesis allows obtaining high SEC value. It opens possibility of manufacturing ion-selective materials with improved sorption activity. These future investigations present especial interest.

Developed material will be interesting in use first of all in hydrometallurgy and in processes of water treatment.

For waste of phosphorus-containing polypeptides are provided possibility conversion in the effective fireproof coverings.

KEYWORDS

- **Fireproof coverings**
- **Fish collagen**
- **Hydrometallurgy**
- **Na-form**
- **Phosphorus-containing polypeptides**

REFERENCES

1. Ana Laura Martinez-Hernandez; Carlos Velasco-Santos; Miguel de Icaza, Victor M. Castano; *e-Polymers* **2003**, *no. 016*, 1.
2. Shapiro D. K. In *Practical Work in Biological Chemistry*; High School: Moscow, 1976.
3. GOST 208512-75.
4. Korostelev P. P. In *Photometric and Complexonometric Analysis in Metallurgy*; Metallurgy: Moscow, 1984.

CHAPTER 4

FREE RADICAL INITIATION IN POLYMERS UNDER THE ACTION OF NITROGEN OXIDES

E. YA. DAVYDOV, I. S. GAPONOVA, T. V. POKHOLOK,
G. B. PARIYSKII, L.A. ZIMINA, M. I. ARTSIS, and G. E. ZAIKOV

CONTENTS

4.1 Introduction.. 42
4.2 Interaction of Nitrogen Oxides with Products of Photolysis and
      Radiolysis of Polymers 43
4.3 Radical Reactions Initiated by Nitrogen Trioxide 46
4.4 Free Radical and Ion-Radical Reactions under Action
      of Nitrogen Dioxide and its Dimers... 48
      4.4.1 Preparation of Spin-Labelled Rubbers 49
      4.4.2 On the Mechanism of Initiation of Radical Reactions by
              Nitrogen Dioxide Dimers 52
4.5 AB Initio Calculations of Energies for Conversions of Nitrogen
      Dioxide Dimmers... 58
4.6 Detecting Radical Cations in Reactions of Electron
      Transfer to NN 60
4.7 Conclusion 62
Keywords 63
References.. 63

4.1   INTRODUCTION

Nitrogen oxides play the important role in various chemical processes proceeding in an atmosphere, and affect on an environment [1–4]. These compounds ejected to atmosphere by the factories and transport in huge amounts creates thus serious problems for fauna and flora as well as synthetic polymeric materials. In this connection, the researches of the mechanism of nitrogen oxide reactions with various organic compounds including polymers are important for definition of stability of these materials in polluted atmosphere. On the other hand, these researches can be used in synthetic chemistry [5–7], in particular for development of methods of the polymer modification, for example, for preparation of spin-labelled macromolecules [8, 9]. The generation of spin labels occurs thus in consecutive reactions including formation and conversion of specific intermediate molecular products and active free radicals. It is necessary to note essential advantages of such way of obtaining spin labels not requiring application of complex synthetic methods based on reactions of stable nitroxyl radicals with functional groups of macroradicals [10]. If the polymers are capable of reacting with nitrogen oxides, the formation of stable radicals in them takes place spontaneously or by thermolysis of molecular products of nitration [11, 12].

Most important for reactions with various organic compounds and polymers are nitrogen oxides of the three types: NO, NO_2, NO_3. All of them represent free radicals with different reactivity [13]. In the present review the features of the mechanism of reactions of these oxides and also NO_2 dimeric forms with a number of polymers and low-molecular compounds are considered. The especial attention is given to the analysis of structure of stable nitrogen-containing radicals and kinetic features of their formation. On the basis of results of the analysis, the conclusions on the mechanism of primary reactions of initiation and intermediate stages of complex radical processes under action of nitrogen oxides are given. The principles of use of nitrogen oxides for grafting spin labels to various polymers are considered.

4.2   INTERACTION OF NITROGEN OXIDES WITH PRODUCTS OF PHOTOLYSIS AND RADIOLYSIS OF POLYMERS

From three nitrogen oxides examined, the radical NO is least reactive. It is not capable of abstracting hydrogen atoms even from least strong tertiary or allyl C–H bonds; the strength of H–NO bond [14] makes only 205 kJ×mol^{-1}. Nitrogen oxide cannot join to isolated double C=C bonds of alkenes [15]. For NO, the recombination with free radicals with formation of effective spin traps (nitroso compounds) is characteristic process. The structure of stable nitrogen-containing radicals formed from nitroso compounds in the subsequent reactions gives information on the mechanism of proceeding radical process in the given reacting system.

During photolysis of polymethylmethacrylate (PMMA) in atmosphere of NO by unfiltered light of a mercury lamp at 298 K, the formation of acylalkylaminoxyl radicals such as $R_1N(O^{\bullet})C(=O)OR_2$ was observed with typical parameters of anisotropic triplet EPR spectrum in a solid phase [16]: $A_{\parallel}^{N} = 2.1 \pm 0.1$ mT and $g_{\parallel} = 2.0027 \pm 0.0005$. If photolysis of the same samples to carry out at 383 K, dialkylaminoxyl radicals $RN(O^{\bullet})R$ in addition to acylalkylaminoxyl radicals are formed with parameters of EPR spectrum: $A_{\parallel}^{N} = 3.2 \pm 0.1$ mT and $g_{\parallel} = 2.0026 \pm 0.0005$. Under action of filtered UV light (260–400 nm) at room temperature there is a third type of stable radicals namely iminoxyls $(R_1R_2)C=NO^{\bullet}$, the triplet EPR spectrum of which in benzene solution of PMMA is characterized by the parameters: $a^N = 2.8$ mT and $g = 2.005$. The occurrence of acylalkylaminoxyl radicals is evidence of eliminating of methoxycarbonyl radicals in a course of the polymer photolysis:

$$\sim(CH_3)C(COOCH_3)CH_2(CH_3)C(COOCH_3)\sim \xrightarrow{h\nu} \sim(CH_3)C(COOCH_3)$$
$$CH_2C^{\bullet}(CH_3)\sim (R^{\bullet}) + {}^{\bullet}COOCH_3 \qquad (1)$$

The subsequent reaction with participation of NO gives acylalkylaminoxyl radicals:

$$^{\bullet}COOCH_3 + NO \rightarrow O=N-COOCH_3 \xrightarrow{+R^{\bullet}} RN(O^{\bullet})COOCH_3 \qquad (2)$$

Dialkylaminoxyl radicals are formed by decomposition of macroradicals $R^{\bullet}$:

$$R^\bullet \xrightarrow{\quad kT \quad} \sim(CH_3)C(COOCH_3)CH_2=C(CH_3)CH_2\sim + {}^\bullet C(CH_3)$$

$$(COOCH_3)\sim \xrightarrow{\quad +NO,R^\bullet \quad} RN(O^\bullet)C(CH_3)(COOCH_3)\sim \qquad (3)$$

Iminoxyl radicals are appeared as a result of generation of macroradicals $\sim(CH_3)C(COOCH_3)C^\bullet H(CH_3)C(COOCH_3)\sim$ ($R_1^\bullet$) which in NO atmosphere are converted into nitroso compounds and further into oximes:

$$R_1^\bullet + NO \rightarrow R_1NO \rightarrow \sim(CH_3)C(COOCH_3)C(=NOH)(CH_3)C(COOCH_3)\sim$$
$$(R_1R_2NOH) \qquad\qquad (4)$$

As a result of the mobile hydrogen atom abstraction from oximes by, for example, methoxycarbonyl radical, iminoxyls are formed:

$$R_1R_2NOH + {}^\bullet COOCH_3 \rightarrow HCOOCH_3 + R_1R_2NO^\bullet \qquad (5)$$

A limiting stage of Eq. (2) is nitrogen oxide diffusion in a polymeric matrix. The rate of Eq. (3) should much more depend on mobility of macromolecular reagents. Therefore, the distinction in composition of radicals in PMMA photolysed at room temperature and 383 K is observed. At room temperature, acylalkylaminoxyl radicals are formed due to accepting low-molecular methoxycarbonyl radicals ${}^\bullet COOCH_3$ by nitroso compounds. At 383K, when molecular mobility essentially grows, dialkylaminoxyl radicals $R_1N(O^\bullet)R_2$ are formed because the meeting two macromolecular particles is provided. The results obtained demonstrate an opportunity of NO use for an elucidation of the polymer photolysis mechanism. With the help of this reactant, it was possible to establish a nature and mechanism of formation of intermediate short-lived radicals in photochemical process using EPR spectra of stable aminoxy radicals.

The application of nitrogen oxide enables to prepare spin-labelled macromolecules in chemically inert and insoluble polymers, for example, in polyperfluoroalkanes. As was shown in the work [16], the radiolysis of oriented films of polytetrafluoroethylene (PTFE) and copolymer of tetrafluoroethylene with hexafluoropropylene initiates reactions with formation of iminoxyl macroradical according to the following scheme:

$$PTFE \xrightarrow{\quad \gamma \quad} \sim CF_2C^\bullet FCF_2 \sim + NO \rightarrow \sim CF_2CF(NO)CF_2 \sim \xrightarrow{\quad \gamma \quad}$$

$$\sim CF_2C^{\bullet}(NO)CF_2 \sim \rightarrow \sim CF_2C(=NO^{\bullet})CF_2 \sim \qquad (6)$$

$$\sim CF_2CF(CF_3)CF_2 \sim \xrightarrow{\gamma,NO} \sim CF_2CF(CF_2NO)CF_2 \sim \xrightarrow{\gamma}$$

$$\sim CF_2CF(C^{\bullet}FNO)CF_2 \sim \rightarrow \sim CF_2CF(CF=NO^{\bullet})CF_2 \sim \qquad (7)$$

The EPR spectrum of iminoxyl radicals in oriented PTFE films represents a triplet (A_{II}^{N}=4.1 ± 0.1 mT) of septets (A_{II}^{F}= 0.5 ±0.1 mT) with g_{II}=2.0029 ±0.0005. The unpaired electron in fluoroiminoxyl radicals interacts with a nitrogen nucleus and four in pairs magnetically equivalent fluorine nucleuses with hyperfine splittings of 1.0 ± 0.1 and 0.5 ± 0.1 mT.

Aminoxyl radicals in polyperfluoroalkanes are not formed in these conditions. However, if to carry out a preliminary γ-irradiation on air, middle and end peroxide macroradicals appear with conversion into fluoroaminoxyl macroradicals ~ $CF_2CF(NO^{\bullet})CF_2$ ~ by the subsequent exposure to NO. Their EPR spectra in oriented films are quintet of triplets with the parameters:A_{II}^{N}=0.46 mT, A_{II}^{F}=1.11 mT, g_{II}=2.006; $A_{\perp}^{N}$=1.12 mT, $A_{\perp}^{F}$=1.61 mT and $g_{\perp}$=2,0071. The following mechanism of formation of aminoxyl radicals in these conditions is offered [17]:

$$\sim CF_2\ CF_2\ C^{\bullet}F\ CF_2 \sim \rightarrow \sim C^{\bullet}F_2 + CF_2=CFCF_2\sim \qquad (8)$$

The end alkyl radical is oxidized into end peroxide radical:

$$\sim C^{\bullet}F_2 + O_2 \rightarrow \sim CF_2OO^{\bullet} \qquad (9)$$

In NO atmosphere, end peroxide radical is converted as follows:

$$\sim CF_2CF_2OO^{\bullet} + NO \rightarrow \sim CF_2CF_2O^{\bullet} + NO_2 \qquad (10)$$

$$\sim CF_2CF_2O^{\bullet} + NO\sim \leftrightarrow CF_2CF_2ONO \qquad (11)$$

$$\sim CF_2CF_2O^{\bullet} \rightarrow \sim C^{\bullet}F_2 + COF_2 \qquad (12)$$

$$\sim C^{\bullet}F_2 + NO \rightarrow \sim CF_2NO \qquad (13)$$

$$\sim CF_2NO + CF_2=CFCF_2\sim \xrightarrow{NO} \sim CF_2N(O^{\bullet})CF_2CF(NO)CF_2\sim \qquad (14)$$

In copolymer of tetrafluoroethylene with hexafluoropropylene, the radicals ~$CF_2N(O^{\bullet})CF_3$ are formed in the same conditions [17].

Thus, spin-labelled macromolecules of fluoroalkyl polymers can be prepared using postradiational free radical reactions in NO atmosphere. The advantage of this method of introduction of spin labels lies in the fact that the radical centre can be located both in the end and in middle of macrochains, that allows basically to obtain the optimum information on molecular dynamics.

4.3 RADICAL REACTIONS INITIATED BY NITROGEN TRIOXIDE

The radicals NO_3 play an essential role in chemical processes proceeding in the top layers of atmosphere [18]. These radicals are formed in reaction of nitrogen dioxide with ozone:

$$NO_2 + O_3 \rightarrow NO_3 + O_2 \tag{15}$$

Under action of daylight, nitrogen trioxide is consumed with releasing atomic oxygen:

$$NO_3 \xrightarrow{\ h\nu\ } NO_2 + O \tag{16}$$

Its disappearance occurs at night in reaction with nitrogen dioxide too:

$$NO_3 + NO_2 \rightarrow N_2O_5 \tag{17}$$

The NO^3 radicals are characterized by high reactivity in reactions with various organic compounds [18–22]. Typical reactions of these radicals are abstraction of hydrogen atoms from C–H and addition to double bonds. Along with these reactions, the radicals NO_3 are decomposed in thermal and photochemical processes. The thermal decomposition of nitrogen tri-oxide generated by pulse radiolysis of concentrated water solutions of a nitric acid takes place with high rate ($k_{208K} = 8 \times 10^3$, s^{-1}) [23]:

$$NO_3 \xrightarrow{\ kT\ } NO_2 + O \tag{18}$$

The radicals NO_3 have three intensive bands of absorption in visible re-gion of an optical spectrum with $\lambda_{max} = 600$, 640 и 675 nm, and in UV region at 340–360 nm [18, 19, 23–27]. The action of light on NO_3 results in dissociation of them by two mechanisms including formation of NO_2 and atomic oxygen similarly to Eq. (18) or nitrogen oxide and molecular oxygen [1]:

$$NO_3 \rightarrow NO + O_2 \qquad (19)$$

The efficiency of the NO_3 conversion by one or either mechanisms is determined by spectral composition of light. The approximate border of wavelengths for photodissociation by Eqs. (18) or (19) lays at 570 nm. Above 570 nm, NO_3 decomposes on NO and O_2 with the very high (~1) quantum yield; the basic products of the photolysis are NO_2 + O below 570 nm [1].

One of the most widely widespread ways of the NO_3 generation is the photolysis of Ce (IV) nitrates in particular ceric ammonium nitrate (CAN) $(NH_4)_2Ce(NO_3)_6$. The absorption spectrum of CAN has a wide and intensive band with a maximum at 305 nm ($\varepsilon = 5890$ l×mol^{-1}×cm^{-1}), which is conditioned by an electron transfer to Ce^{4+} from nitrate anion [26]. Under action of light in the given spectral region there is a photoreduction of CAN [26, 27]:

$$Ce^{4+}NO_3^- \rightarrow Ce^{3+} + NO_3 \qquad (20)$$

Thus, the CAN photolysis gives different active radical particles: NO, NO_2, NO_3 and atomic oxygen. Use of light of various spectral composition allows thus to generate these particles in different ratio.

Atomic oxygen being very active reactant [14] interacts with C–H bonds of organic compounds. Macroradicals formed by action of atoms O on polymers can be converted in the presence of NO into stable aminoxyl radicals. The possibility using these processes for the purposes of chemical modification of polymers is considered by the example of polyvinylpyrrolidone (PVP) in the works [28, 29]. In PVP with CAN (0.05–0.2 mol kg^{-1}) in the course of photolysis by light with $\lambda > 280$ nm, the formation of alkyl macroradicals is registered by EPR as a result of the reaction:

$$\text{(21)}$$

The radicals R_1 are stabilized only at low temperatures (77 K). Photolysis of samples at 298 K by the same light results in the production of stable dialkylaminoxyl radicals characterized by anisotropic EPR triplet spectrum with parameters of $A_{II}^{N} = 3.18$ mT and $g_{II} = 2.0024$. The cross-linkage of PVP macromolecules takes place in these conditions with formation of a gel-fraction as a result recombination of macroradicals R_1 with nitrogen oxide:

$$\text{(22)}$$

As shown in the work [28], the yield of a gel-fraction directly correlates with the amounts of stable radicals, which in the given system represent cross-linkages of macromolecules. Such simple method of cross-linking PVP can be applied for obtaining hydrogels used as specific sorbents [30].

4.4 FREE RADICAL AND ION-RADICAL REACTIONS UNDER ACTION OF NITROGEN DIOXIDE AND ITS DIMERS

Nitrogen dioxide effectively reacts with various low- and high-molecular organic compounds [5–7, 31]. However, it must be emphasized that NO_2 is a free radical of moderate reactivity, and the ONO–H bond strength [14] makes up 320 kJ mol^{-1}. Therefore, radicals NO_2 are capable of initiating free radical reactions by abstraction of hydrogen atoms from least strong, for example, allyl C–H bonds or addition to double C=C bonds [32, 33]:

$$>C=C< \quad + NO_2 \quad \rightleftharpoons \quad >\overset{\bullet}{C}-\overset{|}{C}-NO_2 \qquad (23)$$

This process causes further radical conversions of olefins with formation of dinitro compounds and nitro nitrites:

$$>\overset{\bullet}{C}-\overset{|}{C}-NO_2 \; + NO_2 \quad \longrightarrow \quad \begin{cases} \longrightarrow \; O_2N-\overset{|}{C}-\overset{|}{C}-NO_2 & (24) \\[2ex] \longrightarrow \; ONO-\overset{|}{C}-\overset{|}{C}-NO_2 & (25) \end{cases}$$

$$\sim C(CH_3)=CH\sim \; + NO_2 \quad \rightleftharpoons \quad \sim \overset{\bullet}{C}(CH_3)-CH(NO_2)\sim \quad \longrightarrow \quad destruction \qquad (26)$$

From Jellinek's data [31], butyl rubber destroys under the action of NO_2:
As a consequence of primary reactions of nitrogen dioxide radicals with the isolated double bonds, stable aminoxyl radicals can be generated. Such transformations are characteristic for rubbers.

4.4.1 PREPARATION OF SPIN-LABELLED RUBBERS

The possibility of obtaining spin-labelled rubbers by interaction of their solutions in the inert solvents with a mixture of nitrogen dioxide and oxygen has been demonstrated in the work [34]. Such rubbers can be prepared simply and rapidly by reactions of block polymeric samples with gaseous NO_2 [35]. The experiments were carried on 1,4-*cis*-polyisoprene (PI) and copolymer of ethylene, propylene and dicyclopentadiene. The samples had the form of cylinders of 1.5 cm height and 0.4 cm in diameter. On exposure these polymers to NO_2 (10^{-5}–2.3×10^{-3} mol$\times$l^{-1}) at 293 K, identical EPR spectra were registered. The spectra represent an anisotropic triplet with parameters which are typical for dialkylaminoxyl radicals with A_{II}^N = 3.1 mT and g_{II} = 2.0028 ± 0.0005. The spectra with such parameters testify that the correlation time of rotational mobility τ_c at the given temperature exceeds 10^{-9} s. At increasing temperature up to 373 K, the isotropic triplet signal with a^N = 1.53 ± 0.03 mT and g = 2.0057 ± 0.0005 was observed, that is caused by essential decreasing correlation time (5×10^{-11} < τ_c < 10^{-9} s). The change of τ_c with temperature is described by the relation $\tau_c = \tau_0 \exp(E/RT)$, where $\log\tau_0$ = −14.2, and E is the activation energy of rotational diffusion (34.7 kJ×mol^{-1}).

The study of kinetics of the aminoxyl radical accumulation has shown that concentrations of the radicals pass through a maximum determined by concentrations of NO_2 in a gas phase. So the concentration of radicals peaks in 40 times time more slowly when $[NO_2]$ falls from 2.3×10^{-3} up to 10^{-5} mol×l^{-1}. It is suggested that such kinetic features are due to disappearance of aminoxyl radicals in reactions of their oxidation by NO_2 dimers. The presence of oxygen in a gas mixture results in the long induction period for kinetics of the radical formation, however their maximum concentration in this case are approximately in 3 times higher than on exposure of rubbers to pure NO_2. The scheme of the aminoxyl radical formation includes four basic stages: generation of macroradicals in reaction of NO_2 with rubbers; synthesis of macromolecular, spin trapping of macroradicals by nitroso compounds, and destruction of aminoxyl radicals:

$$\sim CH_2C^\bullet(CH_3) - CH(NO_2) - CH_2 \sim (R_1^\bullet)$$

$$\sim CH_2C(CH_3) = CH - CH_2 \sim + NO_2 \rightleftharpoons CH_2C^\bullet(CH_3) - CH(ONO) - CH_2 \sim (R_2^\bullet) \quad (27)$$

$$HNO_2 + \sim CH_2C(CH_3) = CH - C^\bullet H \sim (R_3^\bullet)$$

$$R_1^\bullet (R_2^\bullet, R_3^\bullet) + NO_2 \rightarrow \text{продукты} \qquad (28)$$

$$2HNO_2 \rightleftharpoons NO + NO_2 + H_2O \qquad (29)$$

$$R_1^\bullet (R_2^\bullet, R_3^\bullet) + NO \rightarrow R_1NO(R_2NO, R_3NO) \qquad (30)$$

$$RNO + R_1^\bullet (R_2^\bullet, R_3^\bullet) \rightarrow (R)_2N-O^\bullet \qquad (31)$$

$$(R)_2N-O^\bullet + R^\bullet \rightarrow (R)_2N-OR \qquad (32)$$

$$(R)_2N-O^\bullet + N_2O_4 \rightarrow [(R)_2N^+ONO_3^-] + NO \qquad (33)$$

As can be seen from this scheme, the accumulation of aminoxyl radicals should be accompanied by cross-linkage of macromolecules. The presence of oxygen inhibits the aminoxyl accumulation as a result of conversions of primary nitroalkyl and allyl macroradicals into peroxide radicals.

In the course of interaction of solid polymers with NO_2 one can expect a non-uniform distribution of formed stable nitrogen-containing radicals in sample owing to diffusion restrictions. A possibility of an examination of the nitration reaction front thus is created in polymeric materials by measurement of spatial distribution of aminoxyl radicals. Such studies can be performed using the method of EPR tomography (or the EPR imaging technique) described in the works [36, 37]. During the reaction of NO_2 with double bonds of rubbers, the spatial distribution of aminoxyls characterizes kinetics of the reaction front progress and thus structural-physical properties of a concrete sample. The EPR tomograms in a non-uniform magnetic field (Fig. 1a and b) were obtained for cylindrical samples PI of 0.4 cm in diameter and height of 1.5 cm. The accumulation of aminoxyl macroradicals up to maximum concentrations in enough thick layers (~1 mm) of samples shows that a spatial grid of cross-linked PI hindered diffusion of NO_2 has not time to be generated. The introduction of O_2 in gaseous mixture narrows front of reaction and reduces rate of the aminoxyl radical formation. This effect is connected with the additional channel of an intensive consumption of intermediate macroradicals $R_1^{\bullet}$ ($R_2^{\bullet}$, $R_3^{\bullet}$) and decreasing their equilibrium concentration. The results obtained by EPR tomography have shown that on exposure of rubbers to NO_2 at enough high concentrations ($10^{-4} - 2.3 \times 10^{-3}$ mol$\times$l^{-1}) the chemical and structural modification occurs in a superficial layer and does not take place in deeper layers.

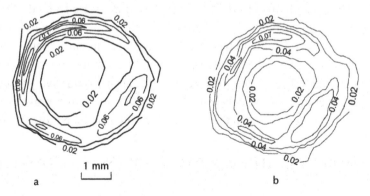

a   b

FIGURE 1 Bulk distribution of dialkylaminoxyl radicals produced upon nitration of PI by NO_2 during 2.5 (a) and 740 hours (b).

4.4.2 ON THE MECHANISM OF INITIATION OF RADICAL REACTIONS BY NITROGEN DIOXIDE DIMERS

As noted above, nitrogen dioxide can initiate free radical reaction in compounds containing least strong C–H bonds or double C=C bonds. However, the effective formation of stable nitrogen-containing radicals was observed also in aromatic polyamidoimides, polycaproamide, polyvinylpyrrolidone (PVP) [8] and also aromatic polyamide (AP) [12]. These facts allow considering other probable mechanisms of the radical processes initiation. The fact is that the basic radical products of interaction of nitrogen oxide with polymers containing amide groups are iminoxyl and acylalkylaminoxyl radicals, which are produced from oximes and acylnitroso compounds [8, 12]. The occurrence of these predecessors of stable radicals is in turn connected with the nitrogen oxide formation.

In this connection, it is necessary to suppose a participation of NO_2 dimeric forms in radical initiation. The main dimers of NO_2 are planar nitrogen tetroxide

$$O_2N\text{-}NO_2 \text{ (PD)} \tag{34}$$

and nitrosyl nitrate

$$ONONO_2 \text{ (NN)} \tag{35}$$

Ab initio calculations [33] show that these dimers are formed with the most probability in NO_2 atmosphere; the form of nitrosyl peroxynitrite ONOONO is too unstable to be considered as efficient participant of reactions, however, it can play a role of intermediate compound at oxidation of nitrogen oxide by oxygen [38]. As NN has strong oxidative properties [39], the generation of radicals can take place by an electron transfer from donor functional groups with the formation of intermediate radical cations [9, 40]:

$$RH + ONONO_2 \rightarrow [R^\bullet H^+(NO \times\times\times ONO_2^-)] \rightarrow R^\bullet + NO + H^+ + ONO_2^- \tag{36}$$

The recombination of radicals with nitric oxide gives nitroso compounds that undergo isomerisation into oximes [41] to produce iminoxyl radicals in the reaction with NO_2:

$$C=NOH + NO_2 \rightarrow C=NO^\bullet + HNO_2 \qquad (37)$$

The tertiary nitroso compounds are effective spin traps and a source of stable aminoxyl radicals:

$$RN=O + R_1^\bullet \rightarrow R(R_1)N-O^\bullet \qquad (38)$$

Thus the mechanism involving Eqs. (36)–(38) formally could explain an appearance of stable radicals in the polymers not containing specific chemical bonds reacting with NO_2 mono radicals. However, there are certain obstacles connected with energetic properties of NO_2 dimers [33] for realizing such mechanism; the energy of syn- and anti-forms of NN exceeds that of PD respectively 29.8 and 18.4 кJ×mol^{-1}; that is the equilibrium

$$O_2N\text{-}NO_2 \rightleftharpoons 2NO_2 \rightleftharpoons ONONO_2 \qquad (39)$$

should be shifted to PD in gas phase.

The diamagnetic (PD) is capable of generating nitrogen-containing radicals in specific reaction with system of the connected double bonds of p – quinines (Q) [42]. On exposure of BQ to nitrogen dioxide, the formation of radicals I of oxyaminoxyl type (R_{oxy}) [43] takes place by the following scheme:

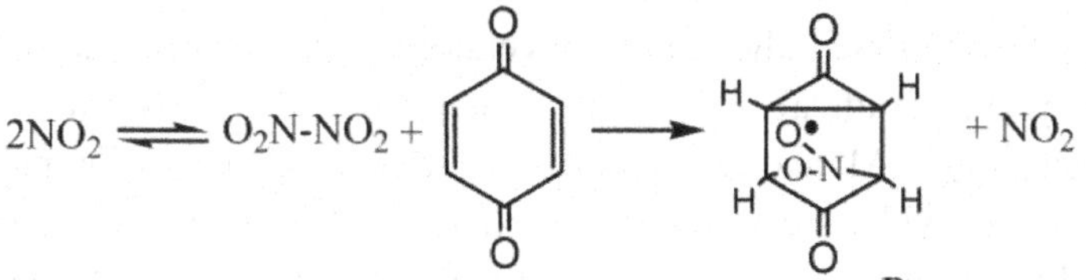

$$2NO_2 \rightleftharpoons O_2N\text{-}NO_2 + \qquad\qquad \longrightarrow \qquad\qquad + NO_2 \qquad (40)$$

$$R_{oxy}$$

The triplet EPR spectrum of radicals R_{oxy} in Q (Fig. 2a) has parameters: $a^N = 2.82$ mT and $g = 2.0053$. The Eq. (40) is confirmed by kinetic data according to which the rate of the radical accumulation is proportional to a square of NO_2 concentration in gas phase.

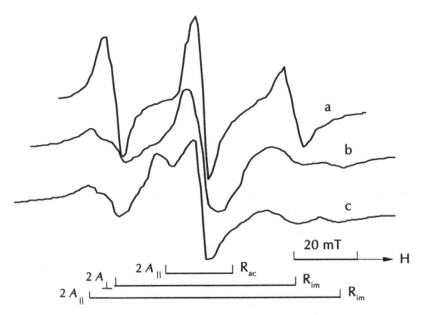

FIGURE 2   EPR spectra of Q after exposure to NO_2 (a); Q + AP (b) and Q+ PVP (c) after preliminary exposure to NO_2 and subsequent pumping-out the composites.

Thus, both dimer forms of NO_2 can be active. On this basis it is possible to suggest that the shift of equilibrium (39) to the formation of NN in PVP and AP is caused by specific donor-acceptor interaction of PD with amide groups, which induce the conversion into NN and ion-radical process by Eq. (36). As the indicator of conversion of PD into NN, the dependence of a yield of radicals R_{oxy} on the contents of AP and PVP was used in composites: Q + AP and Q + PVP. To increase a surface of interaction with nitrogen dioxide, silica gel with particles of 100–160 μ in diameter was added to the composites. Samples for measurement of EPR spectra contained constant quantities of Q (100 mg), SiO_2 (100 mg) and variable quantities of PVP (10–30 mg) or AP (10–50 mg). In addition to R_{oxy}, iminoxyl radicals R_{im} [8] occur in composites of BQ with AP on exposure to nitrogen dioxide. Under the same conditions, the sum of radicals R_{im} and acylalkylaminoxyl radicals R_{ac} [8] along with R_{oxy}, was registered in composites of BQ with PVP. Signals of radicals R_{im} and R_{ac} are masked by an intense signal of radicals R_{oxy} in the EPR spectrum. However on can separate spectra of radicals R_{im} and R_{ac} using the fact that radicals R_{oxy} exist only in NO_2 atmosphere. In view of rather low thermal stability, radicals R_{oxy} quickly disappear at room temperature within several minutes after pumping out nitrogen dioxide from the samples. Remaining spectra of stable radicals R_{im} in AP and the sum of $R_{im} + R_{ac}$ in PVP are shown respectively in Fig. 2b and 2c. They represent anisotropic triplets with $A_\parallel^N$ = 4.1 mT, $g_\parallel$ = 2.0024 and $A_\perp^N$ = 2.6 mT, $g_\perp$ = 2.005 (R_{im}) [12] and with $A_\parallel^N$ = 1.94 mT, $g_\parallel$ = 2.003 (R_{ac}) [8]. Using this procedure, the maximum concentrations of radicals R_{oxy}, R_{im} and R_{ac} were separately determined in composites with the various contents of AP and PVP after exposure to NO_2 within 24 h. The results obtained are shown in Fig. 3a and 3b. As is seen from the figures, the concentration of radicals R_{oxy} accumulated monotonously falls as the relative contents of AP and PVP is increased, while concentrations of radicals R_{im} and $R_{im} + R_{ac}$ vary within 10–20 % of the average value, that is within the accuracy of integration of EPR spectra. This fact is indicative of obvious dependence of the radical R_{oxy} yield on the contents of polymers with amide groups in composites, suggesting that PD is converted under the influence of amide groups into NN that generates stable radicals R_{im} and R_{ac} in the polymeric phases. It is significant that an appreciable decrease of the yield of radicals R_{oxy} was not observed in control experiments when polymers of other chemical structure, for example, acetyl cellulose were used in composites. Therefore one can conclude that amide groups play special role in the process PD $\rightarrow$ NN.

The scheme of R_{im} formation in AP can be presented as follows:

(41)

The structure of radicals R_{im} in AP is confirmed by quantum-chemical calculations of HFI constants [12].

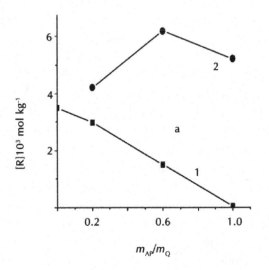

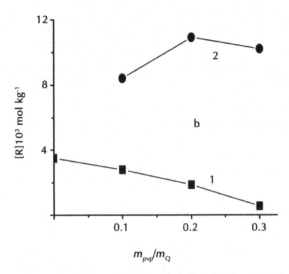

FIGURE 3   Dependence of concentrations of radicals R_{oxy} (1), R_{im} (2) in Q + AP (a) and R_{oxy} (1), R_{im} + R_{ac} (2) in Q + PVP (b) after exposure to NO_2 on weight ratio of Q, AP and PVP.

The formation of radicals R_{im} and R_{ac} in PVP can be described by the following reactions:

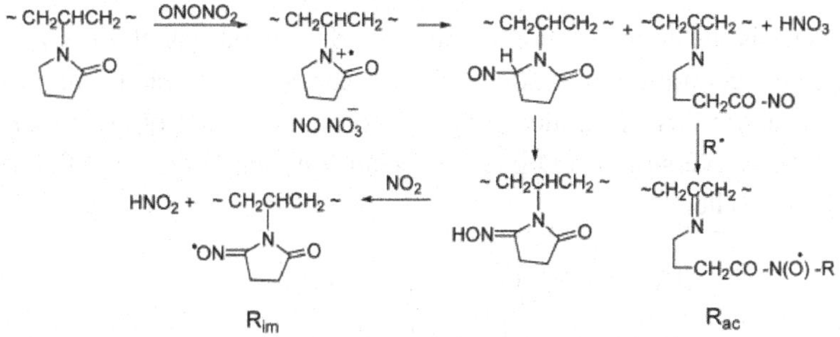

(42)

where $R^\bullet$ appears as a result of the radical cation decomposition:

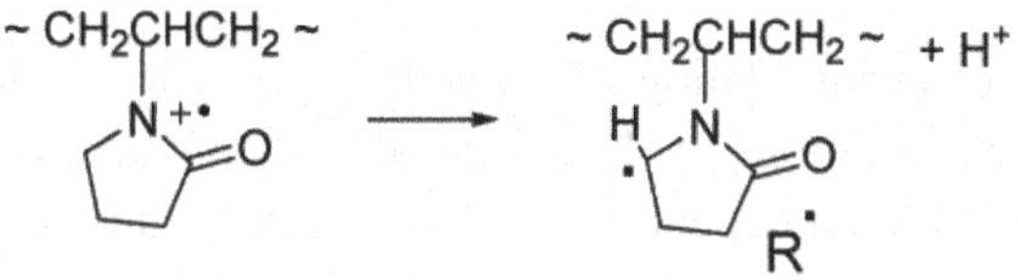

(43)

The decrease of relative yield of radicals R_{im} on addition of polymers with amide groups to composites (Fig. 3) is apparent from the formal kinetic scheme:

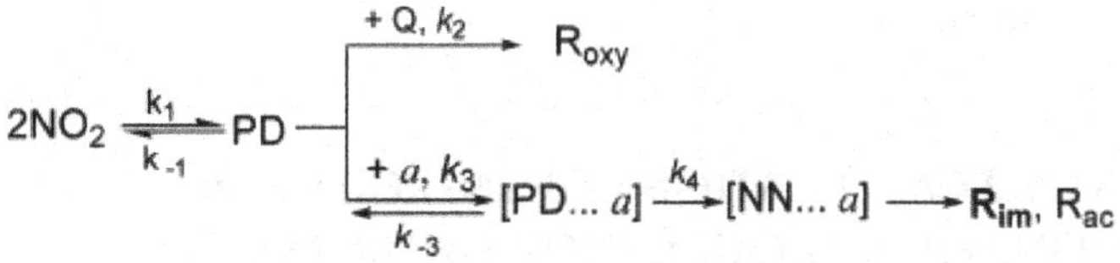

(44)

where a is an amide group. Taking into consideration stationary state for concentrations of PD, NN, [PD×××a], [NN×××a] and invariance of Q contents in composites, the following equations for rates of accumulation of radicals R_{oxy}, R_{im} and R_{ac} can be obtained:

$$\frac{d[R_{oxy}]}{dt} = \frac{k_1 k_2 (k_{-3} + k_4)[NO_2]^2}{(k_{-3} + k_4)(k_{-1} + k_2 + k_3[a]) - k_{-3}k_3[a]}$$

(45)

$$\frac{d[R_{im}],[R_{ac}]}{dt} = \frac{k_1 k_3 k_4 [a][NO_2]^2}{(k_{-3}+k_4)(k_{-1}+k_2+k_3[a])-k_{-3}k_3[a]}, \qquad (46)$$

where $[NO_2]$ is the concentration of nitrogen dioxide in gas phase, $[a]$ is the surface concentration of amide groups. These equations can be simplified if concentrations of amide groups in composites are comparatively high, and the conversion of PD into NN occurs enough effectively, that is $k_3[a] \gg k_{-1}+k_2$. Then

$$\frac{d[R_{ixy}]}{dt} = \frac{k_1 k_2 (k_{-3}+k_4)[NO_2]^2}{k_3 k_4 [a]} \qquad (47)$$

$$\frac{d[R_{in}],[R_{ac}]}{dt} = k_1 [NO_2]^2 \qquad (48)$$

Thus the rate of accumulation of radicals R_{im} and R_{ac} is determined by $[NO_2]$, and concentrations of these radicals accumulated on exposure to nitrogen dioxide do not depend appreciably on AP and PVP contents (Fig. 3a, b (curve 2)). In contrast, the yield of R_{oxy} decreases as polyamides are added to composites and $[a]$ is increased. These plots in character are representative of competitive pathways for PD interactions with Q and amide groups. Note that the yield of R_{oxy} is not changed in the NO_2 atmosphere in composites of Q with other polymers, for instance, acetyl cellulose at any ratio of the components.

4.5   AB INITIO CALCULATIONS OF ENERGIES FOR CONVERSIONS OF NITROGEN DIOXIDE DIMMERS

For validating the mechanism proposed of the conversion of PD into NN, the calculations of energy changes in process of nitrogen dioxide interaction with simplest amide (formamide) have been carried out within the framework of density functional theory by the Gaussian 98 program [44]. The B3LYP restricted method for closed and open shells was used. The intention of the calculations is to correlate energy consumptions for

PD $\rightarrow$ NN with those for other stages of the radical generation process. Energies of the following states according to scheme (44) were calculated:

$$2NO_2 + NH_2COH \tag{49}$$

$$O_2N\text{-}NO_2 + NH_2COH \tag{50}$$

$$ONONO_2 + NH_2COH \tag{51}$$

$$[O_2N\text{-}NO_2 \times\times\times NH_2COH] \tag{52}$$

$$[ONONO_2 \times\times\times NH_2COH] \tag{53}$$

$$N^\bullet HCOH + NO + HNO_3 \tag{54}$$

$$NH_2CO^\bullet + NO + HNO_3 \tag{55}$$

The geometry optimization of all structures was performed applying the basis set 6-31G (d, p). The given process includes intermediate molecular complexes of PD and NN with formamide (52, 53). The changes of minimum energies are shown in Fig. 4. One can see that the formation of PD from NO_2 is energetically advantageous process [33], whereas NN is generated from NO_2 in an endothermic reaction. The complexation of PD with formamide is accompanied by release of energy: $\Delta E = 28$ kJ$\times$mol^{-1}. However, PD in complex (52) is not capable of reacting with formamide and can only leave the reacting cage. At the same time, PD in the complex can be converted approximately with the same energy consumption into NN (53), which further reacts by the electron transfer reactions (54, 55) giving radicals, nitric oxide, nitric acid and significant release of energy (44–57 kJ$\times$mol^{-1}). Such sequence of transformations seems to be more efficient in comparison with a direct interaction of NN and formamide by state (51), as the energy of dimers in complexes (52) and (53) is lower than that of initial state (49).

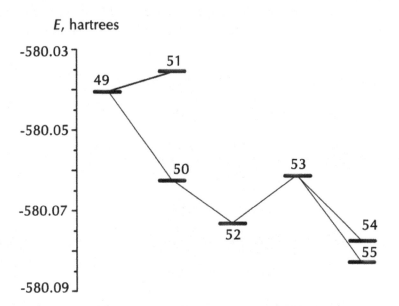

FIGURE 4 Changes of minimum energies calculated for reactions of NO_2 with formamide.

4.6   DETECTING RADICAL CATIONS IN REACTIONS OF ELECTRON TRANSFER TO NN

Registration of radical cations by the EPR method in the presence of nitrogen dioxide could provide direct experimental evidence that the initiation proceeds through scheme (36). However, because of high reactivity and fast decomposition [45], these particles are difficult to detect by this method. Nevertheless, the formation of radical cations can be revealed indirectly in the act of their decomposition with emission of a proton. Pyridine is known to be capable of accepting protons to yield pyridinium cations. Hence, if protons are formed in during decomposition of radical cations by Eq. (36), they can be detected easily from IR spectra typical of pyridinium cations. Note that pyridine can be nitrated only under quite severe conditions. For example, N-nitropyridinium nitrate was obtained only when pyridine was treated with a NO2-ozone mixture in an inert solvent [46]. This is evident from the IR spectrum (Fig. 5) of 1:1 mixture of pyridine and N-methylpyrrolidone (low-molecular analog of PVP) after treating it by nitrogen dioxide. In the spectrum there are two intense bands at 2400–2600 and 2200 cm^{-1} of pyridinium cations [47]. The scheme of reactions in the given system involves the following consecutive stages:

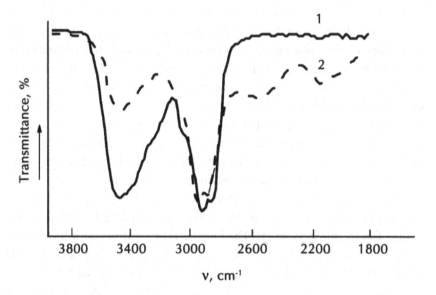

$$CH_3N\text{-}(CH_2)_3\text{-}CO + ONONO_2 \longrightarrow CH_3\overset{+\cdot}{N}\text{-}(CH_2)_3\text{-}CO \longrightarrow$$
$$NO\ NO_3^-$$

$$(56)$$

$$CH_3N\text{-}\overset{\cdot}{C}H\text{-}(CH_2)_2\text{-}CO \longrightarrow CH_3N\text{-}CH(NO)\text{-}(CH_2)_2\text{-}CO \xrightarrow{\ C_5H_5N\ } C_5H_5NH^+\ NO_3^-$$
$$NO\ H^+\ NO_3^- \qquad\qquad HNO_3$$

FIGURE 5 IR spectra of 1:1 N-methylpyrrolidone and pyridine mixture (1) and after exposure of the mixture to NO_2 (2).

Thus pyridine fixes occurrence of radical cations as a result of an electron transfer to NN, and in doing so the ion–radical mechanism (36) is experimentally confirmed. It is possible to believe that such oxidative mechanism of generating radicals is characteristic for compounds with the ionization potential providing an electron transfer from donor groups of molecules to

NN. This regularity can be followed from examples of thermal and photochemical nitration of aromatic compounds under the action of nitrogen dioxide. According UV spectroscopy data, NN can form charge-transfer complexes with methylbenzenes [48, 49]. For the complexes, a bathochromic shift in the corresponding absorption bands was observed with an increase in number of methyl substituents in the benzene ring. Given this, this shift correlates with a decrease in the ionization potential from 8.44 to 7.85 eV on passage from *p*-xylene to hexamethylbenzene.

4.7 CONCLUSION

Nitrogen oxides are the effective initiators of radical reactions in a number polymers with formation the various molecular nitration products and stable nitrogen-containing radicals. Nitrogen oxide does not react directly, but it recombines with free radicals formed in polymers by UV photolysis or γ-radiolysis. The nitro compounds formed in the subsequent reactions are converted into stable radicals. By this way, spin labels can be inserted even to chemically inert polyperfluoroalkanes. However, the chemical structure of macromolecules can essentially change in enough severe conditions of generating radicals. In this connection, nitrogen trioxide obtained by photolysis of the Ce (IV) nitrates is promising in application for spin labels synthesis. Under the action of visible and near UV light on these additives, the radicals and nitrogen oxide are formed simultaneously with transformation finally into spin labels. The light of such spectral composition does not cause undesirable side effects on macromolecules. Nitrogen dioxide is capable of interacting with least strong C–H and double C=C bonds initiating thus radical reactions in the given system. The dimeric forms of nitrogen dioxide actively react by mechanism depending on the chemical structure of those. The dimers in the form of nitrosyl nitrate represent oxidizing agent initiating ion-radical reactions with formation of stable nitrogen-containing radicals. Amide groups can induce a transition of energetically steadier planar dimers of NO_2 into nitrosyl nitrate. This specific ion-radical mechanism determines high activity relative to NO_2 even such stable polymers as aromatic polyamides.

KEYWORDS

- **Aromatic polyamides**
- **Endothermic reaction**
- **Polyperfluoroalkanes**
- **Rubber**

REFERENCES

1. Graham, R. A.;Johnston, H. S. *J. Phys. Chem.*, **1978**, *82* (3), 254.
2. Johnston, H. S.; Graham, R. A. *Canad. J. Chem.*, **1974**, *52* (8), 1415.
3. Stroud, C.; Madronich, S.; Atlas, E.; Ridley, B.; Flocke, F.; Tallot, A. W.; Fried, A.; Wert, B.; Shetter, R.; Lefler, B.; Coffey, M.; Heik, B. *Atmospheric Environment,* **2003**, *37* (24), 3351.
4. Tong, D. Q.; Kang, D.; Aneja, V. P. and Ray, J. D. *Ibid.*, **2005**, *39* (2), 315.
5. Titov, A. I. *Tetrahedron*, **1963**, *19*, 557.
6. Topchiev, A. V. *Nitration of hydrocarbons and other organic compounds,* Moscow, Academy of Sciences of the USSR, 1956.
7. Novikov, S. S.; Shveyhgeymer, G. A.; Sevastyanova, V. V. and Shlyapochnikov, V. A. In *Chemistry of aliphatic alicyclic nitro compounds.* 1974, Moscow, Khimiya.
8. Pariiskii, G. B.; Gaponova, I. S. and Davydov, E. Ya. *Russ. Chem. Rev.*, **2000**, *69* (11), 985.
9. Pariiskii, G. B.; Gaponova, I. S.; Davydov, E. Ya and Pokholok, T. V. In *Aging of Polymers, Polymer Blends and Polymer Composites.* Ed. G. E. Zaikov, A. L. Buchachenko, V. B Ivanov, New York, Nova Science Publishers, 2002.
10. Vasserman, A. M. and Kovarsky, A. L. *Spin labels and Spin Probes in Physical Chemistry of polymers,* Moscow, Nauka, 1986.
11. Gaponova, I. S.; Davydov, E. Ya.; Pariiskii, G. B. and Pustoshny, V. P. *Vysokomol. Soed.*, ser. A, **2001**, *43* (1), 98.
12. Pokholok, T. V.; Gaponova, I. S.; Davydov, E. Ya. and Pariiskii, G. B. *Polym. Degrad. Stability*, **2006**, *91* (10), 2423.
13. Bonner, F. T. and Stedman, G. *In Methods in nitric oxide research*, Ed. M. Feelish and J. S. Stamler, Chichester, Wiley, 1996.
14. Rånby, B. and Rabek, J. F. *Photodegradation, photo-oxidation and photostabilization of polymers,* London, Wiley, 1975.
15. Park, J. S. B. and Walton, J. C. *J. Chem. Soc., Perkin Trans.* **1997**, *2*, 2579.
16. Gaponova, I. S.; Pariiskii, G. B. and Toptygin, D. Ya. *Vysokomolek. Soed.*, **1988**, *30* (2), 262.
17. Gaponova, I. S.; Pariiskii, G. B. and Toptygin, D. Ya. *Khimich. Fizika*, **1997**, *16* (10), 49.
18. Neta, P. and Huie, R. E. *J. Phys. Chem.*, **1986**, *90* (19), 4644.

19. Vencatachalapathy, B. and Ramamurthy, R. *J. Photochem. Photobiology A: Chem.,* **1996**, *93*,1.
20. Japar, S. M. and Niki, H. H.; J. *Phys. Chem.,* **1975**, *79* (16), 1629.
21. Atkinson, R.; Plum, C. N.; Carter, W. P. L.; Winer, A. M. and Pitts, J. N. *J. Phys. Chem., 88,* (6), 1210.
22. Itho, O.; Akiho, S. and Iino, M. *J. Org. Chem.,* **1989**, *54* (10), 2436.
23. Pikaev, A. K.; Sibirskaya, G. K.; Shyrshov, E. M.; Glazunov, P. Ya. and Spicyn, V. I. *Dokl. Akad. Nauk SSSR,* **1974**, *215* (3), 645.
24. Hayon, E. and Saito, E. *J. Chem Phys.,* **1965**, *43* (12), 4314.
25. Dagliotti, L. and Hayon, E. *J. Phys. Chem.,* **1967**, *71* (12), 3802.
26. Glass, P. W. and Martin, T. W. *J. Am. Chem. Soc.,* *92* (17), 5084.
27. Wine, P. H.; Mauldin, R. L. and Thorn, R. P. *J. Phys. Chem.,* **1988**, *92* (5), 1156.
28. Davydov, E. Ya.; Afanas'eva, E. N.; Gaponova, I. S. and Pariiskii, G. B. *Org. Biomol. Chem.,* **2004**, *2* (9), 1339.
29. Davydov, E. Ya.; Gaponova, I. S. and Pariiskii, *Vysokomolek. Soed.,* ser. A, **2003**, *45* (4), 581.
30. Naghash, H. J.; Massah, A. and Erfan, A. *Eur. Polym. J.,* **2002**, *38*, (1), 147.
31. Jellinek, H. H. G. *Aspects of degradation and stabilization of polymers. Chapter 9.* New York, Elsevier, 1978.
32. Giamalva, D. H.; Kenion, G. B. and Pryor, W. A. *J. Am. Chem. Soc.,* **1987**, *109* (23), 7059.
33. Golding, P.; Powell, J. L. and Ridd, J. H. *J. Chem. Soc., Perkin Trans.* **1996**, *2*, . 813.
34. Gyor, M.; Rockenbauer, A. and Tüdos, F. *Tetrahedron Lett.,* **1986**, *27* (39), 4795.
35. Pokholok, T. V. and Pariiskii, G. B. *Vysokomolek. Soed., ser. A,* **1977**, *39* (7), 1152.
36. Smirnov, A. I.; Degtyarev, E. N.; Yakimchenko, O. E. and Lebedev, Ya. S. *Pribory Tekhn. Eksperim,* **1991**, *1*, 195.
37. Degtyarev, E. N.; Polholok, T. V.; Pariiskii, G. B.; Yakimchenko, O. E. *Zh. Fiz. Khim.,* **1994**, *68* (3), 461.
38. McKee, M. L. *J. Am. Chem. Soc.,* **1995**, *117* (8), 1629.
39. White, E. H. *Ibid.,* **1955**, *77*, (20), 6008.
40. Davydov, E. Ya.; Gaponova, I. S.; Pariiskii, G. B. and Pokholok, T. V. *Polm. Sci., ser.A,* **2006**, *48* (4), 375.
41. Feuer, H. *The chemistry of the nitro and nitroso groups,* New York, Wiley, 1969.
42. Davydov, E. Ya.; Gaponova, I. S. and Pariiskii, G. B. *J. Chem. Soc., Perkin Trans. 2,* 2002, 1359.
43. I. Gabr M. C. R. Symons, *J. Chem. Soc., Faraday Trans.,* **1996**, *92* (10), 1767.
44. Frisch, M. J.; Trucks, G. W.; Schlegel, H. B.; Scuseria, G. E.; Robb, M. A.; Cheeseman, J. R. et al. Gaussian 98. Pitsburgh, PA, Gaussian Inc., 1998.
45. Greatorex, D. and Kemp, J. *J. Chem. Soc., Faraday Trans.,* **1972**, *68* (1), 121.
46. Suzuki, H.; Iwaya, M. and Mori, T. *Tetrahedron Lett.,* **1997**, *38* (32), 5647.
47. Bellamy, L. J.; *The infra-red spectra of complex molecules.* London, Methuen, 1957.
48. Bosch; E. and Kochi, J. K. *J. Org. Chem.,* **1994**, 59 (12), 3314.
49. Bockman, T. M.; Karpinski, Z. J.; Sankararaman, S. Kochi, J. K. *J. Am. Chem. Soc.,* **1992**, 114, (6), 1970.

CHAPTER 5

THE STRUCTURE AND PROPERTIES OF BLENDS OF POLY(3-HYDROXYBUTYRATE) WITH AN ETHYLENE-PROPYLENE COPOLYMER

A. A. OL'KHOV, YU. V. TERTYSHNAYA, L. S. SHIBRYAEVA, A. L. IORDANSKII, and G. E. ZAIKOV

CONTENTS

5.1   Introduction... 66

5.2   Experimental... 66

5.3   Results and Discussion 67

Keywords 73

References... 73

5.1 INTRODUCTION

The methods of DSC and IR spectroscopy were used to study various blends of poly(3-hydroxybu-tyrate) with ethylene-propylene copolymer rubber (EP). When the weight fractions of the initial polymers are equal, a phase inversion takes place; as the blends are enriched with EP, the degree of crystallinity of poly(3-hydroxybutyrate) decreases. In blends, the degradation of poly(3-hydroxybutyrate) begins at a lower temperature compared to the pure polymer and the thermooxidative activity of the ethylene-propylene copolymer in the blend decreases in comparison with the pure copolymer.

Composite materials based on biodegradable polymers are currently evoking great scientific and practical interest. Among these polymers is poly(3-hydroxybu-tyrate) (PHB), which belongs to the class of poly(3-hydroxyalkanoates). Because of its good mechanical properties (close to those of PP) and biodegradability, PHB has been intensely studied in the literature [1]. However, because of its significant brittleness and high cost, PHB is virtually always employed in the form of blends with starch, cellulose, PE [2], etc., rather than in pure form.

This work is concerned with the study of the structural features of PHB-EPC blends and their thermal degradation.

5.2 EXPERIMENTAL

The materials used in this study were EPC of CO-059 grade (Dutral, Italy) containing 67.4 mol % ethylene units and 32.6 mol % propylene units. PHB with $M_{ц} = 2.5 \times 105$ (Biomer, Germany) was used in the form of a fine powder. The PHB:EPC ratios were as follows: 100:0, 80:20, 70:30, 50:50, 30:70, 20:80, and 0:100 wt%.

The preliminary mixing of the components was performed using laboratory bending microrolls (brand VK-6) under heating: the microroll diameter was 80 mm, the friction coefficient was 1.4, the low-speed roller revolved at 8 rpm, and the gap between the rolls was 0.05 mm. The blending took place at 150°C for 5 min.

Films were prepared by pressing using a manual heated press at 190°C and at a pressure of 5 MPa; the cooling rate was ~50°C/min.

The thermophysical characteristics of the tested films and the data on their thermal degradation were obtained using a DSM-2M differential scanning calorimeter (the scanning rate was 16 K/min); the sample weight varied from 8 to 15 mg; and the device was calibrated using indium with $Tm = 156.6°C$. To determine the degree of crystallinity, the melting heat of the crystalline PHB (90 J/g) was used [2]. The Tm and Ta values were determined with an accuracy up to 1°C. The degree of crystallinity was calculated with an error up to ±10%. The structure of polymer chains was determined using IR spectroscopy (Specord M-80). The bands used for the analysis were structure-sensitive bands at 720 and 620 cm^{-1}, which belong to EPC and PHB, respectively [3]. The error in the determination of reduced band intensities did not exceed 15%.

5.3   RESULTS AND DISCUSSION

The melting endotherms of PHB, EPC, and their blends are shown in Fig. 1. Apparently, all the first melting thermograms (except for that of EPC) show a single peak characteristic of PHB.

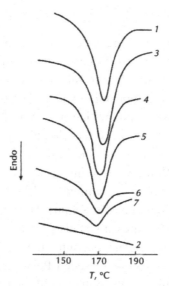

FIGURE 1   The melting endotherms of (1) PHB, (2) EPC, and their blends with compositions (3) 80:20, (4) 70:30, (5) 50:50, (6) 30:70, and (7) 20:80 wt %.

The thermophysical characteristics obtained using DSC for blends of various compositions are listed in Table 1. As is apparent from this table, the melting heat $\triangle H_{ml}$ of PHB during first melting changes just slightly in comparison with the starting polymer. During cooling, only a single peak corresponding to the crystallizing PHB additionally appears.

TABLE 1   The thermophysical properties of PHB-EPC films.

PHB : EPC, wt%	T_m, °C	ΔH^*_{m1}	ΔH^*_{m2}	T_{cr}, °C	Degree of crystallinity**, %
		J/g			
100 : 0	174	88.3	88.9	64	98
80 : 20	173	75.8	76.2	64	84
70 : 30	172	59.5	60.1	60	66
50 : 50	172	56.4	52.1	62	63
30 : 70	172	29.3	20.5	60	33
20 : 80	171	22.3	15.7	–	25
0 : 100	–	–	–	–	–

* Calculated as areas under the melting curves: the first melting and the melting after the recrystallization, respectively.
** Calculated according to the ΔH*ml values.

However, the repeated melting endotherms of some blends (70% PHB + 30% EPC, 50% PHB + 50% EPC) display a low-temperature shoulder. Note that the melting enthalpy significantly changes as one passes from an EPC-enriched blend to a composition where PHB is predominant. When the content of EPC is high, the melting heat AHm2 of the recrystallized PHB significantly decreases. This effect should not be regarded as a consequence of the temperature factor, because the material was heated up to 195°C during the DSM-2M experiment and the films were prepared at 190°C; the scanning rate was significantly lower than the cooling rate during the formation of the films (50 K/min). Thus, the state of the system after remelting during DSC measurements is close to equilibrium.

These results make it possible to assume that the melting heat and the degree of crystallinity of PHB decrease in EPC-enriched blends due to the mutual segmental solubility of the polymers [4] and due to the appearance of an extended interfacial layer. Also note that the degree of crystallinity may decrease because of the slow structural relaxation of the rigid-chain PHB. This, in turn, should affect the nature of interaction between the blend components. However, the absence of significant changes in the Tm and Ta values of PHB in blends indicates that EPC does not participate in nucleation during PHB crystallization and the decrease in the melting enthalpy of PHB is not associated with a decrease in the structural relaxation rate in its phase. Thus, the crystallinity of PHB decreases because of its significant amorphization related to the segmental solubility of blend components and to the presence of the extended interfacial layer.

Figure 2 shows the IR spectra for two blends of different compositions. As is known, the informative structure-sensitive band for PHB is that at 1228 cm⁻¹ [5]. Unfortunately, the intensity of this band cannot be clearly determined in the present case, because it cannot be separated from the EPC structural band at 1242 cm⁻¹ [3]. The bands used for this work were the band at 620 cm⁻¹ (PHB) and the band at 720 cm⁻¹ (EPC) [6], which correspond to vibrations of C–C bonds in methylene sequences (CH₂), where n > 5, occurring in the trans-zigzag conformation. The ratios between the optical densities of the bands at 720 and 620 cm⁻¹ (D_{720}/D_{620}) are transformed in the coordinates of the equation where (5 is the fraction of EPC and W is the quantity characterizing a change in the ratio between structural elements corresponding to regular methylene sequences in EPC and PHB.

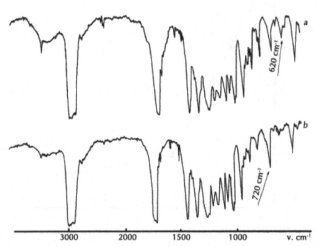

FIGURE 2   The IR spectra of PHB-EPC blends with compositions (a) 80:20 and (b) 20:80 wt %.

Figure 3 demonstrates the value of W plotted as a function of the blend composition. Apparently, this dependence is represented by a straight line in these coordinates but shows an inflection point. The latter provides evidence that phase inversion takes place and that the nature of intermolecular interactions between the polymer and the rubber changes.

$$W = \log [D_{720}\beta/D_{620}(1-\beta) + 2]$$

where (5 is the fraction of EPC and W is the quantity characterizing a change in the ratio between structural elements corresponding to regular methylene sequences in EPC and PHB.

Figure 3 demonstrates the value of W plotted as a function of the blend composition. Apparently, this dependence is represented by a straight line in these coordinates but shows an inflection point. The latter provides evidence that phase inversion takes place and that the nature of intermolecular interactions between the polymer and the rubber changes.

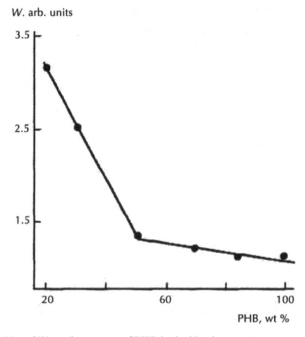

FIGURE 3   Plot of W vs. the content of PHB in the blend.

The phase inversion causes the blends in question to behave in different ways during their thermal degradation. The DSM-2M traces (Fig. 4) were measured in the range 100–500°C. The thermograms of the blends display exothermic peaks of the thermal oxidation of EPC in the range 370–400°C and endothermic peaks of the thermal degradation of PHB at T > 250°C. For the pure PHB and EPC, the aforementioned peaks are observed in the ranges 200–300°C and 360–430°C, respectively. The blend samples studied in this work display two peaks each, thus confirming the existence of two phases. Note that the peak width increases (curves 3, 4 in Fig. 4), and the heat Q of the thermal degradation of PHB changes in all the blends studied here (Table 2). This effect is apparently determined by the blend structure rather than by its composition. In blends, PHB becomes more active compared to the pure polymer and the rate of its thermal degradation increases. The temperature corresponding to the onset of thermal degradation 7^ decreases from 255°C; the value characteristic of the pure PHB, to 180°C (Table 2). The structure of the polymer becomes less perfect in this case; two likely reasons for this are a change in the morphology and the appearance of an extended interfacial layer.

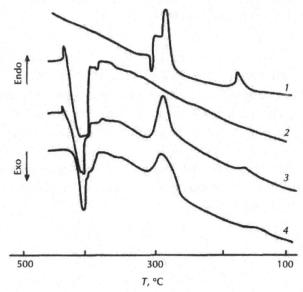

FIGURE 4   The DSC traces of (1) PHB, (2) EPC, and their blends with compositions (3) 70:30 and (4) 30:70 wt %.

As to EPC, it acquires a higher thermal stability in the blends under examination, as indicated by the increase in the temperature corresponding to the onset of its thermal oxidation T° (Table 2). The position of the exothermic peaks on the temperature scale characteristic of EPC indicates that its activity in blends is lower than that in the pure sample. The low-temperature shoulder of the exothermic EPC peak in the range 360–380°C (Fig. 4) decreases with increasing content of PHB. Apparently, this effect is due to a change in the copolymer structure related to the interpenetration of PHB and EPC segments.

TABLE 2   The parameters of the thermal degradation process.

PHB : EPC, wt%	Tod (EPC), °C	Tod (PHB) °C	Q*(PHB), kJ/g
100 : 0	–	255	0.53
70 : 30	370	180	1.38
30 : 70	380	250	0.51
0 : 100	360	–	–

* The specific heat of thermal degradation per g of PHB.

Thus, the existence of two peaks in DSC thermograms of the blends indicates the presence of two phases in the PHB-EPC blends. The phase inversion takes place in the vicinity of the composition with equal component weights. The components influence each other during film formation, and, hence, the appearance of the extended interfacial layer is presumed for samples containing more than 50% EPC. A change in the structure of the blends affects their thermal degradation. The degradation of PHB in blends is more pronounced than that in the pure PHB, but the thermal oxidation of EPC is retarded.

KEYWORDS

- **Exothermic peak**
- **Interfacial layer**
- **Thermal degradation**
- **Trans-zigzag**

REFERENCES

1. Seebach, D.; Brunner, A.; Bachmann, B. M.; Hoffman, T.; Kuhnle, F. N. M.; Lengweier, U. D. In *Biopolymers and Biooligomers of (R)-3-Hydroxyalkanoic Acids: Contribution of Synthetic Organic Chemists*, Edgenos-sische Technicshe Hochschule: Zurich, 1996.
2. Ol'khov, A. A.; Vlasov, S. V.; Shibryaeva, L. S.; Litvi-nov, I. A.; Tarasova, N. A.; Kosenko, R. Yu.; Iordan-Skii, A. L. *Poly. Sci., Ser. A*, **2000**, *42(4)*, 447.
3. Elliot, A. In *Infra-Red Spectra and Structure of Organic Long-Chain Polymers*, Edward Arnold: London, 1969.
4. Lipatov, Yu. S. In *Mezhfaznye yavleniya v polimerakh (Interphase Phenomena in Polymers)*, Naukova Dumka: Kiev, 1980.
5. Labeek, G.; Vorenkamp, E. J.; Schouten, A. J. *Macromolecules*, **1995**, *28(6)*, 2023.
5. Painter, P. C.; Coleman, M. M.; Koenig, J. L. In *The Theory of Vibrational Spectroscopy and Its Application to Polymeric Materials*, Wiley: New York, 1982.

CHAPTER 6

ANALYSIS METHODS OF SOME NANOCOMPOSITES STRUCTURE

KOZLOV GEORGII VLADIMIROVICH, YANOVSKII YURII GRIGOR'EVICH, ZAIKOV GENNADY EFREMOVICH, ELI M. PEARCE, and STEFAN KUBICA

CONTENTS

6.1   Introduction.. 76

6.2   Experimental 78

6.3   Results and Discussion 79

6.4   Conclusion 105

Keywords 105

References.. 106

6.1   INTRODUCTION

The modern methods of experimental and theoretical analysis of polymer materials structure and properties allow not only to confirm earlier propounded hypotheses, but to obtain principally new results. Let us consider some important problems of particulate-filled polymer nanocomposites, the solution of which allows to advance substantially in these materials properties understanding and prediction. Polymer nanocomposites multicomponentness (multiphaseness) requires their structural components quantitative characteristics determination. In this aspect interfacial regions play a particular role, since it has been shown earlier, that they are the same reinforcing element in elastomeric nanocomposites as nanofiller actually [1]. Therefore, the knowledge of interfacial layer dimensional characteristics is necessary for quantitative determination of one of the most important parameters of polymer composites in general their reinforcement degree [2, 3].

The aggregation of the initial nanofiller powder particles in more or less large particles aggregates always occurs in the course of technological process of preparation particulate-filled polymer composites in general [4] and elastomeric nanocomposites in particular [5]. The aggregation process tells on composites (nanocomposites) macroscopic properties [2–4]. For nanocomposites nanofiller aggregation process gains special significance, since its intensity can be the one, that nanofiller particles aggregates size exceeds 100 nm – the value, which is assumed (though conditionally enough [6]) as an upper dimensional limit for nanoparticle. In other words, the aggregation process can result to the situation when primordially supposed nanocomposite ceases to be one. Therefore at present several methods exist, which allow to suppress nanoparticles aggregation process [5, 7]. This also assumes the necessity of the nanoparticles aggregation process quantitative analysis.

It is well known [8, 9], that in particulate-filled elastomeric nanocomposites (rubbers) nanofiller particles form linear spatial structures ("chains"). At the same time in polymer composites, filled with disperse microparticles (microcomposites) particles (aggregates of particles) of filler form a fractal network, which defines polymer matrix structure (analog of fractal lattice in computer simulation) [4]. This results to different mechanisms of polymer matrix structure formation in micro- and nano-composites. If in the first filler particles (aggregates of particles) fractal network availability results to "disturbance" of polymer matrix structure, that is expressed in the increase of its fractal dimension d_f [4], then in case of polymer nanocomposites at nanofiller contents change the value d_f is not changed and equal to matrix polymer structure fractal dimension [10]. As it has been expected, the change of the composites of the indicated classes structure formation mechanism change defines their properties, in particular, reinforcement degree [11, 12]. Therefore, nanofiller structure fractality strict proof and its dimension determination are necessary.

As it is known [13, 14], the scale effects in general are often found at different materials mechanical properties study. The dependence of failure stress on grain size for metals (Holl-Petsch formula) [15] or of effective filling degree on filler particles size in case of polymer composites [16] are examples of such effect. The strong dependence of elasticity modulus on nanofiller particles diameter is observed for particulate-filled elastomeric nanocomposites [5]. Therefore, it is necessary to elucidate the physical grounds of nano- and micromechanical behavior scale effect for polymer nanocomposites.

At present a wide list of disperse material is known, which is able to strengthen elastomeric polymer materials [5]. These materials are very diverse on their surface chemical constitution, but particles small size is a common feature for them. On the basis of this observation the hypothesis was offered, that any solid material would strengthen the rubber at the condition that it is in a very dispersed state and it could be dispersed in polymer matrix. Edwards [5] points out, that filler particles small size is necessary and, probably, the main requirement for reinforcement effect realization in rubbers. Using modern terminology, one can say, that for rubbers reinforcement the nanofiller particles, for which their aggregation process is suppressed as far as possible, would be the most effective ones [3, 12]. Therefore, the theoretical analysis of a nanofiller particles size influence on polymer nanocomposites reinforcement is necessary.

Proceeding from the said above, the present work purpose is the solution of the considered above paramount problems with the help of modern experimental and theoretical techniques on the example of particulate-filled butadiene-styrene rubber.

6.2   EXPERIMENTAL

The made industrially butadiene-styrene rubber of mark SKS-30, which contains 7.0–12.3% cis- and 71.8–72.0% trans-bonds, with density of 920–930 kg/m³ was used as matrix polymer. This rubber is fully amorphous one.

Fullerene-containing mineral shungite of Zazhoginsk's deposit consists of ~30% globular amorphous metastable carbon and ~70% high-disperse silicate particles. Besides, industrially made technical carbon of mark № 220 was used as nanofiller. The technical carbon, nano- and microshugite particles average size makes up 20, 40 and 200 nm, respectively. The indicated filler content is equal to 37 mass %. Nano- and microdimensional disperse shungite particles were prepared from industrially output material by the original technology processing. The size and polydispersity analysis of the received in milling process shungite particles was monitored with the aid of analytical disk centrifuge (CPS Instruments, Inc., USA), allowing to determine with high precision size and distribution by the sizes within the range from 2 nm up to 50 mcm.

Nanostructure was studied on atomic-forced microscopes Nano-DST (Pacific Nanotechnology, USA) and Easy Scan DFM (Nanosurf, Switzerland) by semi-contact method in the force modulation regime. Atomic-force microscopy results were processed with the help of specialized software package SPIP (Scanning Probe Image Processor, Denmark). SPIP is a powerful programmes package for processing of images, obtained on SPM, AFM, STM, scanning electron microscopes, transmission electron microscopes, interferometers, confocal microscopes, profilometers, optical microscopes and so on. The given package possesses the whole functions number, which are necessary at images precise analysis, in a number of which the following ones are included:

- the possibility of three-dimensional reflecting objects obtaining, distortions automatized leveling, including Z-error mistakes removal for examination of separate elements and so on;
- quantitative analysis of particles or grains, more than 40 parameters can be calculated for each found particle or pore: area, perimeter, mean diameter, the ratio of linear sizes of grain width to its height,

distance between grains, coordinates of grain center of mass a.a. can be presented in a diagram form or in a histogram form.

The tests on elastomeric nanocomposites nanomechanical properties were carried out by a nanointentation method [17] on apparatus Nano Test 600 (Micro Materials, Great Britain) in loades wide range from 0.01 mN up to 2.0 mN. Sample indentation was conducted in 10 points with interval of 30 mcm. The load was increased with constant rate up to the greatest given load reaching (for the rate 0.05 mN/s^{-1} mN). The indentation rate was changed in conformity with the greatest load value counting, that loading cycle should take 20 s. The unloading was conducted with the same rate as loading. In the given experiment the "Berkovich indentor" was used with the angle at the top of 65.3° and rounding radius of 200 nm. Indentations were carried out in the checked load regime with preload of 0.001 mN.

For elasticity modulus calculation the obtained in the experiment by nanoindentation course dependences of load on indentation depth (strain) in ten points for each sample at loads of 0.01, 0.02, 0.03, 0.05, 0.10, 0.50, 1.0 and 2.0 mN were processed according to Oliver-Pharr method [18].

6.3 RESULTS AND DISCUSSION

In Fig. 1 the obtained according to the original methodics results of elasticity moduli calculation for nanocomposite butadiene-styrene rubber/nanoshungite components (matrix, nanofiller particle and interfacial layers), received in interpolation process of nanoindentation data, are presented. The processed in polymer nanocomposite SPIP image with shungite nanoparticles allows experimental determination of interfacial layer thickness l_{if}, which is presented in Fig. 1 as steps on elastomeric matrix-nanofiller boundary. The measurements of 34 such steps (interfacial layers) width on the processed in SPIP images of interfacial layer various section gave the mean experimental value $l_{if} = 8.7$ nm. Besides, nanoindentation results (Fig. 1, figures on the right) showed, that interfacial layers elasticity modulus was only by 23–45% lower than nanofiller elasticity modulus, but it was higher than the corresponding parameter of polymer matrix in 6.0–8.5 times. These experimental data confirm, that for the studied nanocomposite interfacial layer is a reinforcing element to the same extent, as nanofiller actually [1, 3, 12].

Let us fulfill further the value l_{if} theoretical estimation according to the two methods and compare these results with the ones obtained experimentally. The first method simulates interfacial layer in polymer composites as a result of interaction of two fractals – polymer matrix and nanofiller surface [19, 20]. In this case there is a sole linear scale l, which defines these fractals interpenetration distance [21]. Since nanofiller elasticity modulus is essentially higher, than the corresponding parameter for rubber (in the considered case – in 11 times, see Fig. 1), then the indicated interaction reduces to nanofiller indentation in polymer matrix and then $l=l_{if}$. In this case it can be written [21]:

$$l_{if} \approx a \left(\frac{R_p}{a} \right)^{2(d-d_{surf})/d}$$ (1)

where a is a lower linear scale of fractal behavior, which is accepted for polymers as equal to statistical segment length l_{st} [22], R_p is a nanofiller particle (more precisely, particles aggregates) radius, which for nanoshungite is equal to ~ 84 nm [23], d is dimension of Euclidean space, in which fractal is considered (it is obvious, that in our case $d=3$), d_{surf} is fractal dimension of nanofiller particles aggregate surface.

The value l_{st} is determined as follows [24]:

$$l_{st} = l_0 C_\infty$$ (2)

where l_0 is the main chain skeletal bond length, which is equal to 0.154 nm for both blocks of butadiene-styrene rubber [25], C is characteristic ratio, which is a polymer chain statistical flexibility indicator [26], and is determined with the help of the equation [22]:

$$T_g = 129 \left(\frac{S}{C_\infty} \right)^{1/2}$$ (3)

where T_g is glass transition temperature, equal to 217 K for butadiene-styrene rubber [3], S is macromolecule cross-sectional area, determined for the mentioned rubber according to the additivity rule from the following considerations. As it is known [27], the macromolecule diameter quadrate values are equal: for polybutadiene – 20.7 Å² and for polystyrene – 69.8 Å². Having calculated cross-sectional area of macromolecule, simulated as a cylinder, for the indicated polymers according to the known geometrical formulas, let us obtain 16.2 and 54.8 Å², respectively. Further, accepting as S the average value of the adduced above areas, let us obtain for butadiene-styrene rubber S=35.5 Å². Then according to the Eq. (3) at the indicated values T_g and S let us obtain C_∞=12.5 and according to the equation (2) – l_{st}=1.932 nm.

The fractal dimension of nanofiller surface d_{surf} was determined with the help of the equation [3]:

$$S_u = 410 R_p^{d_{surf}-d} ,$$ (4)

where S_u is nanoshungite particles specific surface, calculated as follows [28]:

$$S_u = \frac{3}{\rho_n R_p} ,$$ (5)

where ρ_n is the nanofiller particles aggregate density, determined according to the formula [3]:

$$\rho_n = 0.188\left(R_p\right)^{1/3}$$ (6)

The calculation according to the Eqs. (4)–(6) gives d_{surf}=2.44. Further, using the calculated by the indicated mode parameters, let us obtain from the Eq. (1) the theoretical value of interfacial layer thickness l_{if}^T=7.8 nm. This value is close enough to the obtained one experimentally (their discrepancy makes up ~10%).

The second method of value l_{if}^T estimation consists in using of the two following equations [3, 29]:

$$\varphi_{if} = \varphi_n\left(d_{surf} - 2\right)$$ (7)

and

$$\varphi_{if} = \varphi_n\left[\left(\frac{R_p + l_{if}^T}{R_p}\right)^3 - 1\right],$$ (8)

where φ_{if} and φ_n are relative volume fractions of interfacial regions and nanofiller, accordingly.

The combination of the indicated equations allows to receive the following formula for l_{if}^T calculation:

$$l_{if}^T = R_p \left[\left(d_{surf} - 1 \right)^{1/3} - 1 \right] \tag{9}$$

The calculation according to the formula (9) gives for the considered nanocomposite l_{if}^T =10.8 nm, that also corresponds well enough to the experiment (in this case discrepancy between l_{if} and l_{if}^T makes up ~19%).

Let us note in conclusion the important experimental observation, which follows from the processed by programme SPIP results of the studied nanocomposite surface scan (Fig. 1). As one can see, at one nanoshungite particle surface from one to three (in average – two) steps can be observed, structurally identified as interfacial layers. It is significant that these steps width (or l_{if}) is approximately equal to the first (the closest to nanoparticle surface) step width. Therefore, the indicated observation supposes, that in elastomeric nanocomposites at average two interfacial layers are formed: the first – at the expense of nanofiller particle surface with elastomeric matrix interaction, as a result of which molecular mobility in this layer is frozen and its state is glassy-like one, and the second – at the expense of glassy interfacial layer with elastomeric polymer matrix interaction. The most important question from the practical point of view, whether one interfacial layer or both serve as nanocomposite reinforcing element. Let us fulfill the following quantitative estimation for this question solution. The reinforcement degree (E_n/E_m) of polymer nanocomposites is given by the equation [3]:

$$\frac{E_n}{E_m} = 1 + 11 \left(\varphi_n + \varphi_{if} \right)^{1.7}, \tag{10}$$

where E_n and E_m are elasticity moduli of nanocomposite and matrix polymer, accordingly (E_m=1.82 MPa [3]).

According to the Eq. (7) the sum $(\varphi_n + \varphi_{if})$ is equal to:

$$\varphi_n + \varphi_{if} = \varphi_n \left(d_{surf} - 1 \right), \tag{11}$$

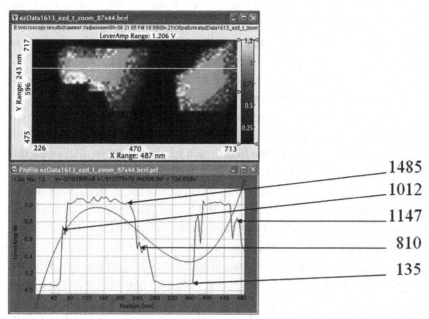

FIGURE 1 The processed in SPIP image of nanocomposite butadiene-styrene rubber/nanoshungite, obtained by force modulation method, and mechanical characteristics of structural components according to the data of nanoindentation (strain 150 nm).

if one interfacial layer (the closest to nanoshungite surface) is a reinforcing element and

$$\varphi_n + 2\varphi_{if} = \varphi_n\left(2d_{surf} - 3\right), \tag{12}$$

if both interfacial layers are a reinforcing element.

In its turn, the value φ_n is determined according to the equation [30]:

$$\varphi_n = \frac{W_n}{\rho_n}, \tag{13}$$

where W_n is nanofiller mass content, ρ_n is its density, determined according to the Eq. (6).

The calculation according to the Eqs. (11) and (12) gave the following E_n/E_m values: 4.60 and 6.65, respectively. Since the experimental value $E_n/E_m = 6.10$ is closer to the value, calculated according to the Eq. (12), then this means that both interfacial layers are a reinforcing element for the studied nanocomposites. Therefore, the coefficient 2 should be introduced in the equations for value l_{if} determination (for example, in the Eq. (1)) in case of nanocomposites with elastomeric matrix. Let us remind, that the Eq. (1) in its initial form was obtained as a relationship with proportionality sign, i.e., without fixed proportionality coefficient [21].

Thus, the used above nanoscopic methodics allow to estimate both interfacial layer structural special features in polymer nanocomposites and its sizes and properties. For the first time it has been shown, that in elastomeric particulate-filled nanocomposites two consecutive interfacial layers are formed, which are a reinforcing element for the indicated nanocomposites. The proposed theoretical methodics of interfacial layer thickness estimation, elaborated within the frameworks of fractal analysis, give well enough correspondence to the experiment.

For theoretical treatment of nanofiller particles aggregate growth processes and final sizes traditional irreversible aggregation models are inapplicable, since it is obvious, that in nanocomposites aggregates a large number of simultaneous growth takes place. Therefore the model of multiple growth, offered in paper [6], was used for nanofiller aggregation description.

In Fig. 2 the images of the studied nanocomposites, obtained in the force modulation regime, and corresponding to them nanoparticles aggregates fractal dimension d_f distributions are adduced. As it follows from the adduced values d_f^{ag} (d_f^{ag} =2.40-2.48), nanofiller particles aggregates in the studied nanocomposites are formed by a mechanism particle-cluster (P-Cl), i.e., they are Witten-Sander clusters [32]. The variant A, was chosen which according to mobile particles are added to the lattice, consisting of a large number of "seeds" with density of c_0 at simulation beginning [31]. Such model generates the structures, which have fractal geometry on length short scales with value $d_f \approx 2.5$ (see Fig. 2) and homogeneous structure on length large scales. A relatively high particles concentration c is required in the model for uninterrupted network formation [31].

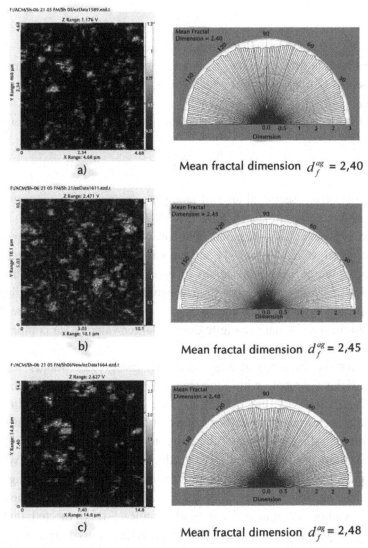

Mean fractal dimension $d_f^{ag} = 2,40$

Mean fractal dimension $d_f^{ag} = 2,45$

Mean fractal dimension $d_f^{ag} = 2,48$

FIGURE 2 The images, obtained on atomic-force microscope in the force modulation regime, for nanocomposites, filled with technical carbon (a), nanoshungite (b), microshungite (c) and corresponding to them fractal dimensions d_f^{ag}.

In case of "seeds" high concentration c_0 for the variant A the following relationship was obtained [31]:

$$R_{\max}^{d_f^{ag}} = N = c/c_0,$$ (14)

where $R_{\max}$ is nanoparticles cluster (aggregate) greatest radius, N is nanoparticles number per one aggregate, c is nanoparticles concentration, c_0 is "seeds" number, which is equal to nanoparticles clusters (aggregates) number.

The value N can be estimated according to the following equation [8]:

$$2R_{\max} = \left(\frac{S_n N}{\pi \eta}\right)^{1/2},$$ (15)

where S_n is cross-sectional area of nanoparticles, of which an aggregate consists, η is a packing coefficient, equal to 0.74 [28].

The experimentally obtained nanoparticles aggregate diameter $2R_{ag}$ was accepted as $2R_{\max}$ (Table 1) and the value S_n was also calculated according to the experimental values of nanoparticles radius r_n (Table 1). In Table 1 the values N for the studied nanofillers, obtained according to the indicated method, were adduced. It is significant that the value N is a maximum one for nanoshungite despite larger values r_n in comparison with technical carbon.

Further the Eq. (14) allows estimating the greatest radius $R_{\max}^T$ of nanoparticles aggregate within the frameworks of the aggregation model [31]. These values $R_{\max}^T$ are adduced in Table 1, from which their reduction in a sequence of technical carbon-nanoshungite-microshungite, that fully contradicts to the experimental data, i.e., to R_{ag} change (Table 1). However, we must not neglect the fact that the Eq. (14) was obtained within the frameworks of computer simulation, where the initial aggregating particles sizes are the same in all cases [31]. For real nanocomposites the values r_n can be distinguished essentially (Table 1). It is expected, that the value R_{ag} or $R_{\max}^T$ will be the higher, the larger is the radius of nanoparticles, forming aggregate, is r_n. Then theoretical value of nanofiller particles cluster (aggregate) radius R_{ag}^T can be determined as follows:

$$R_{ag}^T = k_n r_n N^{1/d_f^{ag}},$$ (16)

where k_n is proportionality coefficient, in the present work accepted empirically equal to 0.9.

The comparison of experimental R_{ag} and calculated according to the Eq. (16) R_{ag}^T values of the studied nanofillers particles aggregates radius shows their good correspondence (the average discrepancy of R_{ag} and R_{ag}^T makes up 11.4%). Therefore, the theoretical model [31] gives a good correspondence to the experiment only in case of consideration of aggregating particles real characteristics and, in the first place, their size.

Let us consider two more important aspects of nanofiller particles aggregation within the frameworks of the model [31]. Some features of the indicated process are defined by nanoparticles diffusion at nanocomposites processing. Specifically, length scale, connected with diffusible nanoparticle, is correlation length ξ of diffusion. By definition, the growth phenomena in sites, remote more than ξ, are statistically independent. Such definition allows connecting the value ξ with the mean distance between nanofiller particles aggregates L_n. The value ξ can be calculated according to the equation [31]:

$$\xi^2 \approx \tilde{n}^{-1} R_{ag}^{d_f^{ag} - d + 2},\tag{17}$$

where c is nanoparticles concentration, which should be accepted equal to nanofiller volume contents φ_n, which is calculated according to the Eqs. (6) and (13).

The values r_n and R_{ag} were obtained experimentally (see histogram of Fig. 3). In Fig. 4 the relation between L_n and ξ is adduced, which, as it is expected, proves to be linear and passing through coordinates origin. This means, that the distance between nanofiller particles aggregates is limited by mean displacement of statistical walks, by which nanoparticles are simulated. The relationship between L_n and ξ can be expressed analytically as follows:

$$L_n \approx 9.6\xi, \text{ nm.}\tag{18}$$

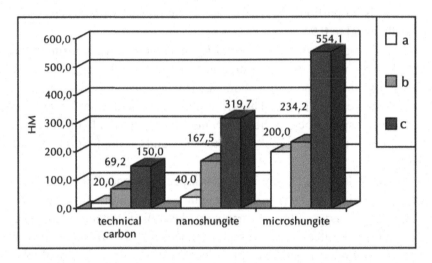

FIGURE 3   The initial particles diameter (a), their aggregates size in nanocomposite (b) and distance between nanoparticles aggregates (c) for nanocomposites, filled with technical carbon, nano- and microshungite.

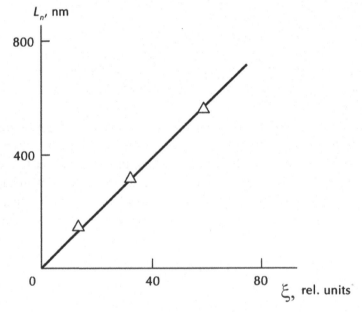

FIGURE 4   The relation between diffusion correlation length ξ and distance between nanoparticles aggregates L_n for considered nanocomposites.

The second important aspect of the model [31] in reference to nanofiller particles aggregation simulation is a finite nonzero initial particles concentration c or φ_n effect, which takes place in any real systems. This effect is realized at the condition $\xi \approx R_{ag}$, that occurs at the critical value $R_{ag}(R_c)$, determined according to the relationship [31]:

$$c \sim R_c^{d_f^{ag} - d}. \tag{19}$$

The Eq. (19) right side represents cluster (particles aggregate) mean density. This equation establishes that fractal growth continues only, until cluster density reduces up to medium density, in which it grows. The calculated according to the Eq. (19) values R_c for the considered nanoparticles are adduced in Table 1, from which follows, that they give reasonable correspondence with this parameter experimental values R_{ag} (the average discrepancy of R_c and R_{ag} makes up 24%).

TABLE 1   The parameters of irreversible aggregation model of nanofiller particles aggregates growth.

Nanofiller	R_{ag}, nm	r_n, nm	N	R_{max}^T, nm	R_{max}^T, nm	R_c, nm
Technical carbon	34.6	10	35.4	34.7	34.7	33.9
Nanoshungite	83.6	20	51.8	45.0	90.0	71.0
Microshungite	117.1	100	4.1	15.8	158.0	255.0

Since the treatment [31] was obtained within the frameworks of a more general model of diffusion-limited aggregation, then its correspondence to the experimental data indicated unequivocally, that aggregation processes in these systems were controlled by diffusion. Therefore, let us consider briefly nanofiller particles diffusion. Statistical walkers diffusion constant ζ can be determined with the aid of the relationship [31]:

$$\xi \approx (\zeta t)^{1/2}, \tag{20}$$

where t is walk duration.

The Eq. (20) supposes (at t=const) ζ increase in a number technical carbon-nanoshungite-microshungite as 196-1069-3434 relative units, i.e., diffusion intensification at diffusible particles size growth. At the same time diffusivity D for these particles can be described by the well-known Einstein's relationship [33]:

$$D = \frac{kT}{6\pi\eta r_n \alpha},$$ (21)

where k is Boltzmann constant, T is temperature, η is medium viscosity, α is numerical coefficient, which further is accepted equal to 1.

In its turn, the value η can be estimated according to the equation [34]:

$$\frac{\eta}{\eta_0} = 1 + \frac{2.5\varphi_n}{1 - \varphi_n},$$ (22)

where η_0 and η are initial polymer and its mixture with nanofiller viscosity, accordingly.

The calculation according to the Eqs. (21) and (22) shows, that within the indicated above nanofillers number the value D changes as 1.32-1.14-0.44 relative units, i.e., reduces in three times, that was expected. This apparent contradiction is due to the choice of the condition t = const (where t is nanocomposite production duration) in the Eq. (20). In real conditions the value t is restricted by nanoparticle contact with growing aggregate and then instead of t the value t/c_0 should be used, where c_0 is the seeds concentration, determined according to the Eq. (14). In this case the value ζ for the indicated nanofillers changes as 0.288-0.118-0.086, i.e., it reduces in 3.3 times that corresponds fully to the calculation according to the Einstein's relationship (the Eq. (21)). This means, that nanoparticles diffusion in polymer matrix obeys classical laws of Newtonian rheology [33].

Thus, the disperse nanofiller particles aggregation in elastomeric matrix can be described theoretically within the frameworks of a modified model of irreversible aggregation particle-cluster. The obligatory consideration of nanofiller initial particles size is a feature of the indicated model application to real systems description. The indicated particles diffusion in polymer matrix obeys classical laws of Newtonian liquids hydrodynamics. The offered approach allows predicting nanoparticles aggregates final parameters as a function of the initial particles size, their contents and other factors number.

At present there are several methods of filler structure (distribution) determination in polymer matrix, both experimental [10, 35] and theoretical [4]. All the indicated methods describe this distribution by fractal dimension D_n of filler particles network. However, correct determination of any object fractal (Hausdorff) dimension includes three obligatory conditions. The first from them is the indicated above determination of fractal dimension numerical magnitude, which should not be equal to object topological dimension. As it is known [36], any real (physical) fractal possesses fractal properties within a certain scales range. Therefore, the second condition is the evidence of object self-similarity in this scales range [37]. And at last, the third condition is the correct choice of measurement scales range itself. As it has been shown in papers [38, 39], the minimum range should exceed at any rate one self-similarity iteration.

The first method of dimension D_n experimental determination uses the following fractal relationship [40, 41]:

$$D_n = \frac{\ln N}{\ln \rho},\tag{23}$$

where N is a number of particles with size ρ.

Particles sizes were established on the basis of atomic-power microscopy data (see Fig. 2). For each from the three studied nanocomposites no less than 200 particles were measured, the sizes of which were united into 10 groups and mean values N and ρ were obtained. The dependences $N(\rho)$ in double logarithmic coordinates were plotted, which proved to be linear and the values D_n were calculated according to their slope (see Fig. 5). It is obvious, that at such approach fractal dimension D_n is determined in two-dimensional Euclidean space, whereas real nanocomposite should be considered in three-dimensional Euclidean space. The following relationship can be used for D_n re-calculation for the case of three-dimensional space [42]:

$$D3 = \frac{d + D2 \pm \left[(d - D2)^2 - 2\right]^{1/2}}{2},\tag{24}$$

where $D3$ and $D2$ are corresponding fractal dimensions in three- and two-dimensional Euclidean spaces, $d = 3$.

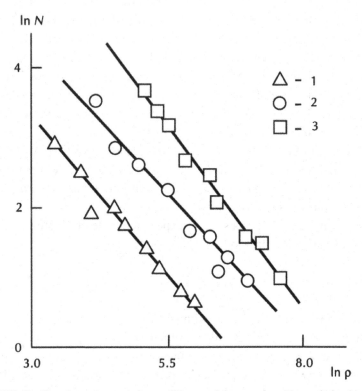

FIGURE 5   The dependences of nanofiller particles number N on their size ρ for nanocomposites BSR/TC (1), BSR/nanoshungite (2) and BSR/microshungite (3).

The calculated according to the indicated method dimensions D_n are adduced in Table 2. As it follows from the data of this table, the values D_n for the studied nanocomposites are varied within the range of 1.10–1.36, i.e., they characterize more or less branched linear formations ("chains") of nanofiller particles (aggregates of particles) in elastomeric nanocomposite structure. Let us remind that for particulate-filled composites polyhydroxiether/graphite the value D_n changes within the range of ~ 2.30–2.80 [4, 10], i.e., for these materials filler particles network is a bulk object, but not a linear one [36].

TABLE 2 The dimensions of nanofiller particles (aggregates of particles) structure in elastomeric nanocomposites.

Nanocomposite	D_n, the equation (23)	D_n, the equation (25)	d_0	d_{surf}	φ_n	D_n, the equation (29)
BSR/TC	1.19	1.17	2.86	2.64	0.48	1.11
BSR/nanoshungite	1.10	1.10	2.81	2.56	0.36	0.78
BSR/microshungite	1.36	1.39	2.41	2.39	0.32	1.47

Another method of D_n experimental determination uses the so-called "quadrates method" [43]. Its essence consists in the following. On the enlarged nanocomposite microphotograph (see Fig. 2) a net of quadrates with quadrate side size α_i, changing from 4.5 up to 24 mm with constant ratio $\alpha_{i+1}/\alpha_i=1.5$, is applied and then quadrates number N_i, in to which nanofiller particles hit (fully or partly), is counted up. Five arbitrary net positions concerning microphotograph were chosen for each measurement. If nanofiller particles network is a fractal, then the following relationship should be fulfilled [43]:

$$N_i \sim S_i^{-D_n/2}, \tag{25}$$

where S_i is quadrate area, which is equal to α_i^2.

In Fig. 6 the dependences of N_i on S_i in double logarithmic coordinates for the three studied nanocomposites, corresponding to the Eq. (25), is adduced. As one can see, these dependences are linear, that allows determining the value D_n from their slope. The determined according to the Eq. (25) values D_n are also adduced in Table 2, from which a good correspondence of dimensions D_n, obtained by the two described above methods, follows (their average discrepancy makes up 2.1% after these dimensions re-calculation for three-dimensional space according to the Eq. (24)).

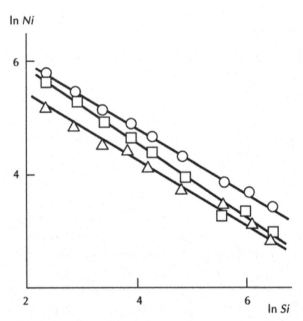

FIGURE 6 The dependences of covering quadrates number Ni on their area Si, corresponding to the relationship (25), in double logarithmic coordinates for nanocomposites on the basis of BSR. The designations are the same, that in Fig. 5.

As it has been shown in the chapter [44], the usage for self-similar fractal objects at the Eq. (25) the condition should be fulfilled:

$$N_i - N_{i-1} \sim S_i^{-D_n}.$$ (26)

In Fig. 7 the dependence, corresponding to the Eq. (26), for the three studied elastomeric nanocomposites is adduced. As one can see, this dependence is linear, passes through coordinates origin, that according to the Eq. (26) is confirmed by nanofiller particles (aggregates of particles) "chains" self-similarity within the selected α_i range. It is obvious, that this self-similarity will be a statistical one [44]. Let us note, that the points, corresponding to α_i=16 mm for nanocomposites butadiene-styrene rubber/ technical carbon (BSR/TC) and butadiene-styrene rubber/microshungite (BSR/microshungite), do not correspond to a common straight line. Accounting for electron microphotographs of Fig. 2 enlargement this gives the self-similarity range for nanofiller "chains" of 464–1472 nm. For nanocomposite butadiene-styrene rubber/nanoshungite (BSR/nanoshungite), which has no points deviating from a straight line of Fig. 7, α_i range makes up 311–1510 nm, that corresponds well enough to the indicated above self-similarity range.

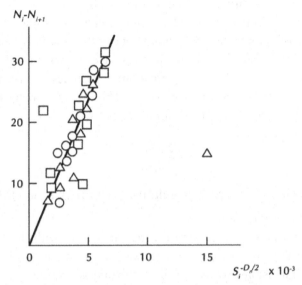

FIGURE 7   The dependences of $(N_i\text{-}N_{i+1})$ on the value $S_i^{-D_n/2}$, corresponding to the relationship (26), for nanocomposites on the basis of BSR. The designations are the same, that in Fig. 5.

In papers of Refs. [38, 39], it has been shown that measurement scales S_i minimum range should contain at least one self-similarity iteration. In this case the condition for ratio of maximum S_{max} and minimum S_{min} areas of covering quadrates should be fulfilled [39]:

$$\frac{S_{max}}{S_{min}} > 2^{2/D_n}. \tag{27}$$

Hence, accounting for the defined above restriction let us obtain $S_{max}/S_{min}=121/20.25=5.975$, that is larger than values $2^{2/D_n}$ for the studied nanocomposites, which are equal to 2.71–3.52. This means, that measurement scales range is chosen correctly.

The self-similarity iterations number μ can be estimated from the inequality [39]:

$$\left(\frac{S_{max}}{S_{min}}\right)^{D_n/2} > 2^\mu . \tag{28}$$

Using the indicated above values of the included in the inequality (28) parameters, $\mu=1.42-1.75$ is obtained for the studied nanocomposites, i.e., in our experiment conditions self-similarity iterations number is larger than unity, that again confirms correctness of the value D_n estimation [35].

And let us consider in conclusion the physical grounds of smaller values D_n for elastomeric nanocomposites in comparison with polymer microcomposites, i.e., the causes of nanofiller particles (aggregates of particles) "chains" formation in the first ones. The value D_n can be determined theoretically according to the equation [4]:

$$\varphi_{if} = \frac{D_n + 2.55d_0 - 7.10}{4.18}, \tag{29}$$

where φ_{if} is interfacial regions relative fraction, d_0 is nanofiller initial particles surface dimension.

The dimension d_0 estimation can be carried out with the help of the Eq. (4) and the value φ_{if} can be calculated according to the Eq. (7). The results of dimension D_n theoretical calculation according to the Eq. (29) are adduced in Table 2, from which a theory and experiment good correspondence follows. The Eq. (29) indicates unequivocally to the cause of a filler in nano- and microcomposites different behavior. The high (close to 3, see Table 2) values d_0 for nanoparticles and relatively small ($d_0=2.17$ for graphite [4]) values d_0 for microparticles at comparable values φ_{if} is such cause for composites of the indicated classes [3, 4].

Hence, the stated above results have shown, that nanofiller particles (aggregates of particles) "chains" in elastomeric nanocomposites are physical fractal within self-similarity (and, hence, fractality [41]) range of ~500–1450 nm. In this range their dimension D_n can be estimated according to the Eqs. (23), (25) and (29). The cited examples demonstrate the necessity of the measurement scales range correct choice. As it has been noted earlier [45], the linearity of the plots, corresponding to the Eqs. (23) and (25), and D_n nonintegral value do not guarantee object self-similarity (and, hence, fractality). The nanofiller particles (aggregates of particles) structure low dimensions are due to the initial nanofiller particles surface high fractal dimension.

In Fig. 8 the histogram is adduced, which shows elasticity modulus E change, obtained in nanoindentation tests, as a function of load on indenter P or nanoindentation depth h. Since for all the three considered nanocomposites the dependences $E(P)$ or $E(h)$ are identical qualitatively, then further the dependence $E(h)$ for nanocomposite BSR/TC was chosen, which reflects the indicated scale effect quantitative aspect in the most clearest way.

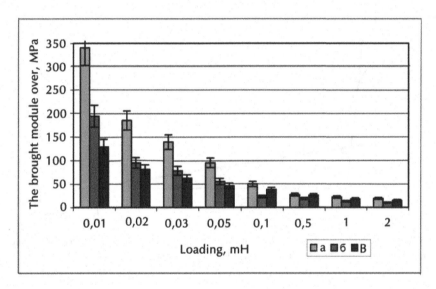

FIGURE 8   The dependences of reduced elasticity modulus on load on indentor for nanocomposites on the basis of butadiene-styrene rubber, filled with technical carbon (a), micro- (b) and nanoshungite (c).

In Fig. 9 the dependence of E on h_{pl} (see Fig. 10) is adduced, which breaks down into two linear parts. Such dependences elasticity modulus – strain are typical for polymer materials in general and are due to intermolecular bonds anharmonicity [46]. In chapter [47] it has been shown that the dependence $E(h_{pl})$ first part at $h_{pl} \leq 500$ nm is not connected with relaxation processes and has a purely elastic origin. The elasticity modulus E on this part changes in proportion to h_{pl} as:

$$E = E_0 + B_0 h_{pl}, \tag{30}$$

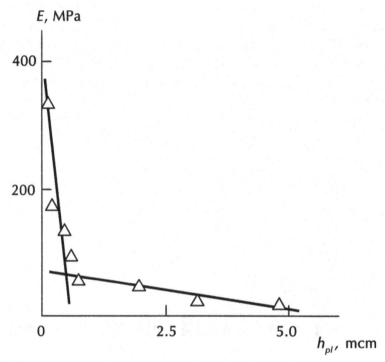

FIGURE 9 The dependence of reduced elasticity modulus E, obtained in nanoindentation experiment, on plastic strain h_{pl} for nanocomposites BSR/TC.

where E_0 is "initial" modulus, i.e. modulus, extrapolated to $h_{pl}=0$, and the coefficient B_0 is a combination of the first and second kind elastic constants. In the considered case $B_0<0$. Further Grüneisen parameter γ_L, characterizing intermolecular bonds anharmonicity level, can be determined [47]:

$$\gamma_L \approx -\frac{1}{6} - \frac{1}{2}\frac{B_0}{E_0}\frac{1}{(1-2\nu)}, \tag{31}$$

where ν is Poisson ratio, accepted for elastomeric materials equal to $\sim$ 0.475 [36].

Calculation according to the equation (31) has given the following values γ_L: 13.6 for the first part and 1.50 – for the second one. Let us note the first from γ_L adduced values is typical for intermolecular bonds, whereas the second value γ_L is much closer to the corresponding value of Grüneisen parameter G for intrachain modes [46].

Poisson's ratio ν can be estimated by γ_L (or G) known values according to the formula [46]:

$$\gamma_L = 0.7\left(\frac{1+\nu}{1-2\nu}\right). \tag{32}$$

The estimations according to the equation (32) gave: for the dependence $E(h_{pl})$ first part $\nu=0.462$, for the second one $-\nu=0.216$. If for the first part the value ν is close to Poisson's ratio magnitude for nonfilled rubber [36], then in the second part case the additional estimation is required. As it is known [48], a polymer composites (nanocomposites) Poisson's ratio value ν_n can be estimated according to the equation:

$$\frac{1}{\nu_n} = \frac{\varphi_n}{\nu_{TC}} + \frac{1-\varphi_n}{\nu_m}, \tag{33}$$

where φ_n is nanofiller volume fraction, ν_{TC} and ν_m are nanofiller (technical carbon) and polymer matrix Poisson's ratio, respectively.

The value ν_m is accepted equal to 0.475 [36] and the magnitude ν_{TC} is estimated as follows [49]. As it is known [50], the nanoparticles TC aggregates fractal dimension d_f^{ag} value is equal to 2.40 and then the value ν_{TC} can be determined according to the equation [50]:

$$d_f^{ag} = (d-1)(1+\nu_{TC}). \tag{34}$$

According to the formula (34) $\nu_{TC}=0.20$ and calculation ν_n according to the Eq. (33) gives the value 0.283, that is close enough to the value $\nu=0.216$ according to the Eq. (32) estimation. The obtained by the indicated methods values ν and ν_n comparison demonstrates, that in the dependence $E(h_{pl})$ ($h_{pl}<0.5$ mcm) the first part in nanoindentation tests only rubber-like polymer matrix ($\nu=\nu_m\approx0.475$) is included and in this dependence the second part – the entire nanocomposite as homogeneous system [51] $-\nu=\nu_n\approx0.22$.

Let us consider further E reduction at h_{pl} growth (Fig. 9) within the frameworks of density fluctuation theory, which value ψ can be estimated as follows [22]:

$$\psi = \frac{\rho_n kT}{K_T},\tag{35}$$

where ρ_n is nanocomposite density, k is Boltzmann constant, T is testing temperature, K_T is isothermal modulus of dilatation, connected with Young's modulus E by the relationship [46]:

$$K_T = \frac{E}{3(1-\nu)}.\tag{36}$$

In Fig. 10 the scheme of volume of the deformed at nanoindentation material V_{def} calculation in case of Berkovich indentor using is adduced and in Fig. 11 the dependence $\psi(V_{def})$ in logarithmic coordinates was shown. As it follows from the data of Fig. 11, the density fluctuation growth is observed at the deformed material volume increase. The plot $\psi(\ln V_{def})$ extrapolation to $\psi=0$ gives $\ln V_{def} \approx 13$ or $V_{def}(V_{def}^{cr})=4.42\approx10^5$ nm^3. Having determined the linear scale l_{cr} of transition to $\psi=0$ as $(V_{def}^{cr})^{1/3}$, let us obtain $l_{cr}=75.9$ nm, that is close to nanosystems dimensional range upper boundary (as it was noted above, conditional enough [6]), which is equal to 100 nm. Thus, the stated above results suppose, that nanosystems are such systems, in which density fluctuations are absent, always taking place in microsystems.

As it follows from the data of Fig. 9, the transition from nano- to microsystems occurs within the range $h_{pl}=408$–726 nm. Both the indicated above values h_{pl} and the corresponding to them values $(V_{def})^{1/3}\approx814$–$1440$ nm can be chosen as the linear length scale l_n, corresponding to this transition. From the comparison of these values l_n with the distance between nanofiller particles aggregates L_n ($L_n=219.2$-788.3 nm for the considered nanocomposites, see Fig. 3) it follows, that for transition from nano- to microsystems l_n should include at least two nanofiller particles aggregates and surrounding them layers of polymer matrix, that is the lowest linear scale of nanocomposite simulation as a homogeneous system. It is easy to see, that nanocomposite structure homogeneity condition is harder than the obtained above from the criterion $\psi=0$. Let us note, that such method, namely, a nanofiller particle and surrounding it polymer matrix layers separation, is widespread at a relationships derivation in microcomposite models.

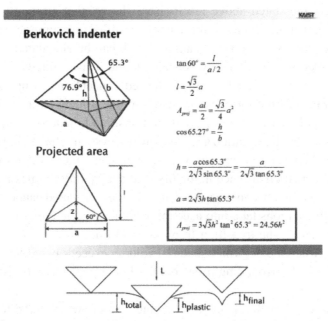

Berkovich indenter

$$\tan 60^\circ = \frac{l}{a/2}$$

$$l = \frac{\sqrt{3}}{2}a$$

$$A_{proj} = \frac{al}{2} = \frac{\sqrt{3}}{4}a^2$$

$$\cos 65.27^\circ = \frac{h}{b}$$

Projected area

$$h = \frac{a\cos 65.3^\circ}{2\sqrt{3}\sin 65.3^\circ} = \frac{a}{2\sqrt{3}\tan 65.3^\circ}$$

$$a = 2\sqrt{3}h\tan 65.3^\circ$$

$$A_{proj} = 3\sqrt{3}h^2 \tan^2 65.3^\circ = 24.56h^2$$

FIGURE 10 The schematic image of Berkovich indentor and nanoindentation process.

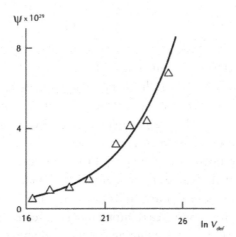

FIGURE 11 The dependence of density fluctuation on volume of deformed in nanoindentation process material V_{def} in logarithmic coordinates for nanocomposites BSR/TC.

It is obvious, that the Eq. (35) is inapplicable to nanosystems, since $\psi \rightarrow 0$ assumes $K_T \rightarrow \infty$ that is physically incorrect. Therefore the value E_0, obtained by the dependence $E(h_{pl})$ extrapolation (see Fig. 9) to $h_{pl}=0$, should be accepted as E for nanosystems [49].

Hence, the stated above results have shown, that elasticity modulus change at nanoindentation for particulate-filled elastomeric nanocomposites is due to a number of causes, which can be elucidated within the frameworks of anharmonicity conception and density fluctuation theory. Application of the first from the indicated conceptions assumes, that in nanocomposites during nanoindentation process local strain is realized, affecting polymer matrix only, and the transition to macrosystems means nanocomposite deformation as homogeneous system. The second from the mentioned conceptions has shown, that nano- and microsystems differ by density fluctuation absence in the first and availability of ones in the second. The last circumstance assumes, that for the considered nanocomposites density fluctuations take into account nanofiller and polymer matrix density difference. The transition from nano- to microsystems is realized in the case, when the deformed material volume exceeds nanofiller particles aggregate and surrounding it layers of polymer matrix combined volume [49].

In the work of Ref. [3], the following formula was offered for elastomeric nanocomposites reinforcement degree E_n/E_m description:

$$\frac{E_n}{E_m} = 15.2\left[1-\left(d-d_{surf}\right)^{1/t}\right], \tag{37}$$

where t is index percolation, equal to 1.7 [28].

From the Eq. (37) it follows, that nanofiller particles (aggregates of particles) surface dimension d_{surf} is the parameter, controlling nanocomposites reinforcement degree [53]. This postulate corresponds to the known principle about numerous division surfaces decisive role in nanomaterials as the basis of their properties change [54]. From the Eqs. (4)–(6) it follows unequivocally, that the value d_{surf} is defined by nanofiller particles (aggregates of particles) size R_p only. In its turn, from the Eq. (37) it follows, that elastomeric nanocomposites reinforcement degree E_n/E_m is defined by the dimension d_{surf} only, or, accounting for the said above, by the size R_p only. This means, that the reinforcement effect is controlled by nanofiller particles (aggregates of particles) sizes only and in virtue of this is the true nanoeffect.

In Fig. 12 the dependence of E_n/E_m on $(d-d_{surf})^{1/1.7}$ is adduced, corresponding to the Eq. (37), for nanocomposites with different elastomeric matrices (natural and butadiene-styrene rubbers, NR and BSR, accordingly) and different nanofillers (technical carbon of different marks, nano- and microshungite). Despite the indicated distinctions in composition, all adduced data are described well by the equation (37).

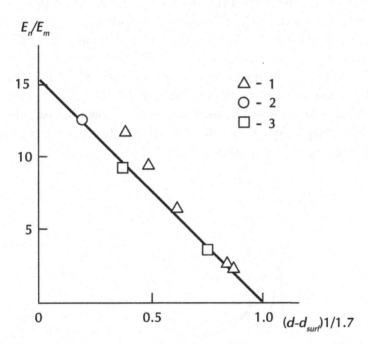

FIGURE 12 The dependence of reinforcement degree E_n/E_m on parameter $(d-d_{surf})^{1/1.7}$ value for nanocomposites NR/TC (1), BSR/TC (2) and BSR/shungite (3).

In Fig. 13 two theoretical dependences of E_n/E_m on nanofiller particles size (diameter D_p), calculated according to the Eqs. (4)–(6) and (37), are adduced. However, at the curve 1 calculation the value D_p for the initial nanofiller particles was used and at the curve 2 calculation – nanofiller particles aggregates size D_p^{ag} (see Fig. 3). As it was expected [5], the growth E_n/E_m at D_p or D_p^{ag} reduction, in addition the calculation with D_p

(nonaggregated nanofiller) using gives higher E_n/E_m values in comparison with the aggregated one (D_p^{ag} using). At $D_p \leq 50$ nm faster growth E_n/E_m at D_p reduction is observed than at $D_p > 50$ nm, that was also expected. In Fig. 13 the critical theoretical value D_p^{cr} for this transition, calculated according to the indicated above general principles [54], is pointed out by a vertical shaded line. In conformity with these principles the nanoparticles size in nanocomposite is determined according to the condition, when division surface fraction in the entire nanomaterial volume makes up about 50% and more. This fraction is estimated approximately by the ratio $3l_{if}/D_p$, where l_{if} is interfacial layer thickness. As it was noted above, the data of Fig. 1 gave the average experimental value $l_{if} \approx 8.7$ nm. Further from the condition $3l_{if}/D_p \approx 0.5$ let us obtain $D_p \approx 52$ nm that is shown in Fig. 13 by a vertical shaded line. As it was expected, the value $D_p \approx 52$ nm is a boundary one for regions of slow ($D_p > 52$ nm) and fast ($D_p \leq 52$ nm) E_n/E_m growth at D_p reduction. In other words, the materials with nanofiller particles size $D_p \leq 52$ nm ("superreinforcing" filler according to the terminology of paper [5]) should be considered true nanocomposites.

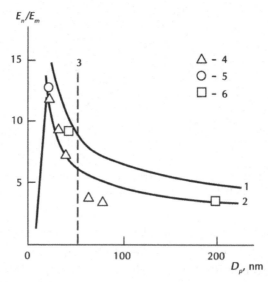

FIGURE 13   The theoretical dependences of reinforcement degree E_n/E_m on nanofiller particles size D_p, calculated according to the equations (4)-(6) and (37), at initial nanoparticles (1) and nanoparticles aggregates (2) size using. 3 – the boundary value D_p, corresponding to true nanocomposite. 4-6 – the experimental data for nanocomposites NR/TC (4), BSR/TC (5) and BSR/shungite (6).

Let us note in conclusion, that although the curves 1 and 2 of Fig. 13 are similar ones, nanofiller particles aggregation, which the curve 2 accounts for, reduces essentially enough nanocomposites reinforcement degree. At the same time the experimental data correspond exactly to the curve 2 that was to be expected in virtue of aggregation processes, which always took place in real composites [4] (nanocomposites [55]). The values d_{surf} obtained according to the Eqs. (4)–(6), correspond well to the determined experimentally ones. So, for nanoshungite and two marks of technical carbon the calculation by the indicated method gives the following d_{surf} values: 2.81, 2.78 and 2.73, whereas experimental values of this parameter are equal to: 2.81, 2.77 and 2.73, i.e., practically a full correspondence of theory and experiment was obtained.

6.4 CONCLUSION

Hence, the stated above results have shown, that the elastomeric reinforcement effect is the true nanoeffect, which is defined by the initial nanofiller particles size only. The indicated particles aggregation, always taking place in real materials, changes reinforcement degree quantitatively only, namely, reduces it. This effect theoretical treatment can be received within the frameworks of fractal analysis. For the considered nanocomposites the nanoparticle size upper limiting value makes up ~52 nm.

KEYWORDS

- Berkovich indentor
- Chains
- Disturbance
- Quantitative analysis
- Seeds

REFERENCES

1. Yanovskii Yu. G.; Kozlov G. V.; Karnet Yu. N. *Mekhanika Kompozitsionnykh Materialov i Konstruktsii*, **2011**, *17*(2), 203–208.
2. Malamatov A. Kh.; Kozlov G. V.; Mikitaev M. A. In *Reinforcement Mechanisms of Polymer Nanocomposites*. Moscow, Publishers of the D.I. Mendeleev RKhTU, 2006; 240 p.
3. Mikitaev A. K.; Kozlov G. V.; Zaikov G. E. In *Polymer Nanocomposites: Variety of Structural Forms and Applications*. Moscow, Nauka, 2009; 278 p.
4. Kozlov G. V.; Yanovskii Yu. G.; Karnet Yu. N. In *Structure and Properties of Particulate-Filled Polymer Composites: the Fractal Analysis*. Moscow, Al'yanstransatom, 2008; 363 p.
5. Edwards D. C. *J. Mater. Sci.*, **1990**, *25(12)*, 4175–4185.
6. Buchachenko A. L. *Uspekhi Khimii*, **2003**, *72(5)*, 419–437.
7. Kozlov G. V.; Yanovskii Yu. G., Burya A. I., Aphashagova Z. Kh. *Mekhanika Kompozitsionnykh Materialov i Konstruktsii*, 2007, v. 13, № 4, p. 479-492.
8. Lipatov Yu. S. In *The Physical Chemistry of Filled Polymers*. Moscow, Khimiya, 1977; 304 p.
9. Bartenev G. M.; Zelenev Yu. V. In *Physics and Mechanics of Polymers*. Moscow, Vysshaya Shkola, 1983; 391 p.
10. Kozlov G. V.; Mikitaev A. K. *Mekhanika Kompozitsionnykh Materialov i Konstruktsii*, **1996**, 2(3–4), 144–157.
11. Kozlov G. V.; Yanovskii Yu. G.; Zaikov G. E. In *Structure and Properties of Particulate-Filled Polymer Composites: the Fractal Analysis*. New York, Nova Science Publishers, Inc., 2010; 282 p.
12. Mikitaev A. K.; Kozlov G. V.; Zaikov G. E. Polymer Nanocomposites: Variety of Structural Forms and Applications. New York, Nova Science Publishers, Inc., 2008; 319 p.
13. McClintok F. A.; Argon A. S. Mechanical Behavior of Materials. Reading, Addison-Wesley Publishing Company, Inc., 1966; 440 p.
14. Kozlov G. V.; Mikitaev A. K. Doklady AN SSSR, **1987**, *294*(5), 129–1131.
15. Honeycombe R. W. K. The Plastic Deformation of Metals. Boston, Edward Arnold (Publishers), Ltd., 1968; 398 p.
16. Dickie R. A. In: *Polymer Blends. V. 1.* Paul D. R.; Newman S., Eds. New York, San-Francisco, London, Academic Press, 1980; p. 386–431.
17. Kornev Yu. V.; Yumashev O. B.; Zhogin V. A.; Karnet Yu. N.; Yanovskii Yu. G. *Kautschuk i Rezina*, **2008**, *6*, 18–23.
18. Oliver W. C.; Pharr G. M. *J. Mater. Res.*, **1992**, *7(6)*, 1564–1583.
19. Kozlov G. V.; Yanovskii Yu. G.; Lipatov Yu. S. Mekhanika Kompozitsionnykh Materialov i Konstruktsii, **2002**, *8(1)*, 111–149.
20. Kozlov G. V., Burya A. I., Lipatov Yu. S. Mekhanika Kompozitnykh Materialov, **2006**, *42(6)*, 797–802.
21. Hentschel H. G. E.; Deutch J. M. *Phys. Rev. A*, **1984**, *29(3)*, 1609–1611.
22. Kozlov G.V.; Ovcharenko E. N.; Mikitaev A. K. *Structure of Polymers Amorphous State*. Moscow, Publishers of the D.I. Mendeleev RKhTU, 2009; 392 p.

23. Yanovskii Yu. G.; Kozlov G. V. *Mater. VII Intern. Sci. Pract. Conf. "New Polymer Composite Materials"*. Nal'chik, KBSU, 2011; 189–194.
24. Wu S. J. *Polymer Sci.: Part B: Polymer Phys.*, **1989**, *27*(4), 723–741.
25. Aharoni S. M. *Macromolecules*, **1983**, *16*(9), 1722–1728.
26. Budtov V. P. The Physical Chemistry of Polymer Solutions. Sankt-Peterburg, Khimiya, 1992; 384 p.
27. Aharoni S. M. *Macromolecules*, **1985**, *18*(12), 2624–2630.
28. Bobryshev A. N.; Kozomazov V. N.; Babin L. O.; Solomatov V. I. *Synergetics of Composite Materials*. Lipetsk, NPO ORIUS, 1994; 154 p.
29. Kozlov G. V., Yanovskii Yu. G., Karnet Yu. N. *Mekhanika Kompozitsionnykh Materialov i Konstruktsii*, **2005**, *11(3)*, 446–456.
30. Sheng N.; Boyce M. C.; Parks D. M.; Rutledge G. C.; Abes J. I.; Cohen R. E. *Polymer*, **2004**, *45*(2), 487–506.
31. Witten T. A.; Meakin P. *Phys. Rev. B*, **1983**, *28*(10), 5632–5642.
32. Witten T.A.; Sander L. M. *Phys. Rev. B*, **1983**, *27*(9), 5686–5697.
33. Happel J., Brenner G. Hydrodynamics at Small Reynolds Numbers. Moscow, Mir, 1976, 418 p.
34. Mills N. J. *J. Appl. Polymer Sci.*, **1971**, *15(11)*, 2791–2805.
35. Kozlov G. V., Yanovskii Yu. G., Mikitaev A. K. *Mekhanika Kompozitnykh Materialov*, **1998**, *34(4)*, 539–544.
36. Balankin A. S. *Synergetics of Deformable Body*. Moscow, Publishers of Ministry Defence SSSR, 1991; 404 p.
37. Hornbogen E. *Intern. Mater. Res.*, **1989**, 34(6), 277–296.
38. Pfeifer. *Appl. Surf. Sci.*, **1984**, 18(1), 146–164.
39. Avnir D.; Farin D.; Pfeifer J. Colloid Interface Sci., 1985,103(1), 112–123.
40. Ishikawa K. J. Mater. Sci. Lett., 1990, 9(4), 400–402.
41. Ivanova V. S.; Balankin A. S.; Bunin I. Zh.; Oksogoev A. A. *Synergetics and Fractals in Material Science*. Moscow, Nauka, 1994; 383 p.
42. Vstovskii G. V.; Kolmakov L. G.; Terent'ev V. E. Metally, **1993**, *4*, 164–178.
43. Hansen J. P.; Skjeitorp A. T. *Phys. Rev. B*, **1988**, *38(4)*, 2635–2638.
44. Pfeifer P.; Avnir D.; Farin D. *J. Stat. Phys.*, **1984**, *36*, *5/6*, 699–716.
45. Farin D.; Peleg S.; Yavin D.; Avnir D. *Langmuir*, **1985**, *1(4)*,399–407.
46. Kozlov G. V.; Sanditov D. S. *Anharmonical Effects and Physical Mechanical Properties of Polymers*. Novosibirsk, Nauka, 1994; 261 p.
47. Bessonov M.I.; Rudakov A. P. *Vysokomolek. Soed. B*, **1971**, *13(7)*, 509–511.
48. Kubat J.; Rigdahl M.; Welander M. *J. Appl. Polymer Sci.*, **1990**, *39(5)*, 1527–1539.
49. Yanovskii Yu. G.; Kozlov G. V.; Kornev Yu. V.; Boiko O. V.;Karnet Yu. N. *Mekhanika Kompozitsionnykh Materialov i Konstruktsii*, **2010**, *16(3)*, 445–453.
50. Yanovskii Yu. G.; Kozlov G. V.; Aloev V. Z. *Mater. Intern. Sci. Pract. Conf. "Modern Problems of APK Innovation Development Theory and Practice"*. Nal'chik, KBSSKhA, 2011,434–437.
51. Chow T. S. Polymer, 1991, 32, 1,29–33.
52. Ahmed S.; Jones F. R. *J. Mater. Sci.*, **1990**, *25(12)*, 4933–4942.
53. Kozlov G. V.; Yanovskii Yu. G.; Aloev V. Z. *Mater. Intern. Sci. Pract. Conf.*, dedicated to FMEP 50-th Anniversary. Nal'chik, KBSSKhA, 2011; 83–89.

54. Andrievskii R. A. Rossiiskii Khimicheskii Zhurnal, **2002**, *46(5)*, 50–56.

55. Kozlov G. V.; Sultonov N. Zh.; Shoranova L. O.; Mikitaev A. K. *Naukoemkie Tekhnologii,* **2011**, *12(3)*, 17–22.

CHAPTER 7

SYNTHESIS, PROPERTIES AND APPLICATIONS OZONE AND ITS COMPOUNDS

S. K. RAKOVSKY, S. KUBICA, and G. E. ZAIKOV

CONTENTS

7.1 Introduction...110
7.2 Ecology 110
    7.2.1 Waste Gases 110
    7.2.2 Waste Water 113
    7.2.3 Natural (Drinking) Water 124
7.2.4 Soils...128
7.3 Industry 130
    7.3.1 Manufacture of Organic Compounds................................130
    7.3.2 Inorganic Productions 142
7.4 Medicine 150
    7.4.1 Therapeutic Application...154
    7.4.2 Toxic Action...165
    7.4.3 Ozone Solutions...170
    7.4.4 Sterilization 172
    7.4.5 Ozonators...172
7.5 Conclusion 173
Keywords...173
References...173

7.1   INTRODUCTION

The ozone application is based on its powerful oxidative action. The ozonolysis of chemical and biochemical compounds, as a rule, takes place with high rates at low temperatures and activation energies [1]. This provides a basis for the development of novel and improved technologies, which find wide application in ecology [2], chemical, pharmaceutical and perfume industries, cosmetics [3, 4], cellulose, paper and sugar industries, flotation, microelectronics, veterinary and human medicine, agriculture, foodstuff industry and many others [5–8]. At present it is very difficult to imagine a number of high technologies without using ozone. Here should be mentioned the technologies for purification of waste gases, water and soils, manufacture of organic, polymer and inorganic materials, disinfecting and cleaning of drinking and process water, disinfecting of plant and animal products, sterilization of medical rooms and instruments, resolving cosmetics problems, sterilization and therapy in veterinary and human medicine, surface cleaning and functionalization in the manufacture of polymer and microelectronics articles, flotation, deodorization and decolorization of gases, liquids and solid substances, oxidation in the intermediate stages of various technologies, etc.

7.2   ECOLOGY

7.2.1   WASTE GASES

More than 100 contaminants of the atmospheric air have been identified. Among them SO_2, CO_2, nitrogen oxides, various hydrocarbons and dust constitute 85%.

The main sources of harmful substances emissions, i.e., dust, SO_2 and CO_2 in the air, are the processing plants for coke, briquettes, coals, the thermal power stations, air, water, and road transport [9]. The exhaust gases contain also, CO, organic and inorganic compounds, etc. [10]. Dust, sulfur dioxide, carbon dioxide, nitrogen oxides and organic compounds, etc. are the main pollutants released in the environment from metallurgy, the manufacture of fertilizers and petrochemistry. The most typical air pollutants emitted by some chemical productions are listed in Table 1.

TABLE 1   Main air contaminants emitted by some chemical productions.

No.	Production	Pollutants	No.	Production	Pollutants
1.	Nitric acid	$NO, NO_x,$ NH_3	12.	Ammonium nitrate	$CO\ NH_3, HNO_2,$ NH_4NO_3 – dust
2.	Sulfuric acid: a) Nitroso b) contact	$NO, NO_x,$ $SO_x, H_2SO_4,$ Fe_2O_3, dust	13	Superphosphate	H_2SO_4, HF, dust
3.	Hydrochloric acid	HCl, Cl_2	14	Ammonia	NH_3, CO
4.	Ocsalic acid	$NO, NO_x,$ $C_2H_2O_2$, dust	15	Calcium chloride	$HCl, H_2SO_4,$ dust
5.	Sulfamidic acid	$NH_3, H_2SO_4,$ $NH(SO_3NH_4)_2,$	16	Chlor	$HCl, Cl_2,$ Hg
6.	Phosphor	P_2O_3, H_3PO_4, HF, $Ca_5F_4(PO_4)_2$, dust	17	Caprolactam	$NO, NO_2, SO_2,$ H_2S, CO
7.	Phosphoric acid	P_2O_3, H_3PO_4, HF, $Ca_5F_4(PO_4)_2$, dust	18	PVC	$Hg, HgCl_2,$ NH_3
8.	Acetic acid	CH_3CHO, CH_3OH	19.	Artificial fibers	H_2S, CS_2
9.	Nitrogen fertilizers	$NO_2, NO, NH_3,$ HF, H_2SO_4, HNO_3	20.	Mineral pigments	$Fe_2O_3, FeSO_4$
10.	Carbamide	NH_3, CO, $(NH_2)_2CO$, dust	21.	Electrolysis of NaCl	Cl_2, NaOH

The technologies with ozone participation are very promising for SO_2 utilization as sulfate, CO as carbonate, nitrogen oxides as nitrates.

Scrubbers used previously for removing only acids, for example HCl and HF, etc., now can be used for SO_2 separation via the addition of an oxidizing agent into the water intended for gas treatment. Ozone as compared with the conventional oxidizers such as hydroperoxide, chlorine, sodium hypochloride, perchlorate, shows appreciably higher oxidizing efficiency [11].

The rate constant of ozone reaction with NO is about 1010 cm^3/(mol.s), and with NO_2 it is $\sim$ 107 cm^3/(mol.s) whereby the oxidation of NO by ozone in the liquid phase is characterized by a higher absorption rate and rise in the concentration of the obtained HNO_3. Moreover, the oxidation of NO can be accomplished in the exhaust gases or in the course of absorption [12].

The CO oxidation by ozone is carried out in the presence of Fe, Ni, Co and Mn oxides. In most cases ozonation is more appropriate than the conventional methods and sometimes it appears the only possible way for its preparation [13].

The purification of exhaust gases from burners working on liquid and gas fuel is accomplished by using oxidation catalysts of Perrovski type oxides – ABO_3 combined with ozonation. Ozone is injected prior the waste gases flow. A may be La, Pr or other alkali earth element; B may be Co, Mn or other transition metal. A may be also partially substituted by Sr, Ca or any other alkali earth element. A catalyst deposited on ceramic support of honeycomb type may contain $SrCo_{0.3}Mn_{0.7}O_3$. It is oxidized 24% CO in the absence of ozone and 80% in ozone atmosphere [14].

The removal of nitrogen oxides (NO_x) from waste gases is carried out through ozone oxidation in charged vertical column. The lower part of the column is loaded by 5–15% $KMnO_4$, and ozone is blown through the lower and upper part of the column and the waste gases are fed to the center of the column charge The inlet concentration of 1000 ppm NO_x in the waste gas is reduced to 50 ppm in the outlet gas flow. Another method for cleaning of exhaust gases from NO_x involves the application of plasma generator and ozonator [15], which practically leads to the complete removal of nitrogen oxides.

Sulfur containing compounds are removed from gas mixtures by silicon oil scrubbing and ozone oxidation of absorbed pollutants. A model system providing that the gas mixture contains methandiol and deimethylsulphide is fed at a rate of 9–120 m^3/h into silicon oil charged scrubber followed by ozone oxidation of the absorbate. The oxidates are extracted with water and the silicon oil is regenerated. Thus the purified gas mixtures practically do not contain any sulfur [16].

A special apparatus is designed for decomposing acyclic halogenated hydrocarbons in gases. It includes a chamber for mixing of the waste gases with ozone coupled with UV-radiation, ozonator, and inlet and outlet units. This method is very appropriate for application in semiconducting industry whereby acyclic and halogenated hydrocarbons are used as cleaning agents [17].

The mechanisms of ozonolysis of volatile organic compounds such as alkenes and dienes are discussed and the products output is determined by matrix isolation FTIR spectroscopy [18].

The synthesis of a material from zeolite via pulverization, granulation and drying at 500–700°C, cooling to 100–200°C, electromagnetic radiation or ozone treatment is described in Ref. [19]. The material thus obtained is suitable for air deodorization, drying and sterilization.

The purification of gases containing condensable organic pollutants can be carried out by gas treatment with finely dispersed carbon, TiO_2, Al_2O_3, Fe_2O_3, SiO_2 and H_2O_2 and ozone as oxidizers [20].

A wet-scrubbing process for removing total reduced sulfur compounds such as H_2S and mercaptans, as well as the accompanying paramagnetic particles from industrial waste gases is proposed [21]. For this purpose a water-absorbing clay containing MnO_2 is used. The collected clay is regenerated by ozone oxidation.

The removal of mercury from waste gases is carried out by catalytic ozonation [22]. The used gases mixed with ozone are blown over a zeolite supported Ni/NiO catalyst. This procedure results in 87% conversion of mercury into mercuric oxide, which is isolated by filtration. Pt and CuO_x/ HgO system is proposed as another catalyst for this process [22].

7.2.2 WASTE WATER

In contrast to the cleaning of natural water by ozonation, which is experimentally confirmed as the most appropriate, the purification of waste water by ozone is still an area of intensive future research. This could be explained by the great diversity of pollutants in the used water and the necessity of specific approach for each definite case.

The recycling of water from cyanide waste water (copper cyanide contaminant) is accomplished by ozone oxidation combined with UV radiation and ion exchange method [23]. Further, the oxidate is passed through two consecutive columns charged with cationite and anionite. The ozonolysis priority over the conventional chlorinating method is demonstrated by the fact that ozonation fails to yield chlorides whose removal requires additional treatment.

The ozonolysis of CN-ions leading to CNO^--ion formation combined with UV-radiation proceeds at 3 fold higher rate as compared with that without radiation employment. Upon exchange in NaOCN, the Na^+ is substituted by H^+ followed by its decomposition to CO_2 and NH^{4+} via hydrolysis in acid medium. The ammonium ion is absorbed by the cation-exchange resin. The solution electroconductivity on the anion-exchange resin becomes lower that 10 muS/cm and cyanides, cyanates or copper ions have not been monitored. On the basis of these experimental results, a recycling method and apparatus for detoxification of solutions containing cyanides is proposed. The recycled water from the cyanide waste water may be re-used as deionization water in gold – plating. The process ion-exchange resins are regenerated after conventional methods. Actually, this method is not accompanied by the formation of any solid pollutants [23].

The waste water from the electrostatic and galvanic coatings containing CN^- -ions and heavy metals is treated consecutively by ozone and CO_2. The cyanides and carbonates being accumulated in the precipitate are already biodegradable [24].

Waste water containing Cr(III) is oxidized by ozone combined with UV-radiation [25]. The radiation reduces the oxidation time about threefold. The oxidate is consecutively passed through columns charged with cationite and anionite resulting in the obtaining of waste water with electroconductivity below 20 mu.S/cm. This deionized water can be successfully used for washing of coated articles. The Cr(IV) concentrated solution from the anionite regeneration can be further treated by appropriate ion-exchange method yielding a high purity Cr(IV) solution [25].

Purification of waste water containing $ClCH_2COOH$ and PhOH is carried out by ozone treatment and UV-radiation in the presence of immobilized photocatalyst. Thus the rate of decomposition is higher than the rate obtained only in the presence of ozone and UV-radiation [26].

Used water containing small amounts of propyleneglycol nitrate or nitrotoluenes is a subject to combined oxidation with ozone and H_2O_2, under pressure and heating to the supercritical point. Thus a nominal conversion of the contaminants higher than 96.7% is achieved [27].

In laboratory experiments for reduction in residual COD in biologically treated paper mill effluents it is subject to ozonation or combined ozonation and UV radiation at various temperatures and pressures [28]. At a ratio of the absorbed ozone to COD less that 2.5:1 g/g, the elimination level with respect to COD and DOC is up to 82% and 64%, respectively. The ozone consumption is essentially higher in the case of the UV combined ozonation at pH>9 and elevated temperature to 40°C.

The decolourization and destruction of waste water containing surface active substances is performed by ozone treatment at ozone concentration of about 80 mg/l. It has been found that for waste water which has undergone partial biological treatment by fluent filters or anaerobe stages up to COD concentration > 500 mg/l and ratio 30D5/COD>0.2 ozonolysis with ozone concentrations up to 1.8 g/l does not change CBR, while after the complete biological treatment at COD<500 mg/l ozonation results in rise of the ratio 30D5/COD from <0.05 to 0.37. This evidences the increase of CBR [28].

The purification of waste water from paper manufacture characterized by enhanced biodegradability, as well as the removal of COD and halogenated compounds is carried out by a method based on ozonation and biofiltration [29].

Waste water from the paper pulp production are clean up by treatment with ozone and activated acid tar. The high values of pH favor the lignin decomposition and carboxylic acids formation since in this case the ·OH – radicals and not ozone are the oxidizing agent. The application of activated acid tar has a positive effect on the dynamics of the microbiological growth, substrate consumption and CBR of organic acids. For example, maleic and oxalic acids are decomposed completely at ozonation of waste water in the presence of activated acid tar. For immobilization of biological culture the use of polyurethane foams appear to be very suitable [30].

The used water in collectors of a plant for paper manufacture are ozonized in a foamy barbotating contactor. The results indicate that ozone is very efficient in the oxidation of coloring and halogenated compounds. Its activity is proportional to the amount of absorbed ozone regardless of the variations in the gas flow rate, the inlet ozone concentration and contactor design. The amount of consumed ozone depends on the operating conditions and waste water characteristics. The absorption rate rises in the presence of stages including ozone decomposition, particularly at higher rates [31].

In biological granular activated carbon (GAC) columns the effect of pretreated ozonation on the biodegradability of atrazine is investigated. The metabolism of isopropyl-^{14}C-atrazine gives higher amount of $^{14}CO_2$ than the ring UL-^{14}C – atrazine which shows higher rate of dealkylation that the process of ring cleavage. The pretreatment with ozone increases the mineralization of the ring UL-^{14}C – atrazine and consequently rises the GAC-columns capacity. A 62% of the inlet atrazine is transformed into $^{14}CO_2$ in columns charged by ozonized atrazine and water. However, in columns with nonozonized atrazine and water only 50% of the inlet atrazine is converted into $^{14}CO_2$ and in columns supplied only with nonozonized atrazine only 38% of it are converted to $^{14}CO_2$ [32].

The waste water from petrochemical industry contains substantial amounts of phenol. The phenols mixtures appear to be more toxic that phenol itself and possess a synergetic effect. Their decomposition to nontoxic products is important and modern problem. The oxidation of phenols such as p-cresol, pyrocatechin, rezorzine and hydroquinone by O_3 in aqueous medium appears to be a promising method for their degradation. At concentrations of 100 mg/l and pH = 11.5 their complete oxidation is performed for 30 min.

The phenol content in used water is reduced from 145–706 mg/l to 2.5 mg/l at ozone consumption of 1.1–2.6 g/l. In real conditions the ozone consumption varies from 5 to 10 g/l per g phenol.

The results from the ozonolysis of waste water from coke processing are shown in Table 2.

TABLE 2　Results from the ozonation of waste water from coke processing. $[O_3]$ = 5 mg/l, flow rate 6 l/min and treatment time – 4 h.

Pollutant, mg/l	Before ozonation	After ozonation
Monophenols	710	0.8
Polyphenols	380	198
Cyanides	3	29
Thiocyanides	384	0
Thiosulfates	538	0
Sulfides	43	2
H_2O_2	0	12
Bases	114	27

The treatment of concentrated aqueous solutions of phenol (1.0g/l) with ozone causes the appearance of a yellowish coloring after absorption of 1 mol O_3 per 1 mol phenol, which gradually fades away. In this case the higher phenols undergo slight oxidation and H_2O_2 is identified in the water after the treatment.

The tests demonstrate that upon varying content of phenols in the used water the ozonation is one of the most promising methods for their removal. The ozonation was found to be applicable for the decomposition of rhodanide in neutral and weak acid medium in the temperature range of 9–25°C. The ozone consumption in this case is 2 mg/mg. The ozone uptake for the oxidation of the cyanide ion (CN^-) is 1.8 mg/mg.

The complex Zn cyanides are oxidized similarly to the simple soluble cyanides. Regardless of the higher resistance of Cu cyanide complexes, the ozone treatment reduces the CN^- content substantially (traces). Copper carbonate is precipitated as a reaction product.

The ozonation can be successfully employed as a method for cleaning the waste water from the manufacture of ammonium nitrate containing CN^-, S^{-2}, CSN^- ions. The purification level is 83, 98 and 95% respectively at O_3 consumption of 0.7 g/l/g water for 10 min contact.

The ozonation of waste water from sulfate-cellulose production results in 60% purified used water which is returned in the process cycle.

The cleaning of waste water from the petrochemical plants by ozone is efficient under wide range of pH (5.8–8.70 and temperatures (5–50°C). Thus the level of petrochemicals in the waste water at pH 5.8, ozone consumption of 0.52 mg/mg and 10 min contact is reduced from 19–33 mg/l to 2 mg/l. Table 3 is shown the efficiency of various treatment procedures for deodorization of petrochemicals contaminated water.

TABLE 3 Result from deodorization of petrochemicals contaminated water (grades/dilution).

Petroleum products	Before treatment	Coagulation and filtration	Charcoal and filtration	Chlorination	Ozonoation	Coagulation, filtration and ozonation
Petrol	5/80	5/60	5/50	5/80	2÷3/10	1/0
Kerosin	5/100	5/80	5/70	5/100	4÷5/25	1/0
Petroleum	5/100	5/60	5/60	5/100	5/20	2/5
Oil	5/10	1/0	1/0	5/10	1÷2/2	1÷0/0

Ozonation is very economical and highly efficient method for decomposing cancerogenic substances such as 3,4-benzopyrene, etc., particularly after the biological purification of waste water.

The application of ozone is also advisable for mercury oxidation in chlorine producing, for decontamination of the used water from the poisonous tetraethyllead, for destroying pesticides traces, etc.

The highest efficiency of cleaning of petrochemicals contaminated water is achieved at the following sequence: primary cleaning (mechanical), secondary (biological) and complete purification (ozonation).

Ozone is also successfully applied for the complete purification of waste water form dyes manufacture. Their color decreases about 10,000 times at exposure to 0.22 g/l ozone and the biological activity of the reaction products is reduced almost to zero. The semi-industrial experiments in CP "Kistenetz" were carried out on the dyes manufactured in the plant:

1. Direct Congo red – $C_{32}H_{26}O_6N_6S_2Na_2$, Mw=698.

2. Direct blue KM – $C_{34}H_{30}O_5N_5S_1Na_1$, Mw=643.

3. Acid blue ZK – $C_{26}H_{20}O_{10}N_3S_3Na_3$, Mw=699.

4. Acid chromium green – $C_{16}H_{11}O_{10}N_4S_2Na_2$, Mw=529.

5. Acid black ATT – $C_{22}H_{16}O_9N_6S_2Na_2$, Mw=618.

6. Acid sulfonic blue – $C_{32}H_{25}O_6N_5S_2Na_2$, Mw=685.

7. Acid chromium black C – $C_{22}H_{19}O_5N_6S_1Na_1$, Mw=502 è $C_{16}H_{13}O_7N_4S_1Na_1$, Mw=428.

8. Direct black Z – $C_{34}H_{26}O_7N_8S_2Na_2$, Mw=768.

9. Acid chromium yelow R – $C_{17}H_{14}O_3N_3S_1Na_1$, Mw=367.

In the process of developing a technology for water purification from various organic contaminants, we have proposed a method and apparatus for quantitative determination of chemical compounds, separated by means of liquid chromatography. Special attention is paid to those compounds susceptible to ozone attack such as olefin hydrocarbons and their derivatives, phenols, amines, thiocarbamates, inorganic compounds, etc. The characterization of the separated compounds is carried out by spectral, massspectral, refractometer, electrochemical analyses, but the detectors in most of these equipments are rather expensive and complex units without sufficient identification capabilities, particularly for the compounds pointed above. In this sense the apparatus designed by us allows to overcome the disadvantages of the known methods and equipments. The scheme of the ozone detector applicable for identification and liquid chromatography separation of chemical compounds is presented in Fig. 1.

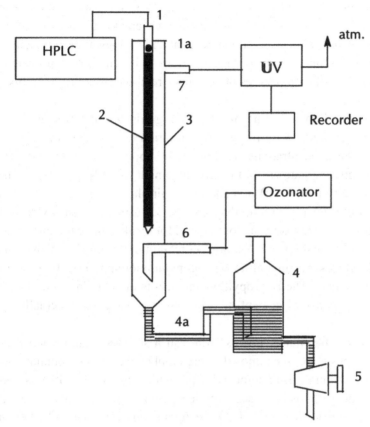

FIGURE 1 Set-up of ozone detector for identification of separated by liquid chromatography chemical compounds. 1 – eluent access; 1a – entry for pouring out the elate as thin film on 2 – solid glass stick; 3 – outer glass jacket; 4 – vessel for eluent accumulation; 4a – hydraulic détente; 5 – tap for pouring out the eluate; 6 – ozone inlet; 7 – ozone outlet.

The ocean power stations operate on the basis of the temperature difference between the upper and lower water layers. Taking advantage of the one-step solution of ozone in depth, the COD is decreased which prevents the befouling in the piping systems found in the upper layers. The ozone-enriched water from the lower layers is supplied to a cyclic filter and is directly injected into upper seawater for control of the sea microorganisms growth [33].

The removal of benzofurenes from ash, clays, soils, water and oils by combustion, ozonolysis of supercritical water, cracking-processes of petroleum, as well as thermal, photochemical and biological decomposition, has been the subject of many investigations [34]. The direct blowing of ozone through water pipers results in slimes removal from the inner pipers walls [35].

The ozone blowing at a rate of 2.87 mg/min through aqueous solutions of 15 pesticides with 10 µg/l concentration leads to their transformation and their concentrations are reduced by about 20% for 60 min. The identified intermediates from the rearrangement of EPN, phenitrothione, malathion, diazinon, izoxathion and chloropiriphos are the corresponding oxones, while for phenthione they are the corresponding sulfoxide, sulfone and sulfooxide oxone analogues, and for disulfothione the intermediate is its sulfo analogue. 2-Amino-4-ethylamino-6-chloro-1,3,5-thriazine is identified as symazin and its 4-isopropylammonium derivative is atrazine intermediate. The isopropylthiolane results in obtaining of its 1-oxo-derivative. Upon replacement of ozone by air changes are actually not registered [36].

At the purification of process water in a soil decontamination plant containing pesticides the removal of organophosphorous compounds such as Thiometon and Disulphoton is of particular importance. In this connection four methods have been applied: (1) ozonolysis at pH = 2.5; (2) ozonolysis at pH = 8; (3) UV/H_2O_2 treatment and (4) oxidation by Fenton reagent [37]. The laboratory experiments were carried out with pure compounds soluted in buffered deoionized water and in process water, which is extracted by solid-liquid extraction of the contaminated soil. The use of ozone in an acid medium turns to be ineffective since the Thimeton reaction practically stops after the formation of PO-derivatives. In all cases when the oxidation is carried out by HO•-radicals, a sufficient removal of the pesticides and their metabolites is achieved. Oxadixyl, a cyclic nitrogen containing compound which is present in high concentrations in the process water is the most stable one, thus being the main soil contaminant. The investigated experimental conditions and the results obtained show that these methods are quite acceptable for universal application [37].

The water-soluble agrochemicals ASULAM and MECOPROP are decomposed for 30 min at ozone blowing (flow rate – 5.1 mg/min and concentration of 10 ppm)) through their aqueous solution (200 ml). The decrease in the flow rate of ozone retards the decomposition process. Similar treatments for 5 h give in ACEPHATE and DICAMBA decomposition to 20%, the rate of decomposition being promoted with H_2O_2 addition [38].

The primary ozonation products of organophosphorous pesticides in water such as diazinone, phenthione (MPP) and ediphenphos (EDDP) [39], are identified by means of GC-MS analysis. The massspectra of the ozonation products of 17 organophosphorous pesticides point oxones as the primary reaction products. This fact is also confirmed by the SO_4^{2-} generation resulting from the ozonation of their thiophosphoryl bonds. Oxones are relatively stable towards ozone attack but they are further hydrolyzed to trialkylphosphate and other hydrolysis products. However, with MPP the thiomethyl radicals are first oxidized to thiophosphoryl bonds giving MPP-sulfoxide, MPP-sulfone, MPP-sulfoxide-oxone are also obtained. Two main oxidation products have been identified at the oxidation of bis-dithio-type ethiones. Phosphate type EDDP is stable towards ozonolysis but its oxidation products are identified after hydrolysis [39].

Pesticides and their degradation products which are not mineralized to CO_2, NH_3, H_2O and inorganic salts can damage and contaminate the water piping. The treatment by photolysis or ozonation may substantially increase the mineralization rate. The photodestructive products of S-triazoles, chloroacetanylide and parakaute are declorinated and/or oxidized. The ozonation of these herbicides results in formation of products whose side alkyl chains are oxidized or removed, and the aromatic ring is oxidized or soluted but dechlorination does not occur. In some cases the medium may cause microbiological effect [40].

A deodorization method for treatment of ill-smelling air containing ammonia, hydrogen sulfide or amines from refrigerator chambers, lavatories, cattle-sheds, etc., involves the ozone contact with air in the presence of porous catalyst of 120 m^2/g specific surface. The used gas passes through a layer of carbon black for adsorption of the residual ozone and the nondegradated smelling substances. The active component of the catalyst may be: transition metal or its oxides supported on porous honeycomb type supports [41].

The contaminated smelling gases in drain and fecal water are dried to 60–30% humidity, followed by ozone treatment on an oxidizing catalyst.

For example, contaminated gas containing 30 ppm methylsulfide is dried to 30% humidity and then is treated with ozone in the presence of honeycomb type catalyst containing 83:12.5 = TiO_2:SiO_2:MnO_2 with a volume rate of 50 000 h^{-1}. In this case the deodorization efficiency is 99% while at 100% humidity it drops to 80% [42].

7.2.3   NATURAL (DRINKING) WATER

The basic sources of drinking water usually contain various organic admixtures in mg/l: carbon – 30, nitrogen – 0.8, fatty acids – 30, highmolecular organic acids – 0.2. naphthenolic acids – 1.5, phenols – 1.2, luminescent substances: neutral resins – 17%, oils, humus – 55%, acid resins – 19%, hydrocarbons, naphthenolic acids – 4–7%.

The coloring of natural water varying from pale yellow to brown is due to the presence of humus substances, Their decolorization is carried out by adsorption on coagulated $Al(OH)_3$ or $Fe(OH)_3$. The increasing demand for fresh water requires the use of more colored water, which however is pre-chlorinated, and then subject to adsorption up to their complete decolorization.

The ozone treatment removes the bad taste and odor of water, the resistant phytoplankton being oxidized to 20%. The ozone consumption for decolorization of natural water varies in broad limits – from 1 to 18 mg/l and depends on the humus composition. The process of humus compounds oxidation by ozone is presented in Scheme 1.

The various biological processes lead to formation of substances causing the bad taste and smell of water – mercaptanes, sulfides, alcohols, carbonyl compounds, acids, terpenes, amines, etc.

The deodorization effect of ozone is preserved in wide range of pH values, temperatures and ion composition. Due to its high oxidation potential, ozone can also degrade the resistant toxic substances. It is known that during the blooming in the water reservoirs are accumulated biologically active substances among, which algotoxine is particularly harmful for the human health. Its complete degradation is achieved by using ozone concentration of 14 mg/l and 50 min contact time.

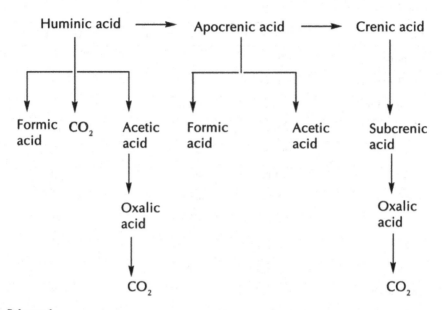

Scheme 1

At ozone concentration of 8 mg/l in water, only 6% of *Scendezmus* algae remain alive after 10–15 min. However, it is to be noted that the ozone action on the various algae is quite different and specific. Thus *Astetionella* type algae are very ozone resistant.

The oxidative, disinfective, decolorizative and deodorizative properties of ozone have found wide application for preparation of drinking water. Many world corporations such as the French Taileygas, Degramont, Ozonia, the German Fisher, AEG, Siemens, the Swiss Braun & Bovery, the American General Electrics, General Motors etc., are among the biggest producers of drinking water. Twenty of the biggest world ozonator stations for drinking water treatment are listed below (Table 4).

TABLE 4 Some of the world biggest ozonation stations for drinking water treatment.

City	State	Water, m³/day	Ozone, kg/h
Moscow	Russia	1,200,000	200
Montreal	Canada	1,200,000	150
Ashford Common	GB	–	146

TABLE 4   *(Continued)*

City	State	Water, m³/day	Ozone, kg/h
Coppermills	GB	–	144
Neuilly-sur-Marne	France	600,000	120
Choisy-le-Roi	France	900,000	120
Masan	Korea	–	112.5
Kiev	Ukrain	400,000	80
Helsinki	Finland	495,000	63
Manchester	GB	480,000	50
Orly	France	320,000	50
Singapore	Singapore	230,000	48
Minsk	Belaruss	200,000	47
Mery-sur-Oise	France	300,000	45
Lodz	Poland	190,000	40
		290,000	
Chiba	Japan	270,000	38
Nizhni Novgord	Russia	350,000	30
Bodenzee	Germany	480,000	30
Nant	France	380,000	30
Amsterdam	Holland	125,000	25
Brussels	Belgium	250,000	24
Quebec	Canada	220,000	19
Wrotzlaw	Poland	180,000	17

Ozonation stations for drinking water preparation work in Portugal, Greece, Bulgaria, USA, etc. all over the world.

Recently water treatment stations in Pancharevo and Bistritsa in Sofia, Bulgaria, of capacity 33 and 66 kg O_3/h, respectively are built.

The technology of drinking water ozonolysis leading to its deodorization, disinfecting and decolorization includes gas units for dry air producing, ozonator, barbators or other mixing units, apparatus for HO•-radicals generation, promoting the ozone formation and filter-catalyst for decomposition of the residual ozone. The preferable materials for barbators are zeolites or ceramics on the base of aluminum and/or silicon with pores diameter of 5-100Å, and the catalysts should contain one or more of the following components: Pt, MnO_2, CuO, Ni_2O_3, Zn, YiO_2 and SiO_2 [43].

For desinfection of drinking water, the ozone-containing air is mixed with water in the water pipe, after which the bubbled water is supplied into a special designed chamber filled by porous ceramics charge for intensification of ozone dissolution into water. Some problems concerning the optimization of the UV stimulated oxidation of organic contaminants in soil water with ozone and hydroperoxide is discussed [44].

The disinfection of drinking water by ozonation can be also carried out at the outlet of various water supplying units before and after consummation, such as taps, fountains, water heaters, dental units [45], etc. In this connection are used small dimensional and low capacity personal ozonators coupled or installed additionally to these equipments.

The attempts for treatment of drinking water with ionization radiation appear to be ineffective mainly because of nitrates and H_2O_2 formation. These problems, however, do not occur when the radiation is performed in the presence of sufficient ozone amount [46]. The addition of ozone before and in the course of radiation converts the radiation into a purely oxidation process, a modification of the so called – Advanced Oxidation Process (AOP). The combined ozone-ionization radiation method is unique since two processes for HO-radicals generation are simultaneously induced. This method yields higher concentrations of HO-radicals as compared with other AOP. This in its turn results in precipitation of smaller amounts at equal ozone concentrations or reduces the ozone consumption at one and the same amount of precipitates. This makes the process particularly suitable for remediation of weakly contaminated soil water. The ozone concentrations at this process are higher than those used for drinking water disinfection. Thus, the drinking water subject to this treatment is disinfected more rapidly. The purification of waste water using this method is more efficient than the UV-radiation treatment only. This combination reduces COD without causing the rise of BOD. In this sense this method appears to be more attractive than the conventional two-stage processes of ozonation and biological treatment. The injection of gaseous ozone into the radiation chamber improves the water flow turbulence and substantially increases the efficiency of contaminants degradation.

The European organization "European regulation" has allowed the use of ozone for iron and manganese removal from mineral water, this initiating studies on secondary products formation, amount and composition, particularly during ozonation of bromine water. This water, usually contain BrO_3^- ions and various halogenated organic compounds (HOC). The present standard for LD of HOC is 100 µg/l [47]. Upon ozonation of bicarbonate mineral water – one sodium-rich water, and the other, calcium rich, it has been established that: (1) the process depends on the initial Fe and Mn concentrations and the Na and Ca concentrations; (2) HOC and bromate ions are secondary products and (3) the specific components characterizing a definite mineral water affect the ozonolysis. This requires definition of the conditions for Fe and Mn removal for each concrete type of mineral water. The removal of iron and manganese is efficient when the amount of ozone is three times greater than the stoichiometric one (0.43 mg O_3/mg Fe and 0.87 mg O_3/mg mn).

The disinfection of swimming pool water is carried out by controlled addition of compounds- bromine and iodine donors (NaBr or NaI) and water oxidation by hypochloride, ozone or potassium peroxymonosulfate [49]. We have designed a compact module system (stationary and mobile), which includes module for air preparation, ozonator, absorption column and unit for the complete decomposition of the residual ozone. Its capacity is 12–15 g/h and it is very suitable for sanitation of drinking water and used water from swimming pool (3.5 l/s capacity).

7.2.4   SOILS

The decontamination of dredging wastes containing spilled petroleum is carried out by using a system for the direct injection of ozone (100–1000 ppm) containing air into pipes at the bottom; means for pumping out the spilled petroleum to a tank for its storage; units for its mixing with ozone for decomposition of organic matter in the tank; units for returning of the processed petroleum for the complete degradation of the residual organic matter by means of aerobic bacteria. The efficiency of organic matter decomposition is by about 20% higher than that by using the conventional methods including only treatment by ozone [49].

Another technology for treatment of contaminated soils [50] involves the ozonolysis of the water slurry of the contaminated soil layer, in the presence of H_2O_2 or in its absence and consecutive biological decomposition. The priority of the proposed technology over the conventional ones is that the removal of organic contaminants is by *on-site* treatment of the soils.

The *on-site* recyclation of petroleum-contaminated soils by ozonation requires the good knowledge of the geological setting and the recycling system. Usually the recycling stages include: (1) system for air evacuation from the soil; (2) means for injection of ozone combined with air evacuation from the soil; (3) development of means for ozone generation; (4) screening of soils contaminants; (5) use of systems for safety control; 6) organization of the cleaning units – pumping and injecting drills; (7) decision for air injection into the soil; (8) methodology for measurement and regulation of ozone concentration; (9) hydraulic purification of the highly contaminated soil areas, etc. [51].

The treatment by ozone can be successfully applied for decomposing mineral oils, polycyclic aromatic compounds (PACs), phenols and some pesticides to biodegradable, nontoxic compounds. It should be noted that although ozone decomposes a great number of microorganisms, the soil microflora can be easily remediated after this treatment [52]. The latter can be carried out on- and out of the contaminated area. A number of problems related to the ozonolysis of organic compounds, the PACs degradation products, the effect of ozone on the soil microflora, the laboratory and pilot results, etc., are discussed.

A method and apparatus for decontamination of wastes such as soil, porous granulated sludge, are suggested. The polluted materials are placed in a special vessel and ozone is bubbled through. The waste gases obtained from the ozone oxidation are further treated with ozone [53].

A method for ozone treatment both *in-situ* and *on-site* has appeared to be applicable for the degradation of organic substances adsorbed on the solid surface [54].

Wet-scrubber method (based on MnO_2) for removing total reduced S-containing compounds such as H_2S, mercaptanes and non-magnetic particles from industrial waste gases using ozone, is suggested [21].

7.3 INDUSTRY

7.3.1 MANUFACTURE OF ORGANIC COMPOUNDS

We will focus our attention to some of the numerous ozone applications in organic polymer and inorganic productions, passivation, cleanup and preparation of surfaces for electronics, superconductors, etc.

The reaction of ozone with olefins is the main reaction for ozonides preparation (1,2,4-trioxalanes), which can be converted into a mixture of carbonyl compounds, the exact composition of which depends on the olefin structure. This reaction is used for synthetic purposes as with small exceptions the C=C bond is cleaved quantitatively under very mild conditions. Usually the ozonolysis is carried out at low temperatures (as a rule at $-70 \div -30$°C) as demonstrated in Chapters 2 and 3. Mostly, ozonolysis of olefins, except in solid phase, is carried out in various solvents such as paraffins, halogenated hydrocarbons, oxygen, sulfur and nitrogen containing hydrocarbons like ahcohols, glycols, aldehydes, ketones, acids, ethers, esters, amines, amides, nitriles, sulfoxides, mineral acids, bases and mixtures from them, etc. In most cases the reaction products are not separated from the reaction mixture except for the purposes of ozonide obtaining, and the oxidates are subject to hydrolysis, reduction or oxidation.

The hydrolysis is applied for preparation of aldehydes [55] and ketones [56]. The reductive decomposition is carried out in the presence of $Zn/AcOH$, SO_3^{-2}, HSO_3^{-1}, $LiAlH_4$, $SnCl_2$, ArP_3, Ph_3P, SO_2, $(CH_3)_2S$, $H_2/$ catalyst. The reduction by metal hydrides leads to alcohol formation in the decomposition products mixture while the use of other reductors yields carbonyl compounds. Actually there are no clear requirements for selection of the most appropriate reductor, but it turned out that dimethylsulf-oxide is probably [57] the most preferable reductor in methanol solution. The ozonolysis in the presence of tetracyanoethylene yields directly carbonyl moieties omitting the reduction step [58]. A novel modification of the conventional methods of ozonolysis is exemplified by the ozonolysis of silica gel adsorbed alkenes (acetylenes) [59] and selective ozonolysis of polyalkenes, controlled with the help of dyes introduced in the reaction mixture [60]. Some examples of ozonolysis using various reductors for ozonides decomposition are given in Table 5.

TABLE 5 Preparation of aldehydes, ketones and alcohols during alkenes ozonolyis with consecutive reduction.

No.	Substrate	Conditions	Product	Yield, %
1.	(cyclohexene)	EtOAc, –20°– –30°C 2. Pd/CaCO₃/H₂	(ring with CHO, CHO)	61
2.	(bicyclic, HO-)	n-C₆H₁₄ Zn/HOAc	(ring, HO-, O, CHO)	59
3.	(pyridine CH=CH₂)	1. MeOH/–40°C 2. Na₂SO₃/H₂O	(pyridine CHO)	65
4.	Me(CH₂)₅CH=CH₂	1. MeOH/–30°÷ -60°C 2. Me₂S	Me(CH₂)₅CHO	75
5.	n-Bu-OOCCH=CHCOO--Bu-n	1. MeOH/–65° ÷ -70°C 2. (MeO)₃P	n-Bu-OOCCHO	78
6.	(bicyclic =CH₂, Me, Me)	1. MeOH/–35° ÷ –75°C 2. NaI/MeOH/HOAc	(bicyclic =O, Me, Me)	75
7.	(cyclopentene Me, Me)	EtOAc/(CN)₂C=C(CN)₂, –70°C	MeCO(CH₂)₃COMe + (CN)₂C—C(CN)₂ (epoxide O)	61
8.	Me(CH₂)₅CH=CH₂	1. n-Pentane/–38° – –42°C 2. LiAlH₄/Et₂O	Me(CH₂)₅CH₂OH	93
9.	H₂C=CH(CH₂)₈COOH	1. MeOH/0°C 2. NaBH₄/NaOH/ EtOH/H₂O	HOCH₂(CH₂)₈COOH	91

Ozonolysis is a powerful chemical method for economical and ecologically pure preparation of various carbonyl compounds [69].

The ozonolysis of cyclic alkenes in protic solvents followed by reductive decomposition of the hydroperoxides formed is a classical method
for dialdehydes synthesis [70]. This method is applied for the preparation
of 3-ethoxycarbonylglutaric dialdehyde by ozonolysis of ethyl 3-cyclo-
pentenecarboxylate. The dialdehyde is an intermediate in the synthesis
of dolazetrone mesitilate, which is an active medical substance in AN-
ZEMET anti-emetik. Zn/AcOH, phosphines, amines and sulfides such as
3, 3'-thiodipropionic acid and its salts are the most suitable reductors for
this system. Their efficiency is comparable to that of methyl sulphide but
without its shortcomings. The polymer immobilized 3,3'-thiodipropionic
acid is also very active in this reduction reaction [70].

The efficient conversion of oximes into the corresponding carbonyl
compounds can be also carried out through ozone oxidation [71].

The application of ozone is a new and convenient way for the preparation of cyanoacetylaldehyde (3-oxopropylonitrile) (1) and its stable
dimethyl acetal (3,3-dimethoxypropylonitrile) (2) used as valuable intermediates for organic synthesis [72]. For this purpose the ozonolysis of
(E)-1,4-dicyano-2-butene or 3-butylonitrile is carried out. Then the oxidates are treated by DMS yielding 1. Further, compound 1 can be used
either directly in the next reactions or is transformed into 2. The 2 output
amounts to from 67 to 71%. The acetal can be again hydrolyzed to 1 by
treatment with ion-exchange resin Amberlyst-15 [72].

The easily available allylphenyl ethers are ozonized at –40°C, treated with DMS giving solutions of the corresponding phenoxyacetalde-
hydes, which are purified by column chromatography. Their reaction with
1-methyl-1-phenylhydrazine yields the corresponding hydrazones [73].

Quinolinealdehyde derivatives can be synthesized by ozonolysis of the
respective quinolineolefins at –60 to –72°C in methanol or ethanol solution
with consecutive reduction with DMS at the same temperatures. Then the
reaction mixture is heated to room temperature for 1 h. The yield of the
products amounts to 29%. [74].

6-Chloro-2-hexanone is prepared from 1-methylcyclopentane via
three-step scheme: (a) ozonolysis of 1-methylcyclopentane in carboyilic
acid solution to 1-methylcyclopentanol; (b) conversion of 1-methylcyclopentanol

into 1-methylcyclopentyl hypochloride using NaClO; (c) cyclization of 1-methylcyclopentyl hypochloride in 6-chloro-2-hexanone [75].

Ozonation is a stage in the synthesis of optically pure (R)-(+)-4-methyl-2 cyclohexen-1-one from (R)-(+)-pulegone and hydroxy ketone from (–)-*cis*-pulegol [76].

Polycarbonyl compounds and aldehyde-acids are the ozonolysis products of (+)-4a-[1-(triethylsiloxy)-ethenyl]-2-carene [77].

Carbonyl oxides obtained from vinyl ethers ozonolysis undergo rapid [3+2] cycloaddition with imines giving the corresponding 1,2,4-dioxazolidines in yields of 14-97% [78].

Carbonyl oxides can be also used for the synthesis of 3-vinyl-1,2,4-trioxalanes (α-vinylozonides) by a [3+2] cycloaddition with α,β-unsaturated aldehydes. Howeve, α,β-unsaturated ketones are practically inactive in this reaction. The reaction of 3-vinyl-1,2,4-trioxalanes with ozone leads to the formation of the corresponding dizonides [79].

The carbonyl oxides prepared from the ozonolysis of enol ethers (for example, 1-metoxy-4,8-dimethylnone-1,7 diene) undergo stereoselective intramolecular cycloaddition with inactivated alkenes yielding bicyclic dioxalanes. The latter are easily converted into β-hydroxycarbonyl species and 1,3 diols by catalytic hydrogenation thus providing a new approach to the synthesis of 1,3-oxygenated products [80].

The ozonolysis of vinyl ethers is discussed in detail in Ref. [81]. The synthesis of 1,2-dioxalanes is carried out by ozonolysis of 1,1-disubstituted nonactivated olefins. Thus the ozonolysis of cyclopropyl-1,1-disubsituted olefins does not produce carbonyl oxides, but formaldehyde oxide. The latter can react with the initial olefin leading to the formation of 2-dioxalanes with 10% yield. At the addition of "foreign" olefins or aldehydes other dioxalanes and normal ozonides can be obtained [82].

The cycloalkenes ozonolysis in the presence of methyl pyruvate results in tri-substituted ozonides formation. The latter contain three reaction centers (peroxides, proton on the ozonides cycle and methoxy-carbonyl group) accessible for various functionalization. The cyclohexene ozonolysis in the presence of methylpyrovate gives ozonide whose treatment with PPh_3 or Et_3N yields $CHO(CH_3)_4CHO$ and $CHO(CH_3)_4COOMe$ (after esterification), respectively. This method proved to be very convenient and practical way for the synthesis of linear compounds containing various terminal groups from symmetrical cycloalkenes, in two steps and with good yields [83].

The direct conversion of olefins into esters is accomplished during mono-, di- and thi-substituted olefins ozonolysis in 2.5 M methanol sodium hydroxide-dichloromethane solution. The methyl esters are obtained in high yields. Thus, 3-benzyloxy-1-nonene (5b) is transformed into 2-benzyloxyoctanoate (7b) in 78% yield [84].

The ozonolysis of acyclic alkenes including terpenes is reviewed in Ref. [85]. The role of ozonolysis as ecological process for the selective and specialized oxidation of petrochemical olefins and cyclic alkenes, for the manufacture of biologically active substances and normal organic compounds is thoroughly discussed in Ref. [86].

In Ref. 87 are summarized the data on ozonolysis of acyclic and cyclic mono-, di- and trienes in the various stages of the synthesis of insect pheromones and juvenoids.

Here should be also mentioned the importance of fulerene (C-60) ozonolysis [70] resulting in the obtaining of mixture containing various oxidation products such as ketone, ester and epoxides species. The intermediate in this reaction can act as oxygen atom carrier thus yielding phenols [88].

N-acylated esters of (cyclohexa-1,4-dienyl)-L-alanine are ozonized aiming at the synthesis of novel unnatural amino acids. The combined reduction and ozonolysis followed by condensation with a suitable nucleophile results in transformation of the aromatic ring of L-Phe to isooxasolyl, N-phenylpyrazolyl and to bicyclic pyrazolo[1,5-a]pyrimidine groups. The preparation of heterocyclic alanine derivatives is reported [89].

Thymidine diphospho-6-deoxy-a-D-ribo-3-hexulose synthesized by D-glucose ozonolysis of methyl-glucophosphate is used as a central intermediate in the biosynthesis of di- and tridioxy sugars [90].

The ozonolysis of vinyl halides followed by reductive regio- and stereo-controlled intramolecular cyclocondensation is a key stage in the synthesis of amino sugars [91].

The ozonolysis of pyrols, oxazoles, imidazoles, and isooxazoles demonstrates another application of this reaction to the organic synthesis. Pyrols are efficient protective groups of the amino functions in the synthesis of α-aminoalcohols, α-aminoketones, α-aminoaldehydes and some peptides. The oxazole ring is also known as a protective group in the peptide synthesis [92].

The 1-substituted imidazoles ozonolysis leads to the formation of the corresponding N-acylamides, which are important amine or acyl derivatives [94].

Some novel tetraacetal oxa-cages and complex tetraquinone oxa-cages are synthesized from alkylfuranes by three-stage reaction. Oxo rings in the tetraacetals are obtained by the ozonolysis of *cis*-endo-1,4-diones (norbornene derivatives) in dichloromethane solution at $-78°C$ and consecutive reduction with DMS. The tetraquinone oxa-cages are obtained in the cis-endo-1,4-diones ozonolysis in dichloromethane solution at $-78°C$ and TEA treatment [94].

A method for cleavage and oxidation of C_{8-30} olefins to compounds containing terminal carboxylic acid groups is reported in Ref. [95]. The process is promoted by oxidation catalysts, such as Cr, Mn, Fe, Co, etc.

The conversion of ethane to methanol and ethanol by ozone sensitized partial oxidation at near atmospheric pressure has been quantitatively studied. The effect of temperature, oxygen concentration in the inlet gas, contact time in the reactor and ozone concentration in oxygen has been evaluated. The selectivity in regard to ethanol, methanol, as well as the combined selectivity towards formaldehyde-acetaldehyde-methanol-ethanol is also discussed [96].

Upon studying the ozone-induced oxidative conversion of methane to methanol and ethane to ethanol it has been established that these reactions do not occur in the absence of ozone which clearly suggests that the partial oxidation is initiated by the oxygen atoms generated from the ozone decomposition [75, 97].

Ortho-selective nitration of acetanilides with nitrogen dioxide in the presence of ozone, at low temperatures results in ortho-nitro derivatives formation in high yields [98].

The ozone mediated reaction of nonactive arenes with nitrogen oxides in the presence of suitable catalysts is reported as a new method for the synthesis of the corresponding nitro derivatives with high yields [99].

The ozone induced reaction of polychloro benzenes and some related halogeno compounds with nitrogen dioxide is a novel non-acid methodology for the selective mono nitration of moderately deactivated aromatic systems [100]. In the presence of ozone and preferably with methanesulfonic acid as a catalyst, the polychloro benzenes undergo mononitration with nitrogen dioxide at low temperatures giving polychloronitrobenzenes in nearly quantitative yields [100].

The ozone mediated reaction of aromatic acetals and acylal with nitrogen dioxide is suggested as a novel methodology for the nuclear nitration of acid sensitive aromatic compounds under neutral conditions [101].

Mineral and acid-free nitro compounds are prepared from CH_2Cl_2 soluted pyridine or its derivatives treated with $NO_2/O_2/O_3$ – mixture for 8 h [102].

The nitration of aromatic compounds with nitrogen oxides in the presence of ozone is a catalytic process [103].

The stereoselective synthesis of vinyl ethers is accomplished by N – (arylidene (or alkylidene) amino) – 2-azetidinones reaction with ozone and $NaBH_4$ treatment resulting in di- and trisubstituted olefins derivatives [78].

The steroeselective synthesis of (2s, 3s) norstatine derivatives is carried out through aldehydes ozonolysis in the presence of lithium methoxyallene [82].

An interesting method of succinic acid preparation from butadiene rubber ozonolysis is suggested [104].

The preparation of α-phenyl ketone, ω-carboxylate-ended telehelic methyl methacrylate oligomers by the ozonolysis of regioregular methyl methacrylate-phenylacetylene copolymers is described in Ref. [105]. The oligomers have a number molecular mass varying from 1600 to 4500 (with respect to the initial copolymers) and polydespersity less than 2.

Ozone is also very convenient agent in the manufacture of organic ceramics [106].

The ability of ozone to destroy the double C=C bonds in organic compounds is the reason for its wide application in the preparation of bifunctional compounds. This is the principle that lies in the organization of the manufacture of dodecanedicarboxylic acid (1) and azelaic acid (2). The initial substrate for the preparation of (1) is cyclododecane which is obtained from the trimerization of butadiene and partial hydrogenation and (2) is prepared on the base of oleic acid:

$$CH_3(CH_2)_7CH=CH(CH_2)_7COOH + O_3 \longrightarrow \underset{\underset{\displaystyle O}{\diagdown \diagup}}{\overset{\overset{\displaystyle O-O}{|\quad\ |}}{CH_3(CH_2)_7CH\qquad CH(CH_2)_7COOH}}$$

$$\underset{\underset{\displaystyle O}{\diagdown \diagup}}{\overset{\overset{\displaystyle O-O}{|\quad\ |}}{CH_3(CH_2)_7CH\qquad CH(CH_2)_7COOH}} \overset{[O]}{\longrightarrow} CH_3(CH_2)_7COOH + HOOC(CH_2)_7COOH$$

The main part of dicarboxylic acids is used in the manufacture of polyester fibers, and the azelaic acid esters (n-hexyl, cyclohexyl-, iso-octyl- and 2-ethylhexyl ester) are excellent plastisizers and synthetic lubricants.

The action of ozone is also used in pharmaceuticals in the preparation of valuable hormone products. Thus the C=C bond at C_{17} in the side chain of stigmasterol is destroyed by ozone yielding progesterone – an initial source for many hormones such as cortisone, male and female sex hormones, etc.

The method of selective decolorization of fabrics containing cellulose materials, such as cotton and oxidizing dyes [107], includes the following steps: application of oxidation blocking agent to the fibers; contact of the fibers with the oxidizing reagent in gas or evaporated state in the presence of moisture till the oxidation and decolorization of the dyer is carried out; interruption of this contact before the beginning of a substantial destruction of the fiber. The oxidizing agents are selected among ozone, chlorine and nitrogen oxides and sulfur dioxide flow. The application of ozone for decolorizing indigo-painted cotton jeans after these procedures results in jeans material, which does not turn yellow for 6 months while the untreated, goes yellow much more rapidly [107].

The changes in the composition and chemistry of UV/ozone modified wool fiber surfaces [108] are investigated by means of photoelectronic spectroscopy (XPS). The oxidation of the disulfide sulfur to sulfone groups ($^-SO_3H$) containing S^{6+} approaches almost 90% conversion. This is much more that the conversion levels by using oxygen plasma. Ozonolysis results in rise of C-O groups content, particularly of the carbonyl ones [108].

The process for producing cellulose fibers and moldings such as fibers, filaments, threads (yarn), films, membranes in the form of flat, pipe and empty fibrous membranes, etc. is carried out using ozone [109]. They are produced by extrusion of cellulose solutions in tertiary amine aminooxides and in some cases in water (particularly N-methylformaline N-oxide and water); regenerating bath and water for washing by treatment with hydrogen peroxide, peracetic acid, ozone or chlorine dioxide for the regeneration of tertiary amines oxides. The introduced compounds can be enzymatically or catalytically destroyed before solvent and water regeneration.

Pitch-based carbon or graphite fibers of high tensile strength are manufactured by primary treatment of the pitch-based fibers with high concentrated ozone for a short period; then the treatment by ozone-free gases follows; in the next step they are subject to carbonization of graphitization to give elliptical fibers of long, wool similar structure. Upon treatment in the absence of ozone the fibers stitched during graphitization [110].

The EPR analysis of thermal decomposition of peroxides in ozonized polypropylene fiber for grafting shows that the decomposition of the peroxide groups begins at about 70°C. The generation of several peroxides radicals is registered; the access of the fiber to the spin sample is enhanced through oxidation. Small amounts of RO_2 – radicals with lifetime of couple of weeks have been identified. The integral intensity of the EPR-signal rises with time and ozonation temperature [111].

Ozonolysis modifies the diffusion pattern of liquid monomers in polypropylene matrix as their addition occurs in the amorphous phase. The appearance of intra-morphological structure and peroxides localization upon ozonolysis of polypropylene (granules and fibers) is monitored using electron spin resonance and transmission electron microscopy [112].

Upon ozonolysis, UV-radiation and plasma treatment polymer peroxides are generated on the surface of films from polypropylene, polyurethanes and polyester fibers [113]. Their thermal and reductive-oxidative decomposition have been studied by means of functional analysis using peroxidase and iodide. However, their disposition in the polymer specimen is quite various thus impeding their analysis and depends on the treatment agent. For example, the treatment with plasma generates easily accessible peroxides in polyurethane films while the UV-radiation and ozonation leads to formation of peroxides uncapable of reacting with aqueous solution of peroxidase. The redox decomposition of the peroxide groups by ferro-ions at 25°C has shown that less than 50% of the peroxides may react with ferro-ions at rate constants similar to those of hydrogen peroxide in aqueous solution. The thermal degradation of peroxides does not follow first order kinetics, most probably because of the generation of various peroxide species, characterized by different rates of decomposition. The lowest rate constant of decomposition observed at 62°C is 3.10^{-3} /min which does not depend on the polymer nature and the method of peroxides generation [113].

Ozone-induced graft polymerization onto polymer surface is an important and convenient method for polymer modification [114].

Ozone-induced graft copolymerization of polyethylene glycol monomethyl ether methacrylate onto poly(etherurethane) improves the hydrophility and water absorption. The autoaccelerated effect in ozone-induced polymerization has been also discussed [115].

The modification of the surface properties of polypropylene and block copolymer is carried out by ozone treatment. Thus, the wettability of the polymers is improved – the contact angle of water becomes $\leq 67°$. The break of the polymer chain and carboxyl groups' formation are accelerated by using high ozone concentrations. The cut-off fragments resulting from the ozone treatment are removed from the surface by ultrasound and organic solvents [116].

The low temperature nonelectrolytic nickel plating onto three types of polypropylene is carried out by substrate pretreatment with ozone. The latter modifies the polymer surface for galvanization while the combination of polar with anchor effect as a result of the ozone etching enhances the adhesive properties of the polymer surface. The washing of the material after ozonolysis is obligatory for ensuring a good adhesion of the material [117].

The surface properties of polypropylene (I), 100 μm films from (I), ethylene-propylene block polymer (II), or ethylene-propylene polymer prepared by the random method (III) can be improved by ozonation at concentrations of 1.38, 0.64 and 0.41 mol%, respectively. The adhesion of dyes, coloring agents, dye layers, on (I) is substantially improved and depends on the substrates in the following order: (II) > (III) > (I). It has been found that for each sample there are optimum conditions of ozonation. Reactive dyes such as epoxy and acrylo urethane resins impart a better adhesion force that the conventional nonreactive acrylic or vinylchloride resin [118].

Freshly extruded, 50 μm C_2H_4-ethyl acrylate-maleic anhydride copolymer film is treated by 500 ml/m² oxygen containing 10 g/m² ozone and is calendered with 200 μm monolayer C_2ClF_3 – polymer film at 15°C to produce moisture-proof packaging material with intra-layer adhesion of 800 g/15 mm and thermosealing force of 3.6 kg/15 mm against 250 and 1.3, respectively, in the absence of ozone treatment [119].

Methylmethacrylate polymers with good thermal degradation resistance are prepared by ozonolysis [120] whereby the end unsaturated groups are converted into carbonyl or carboxyl ones. Ozone/air mixture is blown through a 10 g Acrypet VH solution in CH_2Cl_2 at –78°C followed by exposure only to air for 60 min and after solvent removal the solid residue is dried in vacuum at 297°C for 8 h. The polymer thus obtained is characterized by an initial temperature of thermal degradation of 315°C against 297°C for the polymer without ozone treatment [120].

The surfaces of propene polymer moldings are treated with ozone for improving their hydrophility. The ozone treatment of the surfaces for 8 h reduces the contact angle of water from 100 to 81° [121].

Ozonation of PVC latex is also carried out for removal of vinyl chloride residues. The aqueous dispersions of saturated polymers are treated with ozone and the vinyl monomers are removed [122].

Graft polymerization of vinyl monomers onto Nylon 6 fiber is carried out after ozone oxidation of the fibers or films from Nylon 6 with vinyl monomers such as acrylamide, methylmethacrylate and vinyl acetate. The molecular mass of Nylon 6 decreases slightly upon ozonation. For acrylamide graft polymerization system the preliminary treatment in air or vacuum by γ-rays radiation prior ozonation results in higher graft percent. For methyl methacrylate the apparent graft percent does not rise with the ozonation time. However the apparent graft percent for vinyl acetate is increased with the ozone time treatment [123].

The adhesion of PVC-, fluropolymer, or polyester-coated steel panels is improved by treatment with 5–50% solution of H_2O_2 or ozone [124].

Synthesis of water-soluble telehelic methyl-ketone-ended oligo- N,N-dimethylacrylamides by the ozonolysis of poly(N,N-dimethylacrylamide-stat-2,3-dimethylbutadiene)s is reported in Ref. [125].

The synthesis of telehelic methylmethacrylate and styrene oligomers with fruorophenyl ketone end groups is accomplished by the ozonolysis of copolymers containing 4-flurophenyl butadiene units [126].

Graft polymerization of acrylic acid is carried out into preliminary ozonized siloxane matrixes [127].

The manufacture of base discs for laser recording material is realized by covering of the plastic substrates surface with solid polymer layer possessing directing channels and/or signal holes. The surface of the plastic substrate is preliminary cleaned by UV/ozone exposure before the formation of the hardened polymer layer [128].

The controlled degradation of polymers containing ozonides in the main chain takes place in the presence of: (1) periodate suported on Amberlyst A26 (SPIR); (2) diphenylphosphine deposited on polystyrene (SPR); boron hydride supported on Amberlyst A26 (BER) [129]. Poly (butyl) methacrylate-copolymers are prepared by emulsion polymerization. These materials and the homopolymer poly (butyl) methacrylate are ozonized at various temperatures and treated by any of the reagents described above, thus giving telechelic oligomers in 99% yield. Molecular mass variations with ozonation temperature change have been observed. The end aldehyde groups are registered using ^1H and ^{13}C -NMR spectroscopy; the end hydroxyl groups are observed by means of FTIR and ^{13}C-NMR; the presence of hydroxyl groups is confirmed by tosylate formation; the carboxyl groups are identified by FTIR and ^{13}C-NMR and quantitatively determined by titration; in the SPR-generated oligomers the content of aldehyde groups constitutes about 80% of the end functional groups. The oligomers obtained in BER and SPIR contain 99% hydroxyl and carboxyl groups [129].

Polyethylene fibers are subject to ozone treatment for modifying their surface [130]. The analysis of the surface is carried out by means of X-ray photoelectron (ESCA) and IR (FTIR) spectroscopy. Carbon (C) and oxygen (O) were the main atoms monitored with ESCA (C-1s, O-1s areas) on the treated fibers. The analysis of C-1s peaks (C_1, C_2 and C_3) reveals that the oxidation level depends of the ozonation time. The components of 1s peak (O_0, O_1, O_2) are very useful for carrying out the surface analysis. They demonstrate the presence of carbonyl groups (1740–1700 cm^{-1}) even onto untreated fibers whose intensity rises with treatment time. Ozonolysis is directed from the surface to the fiber bulk. This is confirmed by the great enhancement of the carbonyl bond band after 3 h ozonation. The thermal analysis suggests structural and morphological changes of the fiber when ozonolysi time treatment exceeds 2 h [130].

The processing of fiber-reinforced plastics is performed by blowing an ozone-oxygen mixture (flow rate of 0.4 l/min) for 5 h through a CH_2Cl_2 solution containing glass reinforced fiber particles (1mm diameter) filled with $CaCO_3$ and unsaturated polyester. The fibers emerge on the surface till a fine powder is precipitated [131].

The manufacture of laminates through heat-sealing method involves the application of electric crown, UV-radiation or ozone exposure [132].

The manufacture of pour point depressants for oils, particularly useful for diesel oils, is carried out by oxidative degradation of waste polyethylene and/or polypropylene (I) with ozone at 30–150°C. 1000 g waste (I) is exposed to ozone action at 150°C for 5 h giving the pour point depressant. The addition of 1% depressant to diesel oil reduces the solidification of the oil from –20°C to –35°C and the temperature of filter plugging from –9 to –20°C [133].

7.3.2   INORGANIC PRODUCTIONS

Stainless steel parts are treated by ozone for surface passivation. The parts are heated in oxidative or inert atmosphere at a temperature of condensation lower or equal to –10°C. The unreacted ozone is re-used [134].

The main passivating agent is oxygen in combination at least with ozone. The system is particularly appropriate for passivation of metal (for example, stainless steel, Ti, etc.) equipment, used in chemical plants and exposed to strong corrosion action at high temperatures and pressures [135].

For increasing the corrosion resistance of metals and alloys they were exposed from 1 s to 10 min in cold plasma under pressure 1–103 Pa and 100–5000 V in atmosphere containing O_2, O_3, N_2, H_2, air, CO_2, CO, N-oxides, H_2O (gas), combustible gas and/or inert gas. Thus, 17% C_0-ferrite stainless steel is subject to plasma treatment for 4 min at 103 Pa, 100 mA and 250 V in nitrogen-oxygen mixture with 20% oxygen. The corrosion resistance was evaluated by treatment of the sample with a solution containing 17 ml 28% $FeCl_3$, 2.5 ml HCl, 188.5 ml H_2O and 5 g NaCl. The sample shows relatively good resistance as compared with the untreated one [136].

Electrochemical tests reveal the influence of dissolved ozone on the corrosion behavior of Cu-30 Ni and 304L stainless steel in 0.5N sodium chloride solutions [137]. These experiments include: measurements of the corrosion potential as a function of the time and ozone concentration; cyclic polarization experiments; isopotential measurements of the current density and study of the film components. The results of these experiments show that for Cu-30 Ni and 304L-stainless steel the corrosion potential is shifted to the more noble values (300 mV) at $[O_3]$ < 0.2–0.3 mg/l. At higher concentrations it remains unchanged. The dissolved ozone reduces the corrosion level for Cu-30 Ni – alloys which are evaluated by the substantial decrease in the current density at constant applied potential. The improvement of the corrosion resistance should be related to the decrease of the thickness of the corrosion products film and to the higher oxygen fraction as compared with the chloride in the same film. For stainless steel the differences in the passivating films in ozonized and nonozonised solutions are negligible as it is shown by spectroscopy [137].

Laboratory experiments have been carried out to study the ozone application for acid oxidizing leaching of chalcopyrite with 0.5 M H_2SO_4. For evaluating the reaction mechanism the effect of particles size distribution, stirring and acid concentration, the dissolution reaction and reaction kinetics on the leaching have been investigated. The reaction rate is governed by ozone diffusion to the reaction mixture. Ozone is an efficient oxidizer and the process is most effective at 20°C [138].

The rate of acid leaching of chalcopyrite depends on the use of ozone as an oxidizing agent in sulfuric acid solutions. The leaching of chalcolyrite follows a parabolic law [139]. The use of ozone as an oxidizer provides conditions for the formation of elemental sulfur on the leached surfaces. The rate of leaching is reduced with temperature as the ozone solubility decreases with temperature. The results show that ozone is the best oxidizing agent for acid oxidizing leaching of chalcopyrite and may be applied in pilot plants for regular manufacture.

We have studied the possibility of using ozone for improvement Ag extraction form polymers deposits in a flotation plant in the town of Rudozem. It has been found that upon blowing of ozone (1% vol concentration, flow rate – 300 l/h) through the empty shaft of the flotation machine stirrer (5 m^3 volume) the degree of Ag extraction is increased by 1–2% [139].

The redox leaching of precious metals from manganese-containing ores carried out by other authors show also positive results [140].

Molybdenite flotation from copper/molybdenum concentrates by ozone conditioning results in relatively pure copper-free molybdenum. The process including multi-step ozone flotation proves to be a technical and economical profitable method [141].

The manufacture of potassium permanganate is accomplished by melting of Mn-containing compounds with KOH, dissolution of the melt and the solution oxidation. For reducing the energy consumption $Mn(NO_3)_2$ is used as a Mn-containing compound. It can be easily alloyed with KOH in a 1:5 – 1:10 ratio at 250–300°C. Then the product is dissolved in 20–25% KOH solution, the solid residues are removed, soluted in 3-5% KOH and the solution are oxidized by ozone-air mixture [142].

The manufacture of silicon carbide ceramics is performed as melted organosilicon polymer is oxidized with 0.001% vol. ozone. Polycarboxylstyrene is the preferred polymer. The ceramics obtained is characterized by high thermal resistance and acceptable physical properties [106].

Mixtures containing In- and Sn-compounds are molded and sintered in a furnace in air atmosphere containing ≥ 1000 ppm ozone. The ITO ceramics thus obtained are characterized by high density at low temperature sintering for a short time [143, 144].

A method for oxidizing carbonaceous material, and especially for bleaching gray kaolin for subsequent use as coating or filler for paper in the presence of ozone is reported in Ref. 145.

The manufacture of mercury(I)chloride includes the reaction of Hg with hydrochloric acid in the presence of water, subsequent removal and drying at 95–105°C. Ozone (0.1–0.1 g/g product) is bubbled through the reacting mass to increase the product yield and quality [146].

Arsenic acid is prepared from $(As_2Cl_6)_2$-ions by ozonolysis [147].

The removal of color and organic matter in industrial phosphoric acid by ozone and the influence on activated carbon treatment is described in

Ref. [148]. Industrial phosphonic acid containing 42–45% P_2O_5 and 220–300 mg/l organic matter (OM) is subject to combined treatment with ozone and activated carbon. The independent ozonation results in removal of the initial dark color of the acid and the organic matter. It is only through absorption on activated carbon that the level of OM could be reduced to 80% per 25 g/kg P_2O_5. The ozonation prior adsorption enhances the efficiency of the activated carbon effect and decreases its specific consumption [148].

The fabrication of high-Tc superconductors is carried out using ozone-assisted molecular beam epitaxy (MBE). It includes the simultaneous evaporation of the elementary components and application of ozone as a reactive oxygen source. The ozone priority over the other oxygen forms is that it is rather stable and could be produced and supplied to the substrate in a very pure state using simplified apparatus ensuring a well defined flow of oxidizing gas. Upon application of ozone the untempering through a thermal treatment for removing the inner stresses could be ignored. In order to prepare films with high temperatures of superconductivity the growth should be carried out at relatively low stresses in the system. In addition, the surface during film growth can be analyzed by reflectance high-energy electron diffraction. The most recent improvements in ozone-assisted MBE for family of $YBa_2Cu_3O_7$-delta films are described. The results show that this technique is very appropriate for growth of high quality superconductive films and could be ideal for the manufacture of such structures [149].

The growth of superconductive oxides under vacuum conditions compatible with MBE requires the use of activated oxygen. The atomic oxygen or ozone appear to be such species [150]. The characteristics of a radio-frequency plasma source for molecular beam epitaxial growth of high-Tc superconductor films (200 Å) from $DyBa_2Co_3O_7$ on $SrTiO_3$ are described and discussed in Ref. [150].

An apparatus for the preparation of pure ozone vapor for use *in situ* growth of superconducting oxide thin films are designed [151]. Pure condensed ozone is produced from distillation of diluted ozone-oxygen mixture at 77K. The condensed ozone is heated until the pressure of its vapors approaches the necessary for an adequate flow of ozone- gas to the chamber of thin films growth. The thin films from $YBa_2Cu_3O_7$ with zero resistance at temperatures of about 85K grow at ozone pressure in the chamber of 2.10^{-3} Torr even without subsequent untempering. It should be noted that in contrast to other highly reactive oxygen species ozone can be prepared and stored in very pure form which makes it very convenient for studies on the kinetics of growth and oxidation of films with a well defined gas [151].

A patent for preparation of oxides superconductors $M-M^1-M^2-M^3$ includes ozone oxidation where: M = elements of III B groups such as: Y, Sc, La, Yb, Er, Ho or Dy; M^1 = elements from II A group, such as: B, Sr, Ba or Ca; M^2 – Cu and one or more elements from I B group such as Ag or Au; M^3 = O and on or more elements belonging to VI A group like S or Se and/or elements from VII A group as F, Cl or Br. The superconductors thus prepared have high critical temperature and high critical current densities [152].

The manufacture of bismuth-, copper-alkaline earth oxide high-temperature superconductors is carried out through calcination and/or sintering in ozone-containing atmosphere with $Bi(OH)_2$, $Ca(NO_3)_2$ and $Sr(NO_3)_2$ and subsequent treatment by $CuCO_3(OH)_2$; the components are dried, calcinated in ozone containing air, molded and dried in air atmosphere. Their temperature of superconductivity is 107K and the critical current density is 405 A/cm^2 [153].

A low temperature method for preparation of superconductive ceramic oxides is described in Ref. [154]. It includes the treatment of the substrate-heated surface by ozone to the complete evaporation of the other components thus forming a superconductive ceramic oxide.

The manufacture of rare earth barium copper oxide high-temperature superconductor ceramics is carried out by sintering at 930–1000°C in ozone atmosphere containing oxygen. Y_2O_3, $BaCO_3$ and CuO are mixed, calcinated, pulverized, pressed and sintered at 950°C in the presence of ozone-oxygen mixture after which they are gradually cooled to the critical superconductive temperature 94K [155].

A review devoted to the growth of co-evaporated superconducting yttrium barium copper oxide (Y Ba_2CuO_7) thin films oxidized by pure ozone are presented in [156].

The removal of organic pollutants from the surface of supports for microelectronics purposes is conducted by UV/ozone treatment [157].

Low temperature silicon surface cleaning is carried out by fluoric acid (HF) etching, washing, cleaning with deionized water, N-gas blowing, UV-ozone treatment. This is a treatment preceding the process of silicon epitaxy [158–160].

The method for total room temperature wet cleaning of silicon surface comprises the use of HF, H_2O_2 and ozonized water and is with 5% more economical than that the standard procedure of wet cleaning [161].

The removal of resist films supported on semiconductive substances is accomplished by using ozone [162].

The application of ozone for reducing the temperature and energy in the process of very large scale integration (VLSI) is described in [163, 164]. The main problems under discussion are as follows: (1) cleaning of Si-support with ozone by two methods: dry process combined with UV and wet one with ozonized water; (2) stimulation of Si thermal oxidation with ozone; (3) deposition of SiO_2 under atmospheric pressure and low temperature via chemicals evaporation using tetraethylorthosilicate and ozone; (4) deposition of Ta_2O_3 at atmospheric pressure and low temperature applying $TaCl_5$ and ozone; (5) tempering with ozone and UV-radiation of Ta_2O_5 films used for dielectric for memory units , this treatment reduces the current permeability into the film; (6) etching of photoresistant materials by $O_2/O_3/CF_4$ and the effect of ozonator charge and injection of CF_4 [163, 164].

The cleaning of synchrotonic radiation optics with photogenerated reactants has a number of priorities over the methods of discharge cleaning [165]. Upon discharge cleaning, the discharge particles should react with the surface contaminants until its shielding by the rough discharge elements, which may pollute or destroy the surface. Contrary, if the particles can be photon generated near the surface, the problem with the protection drops off and in some cases the cleaning can be more efficient. An estimation of the various methods for cleaning was made comparing the rates of polymethylmethacrylate films removal. A number of various light and geometry sources have been tested. The highest rate of cleaning was achieved upon using UV/O_3 cleaning method at atmospheric pressure. This method has been widely applied for cleaning semiconductive surfaces from hydrocarbon contaminants. It is noted that it is also effective in removing graphite-like pollutants from synchrotonic radiation optics. It proves to be more simple, economical and selective method as compared with other cleaning method. Although it requires the drilling of a hole in the vacuum chamber, the cleaning of the optics can be carried out without dissemble which saves much time. This method is successfully applied for cleaning grates and reflectors in several beams [165].

Native oxide growth and organic impurity removal on silicon surface is carried out by ozone-injected ultrapure water [166]. In order to manufacture high-efficient and reliable ULSI-units the further integration and minimization is in progress. The cleaning with $H_2SO_4/H_2O_2/H_2O$ is accompanied by serious problems: a great amount of chemical wastes is obtained which must be suitably treated. The cleaning technology with ultrapure water includes ozone injection in concentrations of 1–2 ppm. This method is very efficient in removing the organic impurities from the surface for a short time and at room temperature. The process wastes can be treated and recycled. The ozonized water components can be easily controlled [166].

The reaction mechanism of chemical vapor deposition using tetraethylorthosilicate and ozone at atmospheric pressure is reported and discussed in Ref. [167].

Covering of semiconductor devices with silica films is carried out via CVD using $Si(OEt)_4$ and ozone. The deposition process is followed by heating in oxygen atmosphere with simultaneous UV radiation. The SiO_2-film thus formed can improve the water resistance of semiconductive devices [168].

The fabrication of nondoped silicate glass film with flat structure by O_3-TEOS deposition includes three stages: (1) substrate treatment; (2) formation of Al conducting layer on it using oxygen plasma through heating; (3) formation of SiO_2-film on the substrate by plasma using ozone and TEOS [169].

In Ref. [170] are reviewed the future trends for interlayer dielectric films production and their formation technologies in ULSI multilevel interconnections. The properties of the interlayer dielectric films and their preparation technologies should satisfy the following three requirements: (1) available for disposition (setting) on large surface; (2) low dielectric constant; (3) low deposition temperature. Two techniques have been developed for the selective deposition of SiO_2-films which is the best way for achieving a complete planarization of the interlayer surface of dielectric films: (1) low temperature liquid phase deposition using a saturated aqueous solution of hydrofruorosilicon acid H_2SiF_6; (2) half-selective SiO_2-films deposition at 390°C with $Si(OC_2H_5)_3$ and ozone and preliminary treatment with tetraflurocarbonic (CF_4) plasma on TiW or TiN surfaces [170].

Lead zirconate and titanate thin films are successfully prepared by reactive evaporation. The elements Pb, Zr and Ti are evaporated in ozone-oxygen mixture. The films obtained have equal thickness and composition per a large surface (in the range of ±2% from the 4-inch support) [171].

Ceramic coatings on substrates are formed in the presence of ozone [172]; the substrates (silicon checks) are covered (dipped) in a solution of one or more (partially) hydrolyzed pre-ceramic silicon alholates with general formula $R_xSi(OR)_{4-x}$ ($R = C_{1-20}$ alkyl, aryl, alkenyl or alkinyl; x = 0–2). Further, the solvent is evaporated to form a coating, which is subsequently heated in the presence of ozone up to 40–100°C thus converting into a ceramic coating. It in its turn may be covered also by a protective layer containing Si, or Si and C, or Si and N, or Si, C and N, or SO_2 and oxide. These coatings which are abrasive-, corrosive and thermo-resistant have also a small number of defects and suppress the diffusion of ionic contaminants such as Na and Cl ions and are particularly convenient for electronic units [172].

Method of forming zinc oxide light-shielding film for liquid crystals shields includes the injection of vapors of alkyl zinc compound with ozone or atomic oxygen in the activation chamber after which they pass through a chamber for ZnO film deposition at low temperature heating at about 200°C. The method is applicable for large-scale production of these films [173].

An evidence for a new passivating indium rich phosphate prepared by UV/O_3 oxidation of indium phosphide, InP, is provided in Ref. [174]. The phosphate does not exist as crystal compound and its composition is $InP_{0.5}O_{2.75}$. The passivating ability of the latter with respect to InP surface is discussed.

For improving the light-absorption characteristics of oxide optical crystals they are heated in to ozone-oxygen atmosphere [175, 176]. Thus the light absorption from the optical crystal in a wave range different from that with which the crystal affects its own absorption is reduced to the most possible level. Devices using such oxide optical crystals with improved absorption characteristics work with high efficiency as optical amplifiers, optical isolators, optical recording medium, and optical generators.

The manufacture of solar cell modules with transparent conducting film covered by amorphous Si layer and electrode on the backside linked to the transparent isolator layer is described in Ref. [177]. The preparation includes the application of a laser beam for electrode molding and exposure in oxidizing atmosphere containing 0.5–5% O_3 or ≥10% O [177].

A method for strong oxidation in ozone atmosphere is proposed for surface activation of photoconductive PbS films [178].

Adhesion-producing materials for electroless plating and printing circuits contain particles of thermostable material slightly soluble in the oxidizing agent; the particles are dispersed in the thermostable resin, which becomes almost insoluble in the oxidizing agent at hardening. These materials find application for printing circuits and are treated by oxidizing agents (for example, chromic acid, chromate, permanganate or ozone) to create a concave material surface [179].

7.4 MEDICINE

Upon exposure of living organisms to ozone, depending on doses, the different vital systems can be suppressed or stimulated to a certain extent.

For the first time already in 1898 Binz [180] showed interest in the effect of ozone on the blood and plasma and found that ozone exhibits a somnolent effect. Brinkmson and Lambert [181] proved that ozone inactivates the cell enzymes and decreases the oxygen exchange between blood and tissues.

A thorough investigation of the effect of ozone on the human organism should allow extending the range of its application in medicine. The operation mechanism of ozone at small (therapeutic) concentrations, up to 100 $\mu g/m^3$, in the human cell has been already studied. Ozone enhances the cell membrane resistance to damaging effects. Human blood, treated with ozone, is able to keep erythrocyte membrane intact for 5 to 15 days. A method for preparation of nonhemolyzed plasma has been developed which facilitates laboratory diagnostic investigations and preserves the blood without inducing hemolysis [182]. *In vivo* and *in vitro* experiments have shown that ozone stimulates erythrocytes gas-transportation function, which provides a continuous optimum tissue oxygenation. Besides, ozone strengthens leukocytes antibacterial function, the high activity of the mieloperoxidase enzyme system being preserved for 15 days. Also, ozone could find application in disinfection of biological materials, blood in particular. Encouraging results of HIV inactivation by means of ozonolysis have been reported in Refs. [183–189]. Blood, plasma, infected with retrovirus, whose transcriptase (RT) activity was 1.4×10^5 cmp/ml (enzyme for retrovirus replication) was separated into three fractions – A, B and C. Fraction A and fraction B (after cryoprecipitation) were exposed to ozone, 12 $\mu g/m^3$, at 0°C for 30 min. Fraction C was not treated with ozone. Subsequently, after 13 days of incubation, the following changes in RT activity were observed: fraction A = 130,000; B = 0 and C = 1.75,000.

In acute massive blood losses the adaptive potentialities of the human organism are increased. Besides, the operating strength of the intracellular mechanism of hemoglobin antioxidative resistance is enhanced, as deduced from superoxide dismutase activation in erythrocytes. Phrophylaxis in hemorrhages by means of ozone inhalation is due to the activation of biological processes, which provide erythrocytes gas-transportation function. The application of ozone in medicine opens prospects to invent means for correcting the respiratory systems by changing the molecular regulation mechanism of hemoglobin affinity to oxygen in various hypoxic conditions. Methods can be developed to preserve blood cell vitality and dynamic activity during conservation and transfusion. There is growing interest in looking for techniques for ozone use in cardiovascular diseases. Combined effects and hyperbaric oxygenation in conditions related to hypoxia have also been studied.

The ability of ozone to oxidize toxic products formed under various pathological conditions has attracted clinicians attention. Ozone either directly oxidizes double bonds (in unsaturated compounds) or, by activating free radical processes, initiates a danger of enhancing the spontaneous process of lipid peroxide oxidation. Kontorshticova et al. [190] studied the lipid and protein spectra of blood plasma exposed to ozone. Blood, taken from a dog 15 min after suffering from hemorrhagic shock, was exposed for 5 min to oxygen-ozone mixture at a rate of 1.0 l/min and ozone concentration of 0.048 mg/l/ The total protein content and separate protein fractions were analyzed. The fatty acids content was determined by gaschromatography. The protein and lipid spectra reflect processes, which take place in hypoxia and subsequent oxygenation in the organism of test animals. These are: (a) elimination of lipid depots of nonesterified fatty acids; (b) nonoxidized or partially oxidized products of lipid and protein nature; (c) blood saturation with oxygen and activation of oxygenation processes. An increase by 20% in α_2-globulin content zeruloplasmin and haptoglobin, which possess antioxidative properties, is observed in the protein spectrum. The content of fatty acids is decreased by about twofold. The total protein and β_2-globulin concentration also decrease. These changes are consistent with a direct effect of ozone on peptide bonds in proteins and on double bonds in unsaturated fatty acids.

We have studied the effect of ozone [191] on POL of rabbit brain and liver. In this connection the rabbits were exposed to ozone action in a special chamber for 2, 3, 10 and 15 h. The ozone concentration in the gas phase was varied in the range 50–250 ppm. We have found a substantial enhancement of the content of endogenous POL fluorescence products (FP) and TBA-reacting products (TBA-rp) after 2 h exposure to 250 ppm ozone concentration. However, after 4 h exposure to the same ozone concentration the rabbits die. When the concentration of exposed ozone was 50 ppm the increase in FP content in the rabbits brain is increased by about 1.8 times after 10 h, and in the liver – by about fold 2 compared with that of the control test. The level of TBA-rp was 3.1 and 3-fold higher, respectively. It has been observed that the amount of FP in the brain between the 10th and 15th h does not change, but the FP content in the liver microsomes rises 2.6 times and that of TBA-rp – 3.6 fold. A series of test animals, which were preliminary injected with 56 mg/kg potassium salt of DL-α-tocopherol 10 h before the test, were also studied under the same conditions. The results from FP determination after 10 and 15 h of exposure reveal that their amount in the brain has increased by 1.09 and 1.3 times, respectively and in the liver microsomes – 1.5 and 1.9 fold. Simultaneously there is no marked difference in the DL-α-tocopherol content before and after exposure of the test animals as demonstrated by the HPLC data. These facts indicate that the brain is either more stable or more protected in regard to ozone action since DL-α-tocopherol protects first of all the biomembranes.

The effect of tocopherol and more than 20 its homologues and derivatives was investigated in model reaction of cumene oxidation and in enzymatic and enzyme-free POL in rabbits brain and liver microsomes [192–196]. It has been shown that the inhibitor efficiency depends rather on their diffusion coefficients through the membrane biolayer as well as their mobility than on the chemical nature of the substituents. A correlation between the rate of POL inhibition and the suppression of the luminol-dependent chemiluminescence in biomembranes has been established. However, no relationship was found between the action of 2,6-di-*tert*-butylphenol inhibitor and its derivatives and the chemiluminescence produced during the interaction between superoxide anion and luminol.

Fairly high doses of ozone can be employed to saturate blood with oxygen and deactivate nonoxidized and partially oxidized products. Therapeutic doses of ozone, 0.048 mg/l, are applied to correct pulmonary insufficiency in postreamination periods.

High concentrations of ozone, 0.2 mg/l, can be used to model pathological conditions, which allow studying morphological changes in pulmonary tissue. The effect of ozone upon varying the mode of administration in organism has also been investigated [197].

An isotonic NaCl solution saturated with ozone was injected intracutaneously or a saturated aqueous solution was administered orally to rabbits. In animals, injected with ozonized water, the red blood cell level remained stable in the course of experiment (10 days). An increase in leucocyte level (2 fold) was observed at the expense of monocytes. Changes of the granulocytes were not pronounced. The functional activity of neutrophilic cells, evaluated by phagocytosis activity is increased 1.5 times. On intracutaneous administration of ozonized isotonic NaCl solution, the leucocyte level was diminished after the first hour, enhanced after the third hour and remained high by the end of experiment with respect to the initial level. By comparing the two methods, it was established that the toxic effect was more pronounced with the intracutaneous administration than the oral one. Thus, for example, the activity of the process of lipid peroxide oxidation, evaluated through malonic dialdehyde determination in erythrocytes, increased by 3-fold upon intracutaneous injection and did not change after oral administration. On exposure of blood from healthy donors aged 25–35 to 3 mg/l ozone the activity of the aspargataminotranspherase was increased whereas that of lactic dehydrogenase attained a maximum concentration and then fell down to the initial one.

Ozone therapy, being an effective means to treat a number of diseases, attracts the attention of practicing clinicians, dermatologists in particular [198]. Attempts have been made to treat various dermatoses; psoriasis, herpes, microbial eczemas, etc. [199]. The obtained results were positive. In this connection a joint project with the dermatology department at the Sofia Medical Academy is developed. Its object is to study the effect of ozone on dermatologically injured skin and development of ozone generating units.

The effect of ozone has focused scientist's attention for more than two decades. Of particular interest is the oxidative effect of ozone on cells, especially tumor cells [200–202]. Cells from alveolar adenocarcinomas, breast adenocarcinomas and uterine carcinosarcoma and normal cells from lung diploid fibroplast as a reference were incubated in a special chamber. They were exposed to ozone (0.3 to 0.7 ppm) for 8 days. The number of cells was determined every 48 h. Ozone levels from 0.3 to 0.5 ppm inhibited the accumulation of tumor cells by 40–60% and did not affect the normal cells. The inhibition was 90% at a concentration of 0.8 ppm.

Many viruses are sensitive to ozone, for example *vesicular stomatitis, encefalomyocarditis, poliovirus type 2 and 3, coxsacknevirus, echovirus* and *adenovirus* [203–205]. The authors have studied the inactivation of rotaviruses HRV type 2 from humans and SA-11 from apes for various exposures of ozone (0.05, 0.1, 0.2 mg/l) at 4°C. HRV are considerably more sensitive to ozone than SA-11. In this connection ozone might find therapeutic application in oncology.

7.4.1 THERAPEUTIC APPLICATION

As a therapeutic means the so-called medical ozone is employed in concentrations from 0.05 to 5%. The types of application and the relevant concentration range of medical ozone/oxygen mixtures are specific for each pathological change (Table 6).

TABLE 6   Therapeutic doses.

Types of applications and appropriate range of doses of ozone/oxygen mixtures for medical purposes
Colonic insufficiency and external exposure in closed system ↓

Intravenous	
Intramuscular injection ↓	
Hypodermic	
Blood treatment ↓	Production of ozonized
Stimulating treatment ↓	water
Circulatory ↓	Cleaning of wounds and disinfection

```
0     10    20    30    40    50    60    70    80    90    100.
```
Ozone concentration, μg/ml.

As seen the ozone concentrations administered by injection reach 40 mg/l. Upon extracorporeal blood treatment, stimulation and circulatory treatment ozone is employed in concentrations up to 30 mg/l whereas for colonic insufflations, external gas application in closed system, cleansing of wounds and in ozonized water production it can be used up to 100 mg/l level.

Ozone therapy is successfully applied in a wide variety of diseases as shown in Table 7.

TABLE 7   Some examples of diseases and medicine fields of ozone application.

Disease	Medical field
Anal fissure	Dermatology / proctology
Acne	Dermatology
Abcess	Surgery / dermatology
Allergies	General medicine, dermatology, allergology
Virus infections	AIDS, various special fields
Arterial circulatory disturbances	Surgery, vascular surgery
Fistulare	Proctology, drmatology, urology, gynecology
Mucous colitis	Proctology, gastroenterology

TABLE 7   *(Continued)*

Disease	Medical field
Cystitis	Urology
Cerebral sclerosis	Gerontology, internal medicine and neurology
Decubitus (ulcers)	Surgery, gerontology and dermatology
Furunculosis	Dermatology
Gangrene	Surgery
Ulcus cruris	Dermatology
Hepatitis	General medicine, internal medicine
Hypercholesterolaemia	Internal medicine
Herpes (AIDS)	Dermatology, internal medicine
Climacterium/menopause	Gynecology
Cirrhosis of the liver	Internal medicine
Raynaud's disease	Surgery, vascular surgery
Constipation	Internal medicine
Radiation scars (following	Radiology, dermatology irradiation/treatment)
Osteomyelitis	Surgery
Oncological additives	Oncology and relevant fields
Parkinson's disease	Neurology
Mycosis/fungus infections	General medicine, dermatology
Polyarthritis	Internal medicine, neurology
Spondylitis	Orthopedics, surgery
Stomatitis	Odontology, dentistry
Sudeck's disease	Surgery, orthopedics
Thrombophlebitis	Internal medicine, dermatology
Wound heling disturbances	Surgery

As a rule, the ozone-oxygen mixture is employed in cases of arterial circulatory disturbances. This method was first described by Dr. Lacoste in 1951 and since then has attracted the attention of many scientists. Rokitansky [207] was the first who made a statistical evaluation, according to which four out of ten patients with arterial-metabolic circulatory disturbances treated with ozone did not need operation the results of other authors confirm his conclusions [208, 209].

Rectal application method (called ozone enema) was reported in 1935 and its successful application has been proved in cases of *mucous colitis* and *fistulae*. It was established that the oxygen content in the blood was increased [210, 211].

Ozone has shown an antalgic effect, this property being used through injections in case of pains of the spinal column [21]. Care should be taken of sensitive patients, since the initial pain on the injection is considerable. Usually 30–50 cm³ ozone are injected intracutaneously in the lumber and sciatic region to relieve pain, which is observed after 1–2 h and may continue for a couple of days.

Wolff [213] reported that as early as 1915 ozone was provided for local gas treatment in cases of *fistulae, dicubitus (ulcers), ulcus cruries, osteomyelitis* as well as badly healing wounds. For that purpose plastic bags filled with gas mixtures were employed. This therapy became particularly useful for inaccessible ulcers, fistulae and damage from X-radiation.

Payr [214] applied subcutaneous injection of ozone gas in cases of venous circulatory disturbances and varicosis. For the same cases, he also used intravenous injections. He also reported that Dr Fisch had tested this method on himself without side effects.

The intraarterial application of ozone therapy is promising for orthopedic practice and for accident surgery. Intramuscular injection of oxygen-ozone mixture is mainly applied in some cases with obstinate carcinoma patients or as a supportive therapy together with the so-called "minor" autohemotherapy. This is a successful therapy: painless, harmless and widely applied in inflammatory processes.

Dorstewitz [215] employed autohemotherapy to treat hepatitis. 50–100 ml blood from the patient are treated with ozone and then reinjected. A 20 patients with herpes simplex and herpes zoster were treated in the same way for 5–10 days. A complete healing of skin lesions was observed. In one of the cases of extensive herpes in the trigeminal region with keratitis, healing was achieved within 10 days, although a persistent opacity of the keratic tissue remained. In six cases of herpes simplex, upon double treatment with ozone, a positive effect was observed after 28 h. It is desirable to start therapy at an early stage of illness.

Moreover, the so-called hematological ozone therapy (HOT) has found wide application [216]. The therapy has been employed since 1947. In a special hyperbaric chamber the air is enriched with ozone by means of UV-Cs lamp. It has been suggested that singlet oxygen is formed under these conditions. 60–70 ml blood from patients are placed in the HOT apparatus and enriched with oxygen-ozone. The blood thus treated is re-injected intravenously or intramuscularly into the patients. HOT therapy is applied in the following affections: chronic ulcers and gastritis, pro-phylaxis in recoveries from myocarditis; lipid metabolism disturbances hypercholesterinae; chronic liver trouble and recoveries from infectious hepatitis; prophylaxis in acute hepatitis, donor blood preliminary treated with ozone-mixture migraine, chronic nephritis with increased creatine and urea levels; acute artery retinae occlusions, chronic polyarthritis; scab treatment after burning; acute bleeding of the eye. Most likely, HOT ther-apy enhances prostoglandine synthesis. Following HOT therapy, the cho-lesterol level of 32 patients was examined (322 determination tests). It was found that the cholesterol level dropped at an average of 20% with respect to the initial values. It was also demonstrated that this therapy caused no toxic effect on the marrow by blocking the enzyme system.

Major autohemotherapy is applied if the artery in question is not acces-sible to palpation. However, in cases of milder circulatory disturbances, intramuscular and/or subcutaneous injections may be sufficient.

Rokitansky [217] widely recommended an external ozone gas therapy in the advanced stages of gangrene, which prevents from further inflam-mation. He has indicated that as a result of intravascular therapy applied in gangrene of the toes of the foot, the average rate of amputations at the upper thigh is reduced from 15 to 8% in stage III. In some cases during the postoperational period, an extensive formation of necrosis with purulent secretion might start to develop. However, by making use of intraarte-rial ozone injection and local ozone-oxygen gas treatment, it became pos-sible to improve circulation and oxygenation of the tissue. Ozone therapy was also used to treat gas gangrene. Guinea-pigs were treated every day with 450 µg ozone. The death-rate decreased from 97 to 72% [218]. Pe-ripheral artery circulatory disturbances are treated with intraarterial ozone injections at a maximum concentration of 33 µg/l. In autohemotherapy the concentrations should not exceed 40 µg/l because of risks to develop hemolysis [219].

In cases where the microorganisms come into direct contact with ozone in a sufficient quantity, killing of the microbes most certainly take place via the oxidative decomposition of the capsid/protein shell of the virus, thus exposing their DNMA and RNA components to open attack; the presence of DNA and A catabolism products could be demonstrated empirically [220].

In the treatment of virally produced diseases such as herpes or hepatitis, however, the circumstances are quite different. In this case, a completely different method of application must be applied, which is safer for the patient, the choice here generally being in favor of "major autochaemotherapy" whereby the patients own blood receives an admixture of ozone upon withdrawal from the body and is immediately put back into system whereas during ozone therapy session lasting for 5 min a wound receives almost 800 mg of ozone, as acute case of virus infection needs only 10 mg of ozone in autohaemotherapy. It is for that reason that, as an active mechanism, a killing off of the virus does not come into question, such as is the case for wound treatment: instead there are two processes to be discussed here which probably take place simultaneously: (1) virus inactivation through ozone or ozone peroxides and (2) a peroxide intolerance of those body cells infected by viruses. Therefore, this means that in the ozone/oxygen therapy, not only the reaction of ozone itself, but also those of its reaction products must come under discussion. Ozone reacts with unsaturated fatty acids in the cell forming ozonides and peroxides. The intervention of ozone and/or its peroxides ought to take place at the acceptors of the free virion, whereby the virus-to-cell contact, and thus the entire reproductive cycle is interrupted. The free pair of electrons on the nitrogen of the N-acetylglucosamine probably represents the point of attack for the electrophilic ozone molecule or its peroxides, so that the virus is blocked for further reaction by oxidation, making it effectively inactive.

The infected cell, however, produces hydrogen peroxide as a defensive function. Apart form this, the electrophilic reaction of ozone can also take place with unsaturated fatty acids, as a membrane constituent of the virally infected body cell, and injected peroxides into the cell. The enhanced quantity of peroxide in the cell affect the cell processes through two ways: (1) the peroxides destroy the membranes of the human cell before reproduction cycle of the infected viruses has been completed or (2) produce a synergetic effect, destroying the microorganisms which have penetrated the cell. Intravascular ozone/oxygen injection using small quantities of ozone constitutes a direct intervention in the metabolic processes, particularly in that of erythrocytes, participating in the normal glucose and pentose phosphate cycle. The final product is 1,3-diphosphoglycerate, which is of decisive importance for the deoxygenation of hemoglobin.

Under the effect of peroxides, obtained as a result of ozone reaction with unsaturated fatty acids in the erythrocytes membranes, the detoxifying mechanism of these substances is immediately put into action via the glutathione system. In order to maintain the glutathione redox system, the pentose/phosphate path must be accelerated in order to cover the ADPH debit as a reducer of GSSG. This means, simultaneously, an increased breaking down of sugar. The decisive product of the accelerated glucose metabolism is the presence of 2,3-diphosphoglycerate (2,3DPG), which represents a key compound in the curative effect of ozone:

$$HbO_2 + 2,3\ DPG \Leftrightarrow Hb^{\cdot}2,3\ DPG + O_2$$

Every increase in 2,3-DPG levels facilitates the release of oxygen through a shifting of the HbO/Hb balance in favor of deoxygenated hemoglobin. As hemoglobin is very stable, particularly in the case of diabetics, ozone therapy is particularly indicated for this decrease and brings about an improvement in the peripheral oxygen supply. It is possible, by measurement of the arterial and venous partial pressure of O_2, to observe *in vivo* the change in the oxygen situation during the course of treatment. However, not only the increase in arterial pO_2 but primarily the reduction in venous pO_2 is decisive for an increased supply of oxygen, which is equal to an increase in the arterial/venous pO_2 difference, even up to normal values of $pO_2 = 60$ mm Hg.

The immunoglobulins are proteins produced from lymphoid cell, which are unified according to similar general principles of their structural organization. The functional integrity of various classes of immunoglobulins is reduced to their participation in homeostasis, the most important factor of immunologic defense systems. Immunologic processes are associated with the presence of antibodies. Five classes of immunoglobulins are known: IgG, IgA, IgM, IgD and IgE. Changes of normal levels cause certain troubles, for example, increased levels of IgG lead to chronic liver deseases. The effect of ozone on the immune system was investigated by monitoring changes of immunoglobulins levels in the blood [221].

Nineteen patients with cervix carcinoma and 20 patients with ovarial carcinoma were entered into a study. They were injected intravenously with ozone-oxygen mixture (540 µg ozone) for 5 days. However, the therapeutic effect was not confirmed.

Thirty four patients with chronic rheumatic troubles underwent autohemotherapy through intramuscular injections with 5 ml blood from patient enriched with 300–350 µg ozone.

A 1000 µg ozone were administered intravenously into 47 patients every week for 6 months (50–60 ml blood from patient enriched with ozone). Statistical data processing indicated insignificant changes of the IgA, IgG, IgC and IgM levels. Therefore, no immunodeficient effects were observed. Certain changes of IgG levels can be evaluated as immunostimulating.

The inhibiting effect of ozone on tumor tissue has been investigated by Sweet [222]. Zanker confirmed this effect, his attention being attracted to the effect of ozone on tumor metabolism.

Steiner et al. [223] studied the changes in the level of: DPG, DPN, F-6-PK, GAPDH, HBDH, ICDH, LDH and NAD in carcinomatous and healthy ovarial tissues (*in vivo*) under the influence of ozone. Cytostatics like Adriblastin and Xoloxan were also used for reference purposes. The lysolecithin levels decreased in the tumor tissue but did not change in the healthy one, which indicates that ozone has no influence on the tumor metabolism. However, the ozone therapy has a positive effect in employing ozone in tumor tissue.

Brinkman and Lamberts [224] were the first who reported on the effect of ozone on erythrocyte (RBC's) behavior. These authors exposed volunteer patients to the influence of ozone (1 ppm) for 10 min and then measured the oxygen evolved from hemoglobin. Upon prolonged ozone exposure oxygen is not bound with hemoglobin, which causes blood-vessel occlusions in fingers and toes. Other authors [225] have observed changes in red blood cell biochemistry, the RBC's being taken from mice and apes. Brikley et al. [226] have investigated the morphological and biochemical changes in human red blood cells for ozone concentration of 0.5 ppm. This study indicated a 20% increase of erythrocyte shortness, a 20% drop in acetylcholine esterase activity, a 15% fall in glutathione levels (GSH) and a comparatively slight increase in the activity of lactic dehydrogenase (LDH). Some authors [227, 228, 230] have suggested that, due to inhaled ozone, the amount of oxygen free radicals is increased. As a result of this, hydrogen peroxide, which might be a notable oxidative metabolite, also increases its levels. The catalase activity in red blood cells is diminished by 50% and the intracellular hydrogen peroxide is formed upon ozone exposure 6-6 ppm.

Changes in erythrocyte membranes stability have been observed [229]. Ozone concentrations of 0.25 ppm from 1 to 6 h change the erythrocyte spherical shape, which may change the normal capillary permeability. Schulz et al. [230] have found that exposures of 0.7 ppm ozone for 2 h increase the methemoglobin levels and decrease the total number of red blood cells. In some mice and Guinea-pigs the hemoglobin and hematocrit levels were increased at high concentrations of ozone as compensating response against the decreasing number of erythrocytes.

Many authors [231–233] have investigated the ability of vitamin E to protect erythrocytes and lung tissue from the influence of ozone because, being a biological antioxidant, ozone interferes with the metabolism and protects other vitamins, hormones and enzymes from oxidation. Chow and Kanes [231] found that under vitamin E deficiency in mice blood during a 7-day period with 0.8 ppm ozone, the activity of glutathione peroxidase, pyruvate kinase and LDH in the erythrocytes was enhanced while the total glutathione level was decreased. The latter maintains the normal functional state of erythrocytes, low redox potential of the cell constituents that preserves the thiol groups of a number of enzymatic proteins in reduced state. The principal function of glutathione is to eliminate hydrogen peroxide formed during direct oxidation of certain drugs. The toxic effect of hydrogen peroxide is due to its ability to oxidize hemoglobin to methemoglobin and to form hydroperoxides with unsaturated fatty acids, which lead to changes in the lipids of the cell membranes.

The effect of ozone on erythrocyte membranes has been investigated with respect to the SH-groups in some enzymes. At an ozone concentration of 40 mmol/min over 5 ml membrane suspension, the activity of ATPase is totally lost. The latter can be regenerated by means of phosphatidilserine

[234]. Peterson [235] observed changes in T-lymphocyte and B-lympho-cyte levels for an ozone concentration of 784 $\mu g/m^3$ for 4 h.

Canada and Airriess [236] studied changes in the concentration of vi-tamin E, vitamin A and vitamin C (antioxidants) in the human serum after exposure to ozone concentration of 0.37 ppm for 2 h. These changes were insignificant. The alkali phosphotase in the serum was increased at ozone concentrations of 10 ppm for 1 h. This was explained by the fact that the blood tissues let the enzymes pass because of a visible danger to lungs at these concentrations. Other authors [57, 58, 237, 238] have followed the concentration levels of prostoglandines PGFa and PGEa during expo-sure of mice to ozone at concentrations of 4 ppm for 8 h. They observed an increase by 186 and 200%, respectively, which was attributed to ara-chidonate oxidative induction by plasma membranes. Veninga and Wage-naar [239] established that creatine kinase (CPK) levels were increased after exposure to 0.02 ppm ozone for 2 h but, unexpectedly there were no changes at higher ozone concentrations. Probably, this effect is associated with some adaptation of the organism. There was response in serum lipids [240] at 1.1 ppm ozone for 24 h; the activity of lecithincholesterol trans-ferase, free cholesterol, and lipoproteins was increased, however, there was a descending tendency in triglyceride levels. The bacterial effect of ozonized blood serum on gram-negative bacteria (*Klebsiella, Pseudomo-nas* and *Salmonella*) has also been studied. This effect is dependent on the serum ozonation time, type of bacteria and bacteria membrane structure [241]. Prolonged exposures to ozone (0.4 to 1 ppm, 6 h daily) caused a drop in serum albumen levels and increased the α-globulin and β-globulin levels. The total protein level was not changed [242].

It has been established that ozone affects the central nervous system (CNS) [243]. Exposures to > 0.6 ppm ozone concentrations cause a head-ache and lethargic states. Changes in the visual sharpness (0.2 ppm) and eye muscle may occur [244]. A fall in the cerebral levels of catechol-*o*-methyl tranferase on exposure of dogs to 1 ppm of ozone (8, 12 or 18 h) was observed [245]. The reduction of enzyme activity was in a good cor-relation with a decrease in catechol amounts in the brain.

Takanashi and Miura have reported a considerable fall in the activity of metabolite enzymes in the liver: benzopyrene hydroxylase, 7-ethoxyc-umarine-*o*-diethylase and aniline hydroxylase. No effects were observed on p-nitroanisole-N-dimethylase on exposure to ozone of 0.8 ppm for 7 h. A 20% drop in cytochrome P-450, NADPH cytochrome, P-450 reductase, and cytochrome B5, included in mitochondria and/or ergastoplasma membranes, was established. Fluorescent pigments in the liver, being an end product in lipid oxidation, were increased upon exposure to 1.5 ppm ozone for 30-60 min. The effect of ozone concentration 1.0, 0.5 and $0.2.10^{-5}$ M has been investigated on mice for exposure time 2, 3 and 15 h [249, 250]. The quantity of cytochrome P-450 in liver microsomes at 1.10^{-5} M and 2 h exposure is 73.4% with respect to the control level and B5 – 93.3%, respectively. After 3 h under the same conditions these parameters are already 75 and 69.6%, respectively and after 4 h the mice die. The cytochrome level at ozone concentration of $0.2.10^{-5}$ m and exposure time of 15 h is 91.3 and 93.2%, respectively. Simultaneously 10 h before the exposure of mice to ozone atmosphere the potassium salt of α-tocopherol is administered intraperitonitously – 56 mg/kg. This leads to increase in P-450 level to 158.9% and that of B5 to 136.6%. In the control mice this leads to increase of their levels with 135 and 116%, respectively. These data allow to assume that tocopherol forms adducts with cytochromes and ozone stimulates the adduct formation.

Zelac and Cromroy [251] have established that ozone is a better mutant than radiation. Humans exposed to 0.5 ppm ozone for 8 to 10 h manifested chromatid and chromosomal aberrations. No increase in chromosomal aberration was observed at 0.4 ppm ozone for 4 h.

In tracheotomy with dogs, with acquired interrupted respiration accompanied by hypertension and bradicardia during exposure to ozone (50 ppm, 4 min), the symptoms decreased due to stimulated *parasympathetic vagus nerves* [252, 253]. Exposure of rabbits to ozone (0.2 ppm, 5 h) gave rise to pathological changes without myocardial damage but cell membranes did not remain intact.

Ozone causes a multiplex effect on the endocrine system [254]. The authors report an inhibiting effect on [131]I (ozone 4 ppm, 5 h), which was observed even 12 days after exposure. Plasma thyrotropin was reduced by 50%. The thyroid gland increased its size. The authors suggest that decreased thyroid hormone levels appear to be prophylactic against ozone damage.

Gordon et al. [255] have reported that upon inhalation of ozone (0.8 ppm, 1 h) the activity of choline esterase was diminished not only in the blood, but also in the diaphragms of Guinea-pigs by 14% and lung tissue (16%). The investigations [256] of the effect of ozone on the immune system show that exposures of 0.6 ppm for 2 h did not affect the number of T-lymphocytes. However, an exaggerated spleen was observed during ozone exposure of 0.31 ppm for 103 h. Kidneys manifested dilated tubules.

7.4.2 TOXIC ACTION

Mechanisms, according to which ozone damages biological tissue, are not clearly defined. However, several processes should be considered which lead to disturbances in normal cell functioning. Ozone oxidizes polyunsaturated fatty acids (PUFA) in the cell membranes according to Criege's mechanism [257]. Also, it contributes to the increase of malonic dialdehyde content and interacts with natural antioxidants (vitamin C, vitamin A and vitamin E). On the other hand, ozone reacts directly with aminoacids in tissues, proteins, and small peptides [240]. It inactivates the following enzymes: lysozymes, betaglucoronidase and acid phophatase, while urease is resistant to ozone.

Very likely, the cellular regulation of the electrolytes becomes worse, and in consequence of this, the enzymes become inactivated and the mitochondrial metabolism is changed. Normal cell functioning is disturbed which might lead to cell destruction [258].

Lung tissue, being vulnerable to ozone attack, has been the object of many investigations. Biochemical studies indicated changes of enzyme levels, the enzymes being an integral part of the intracellular protection from disturbances. These enzymes bind with free radicals, lipid peroxides, and hydrogen peroxide, which are factors for changes of tissues. Chow and Hussain [259] have reported enhanced activity of glutathione peroxidase in mice inhaled with 0.8 ppm ozone for 3 days. Other natural antioxidants, such as superoxide dismutase, catalase, vitamin E, and vitamin C, destroy the free radicals, thus contributing to reduce the risk of permanent tissue damage due to ozone. For example, bronchia constriction, caused by ozone, fades away or disappears after treatment with vitamin C or with combination between vitamin E and vitamin C (synergistic effect). Complementary biochemical investigations of ozone toxicity involve measurements of the enhanced synthesis of collagen, which induces fibrosal changes in lung tissue [260, 261].

Exposure of mice to small doses of ozone [262] gave rise to changes in two enzyme systems: 1) a decrease in the hepatic ascorbic acid (HAA) concentrations and 2) a change in the creatine phosphokinase (CPK). HAA was considerably changed for 30 and 120 min after cessation of the ozone exposure whereas CPK demonstrated changes only 15 min after that. Compared to reference animal, no significant changes were observed after 24 h for HAA and 30 min for CPK. Plasma histamine lactodehydrogenase was not affected with various doses of ozone. It is assumed that this is due to the ability of organisms to adapt themselves to the harmful agents, which, on the other hand, stimulates metabolite processes and enzyme activity. At higher doses of ozone the enzyme activity gradually falls down, probably due to some compensatory balanced conditions. At ozone concentrations of 0.8 ppm (8 h) the activity of the total lactate dehydrogenase (LDH) in the lungs was enhanced but not in the plasma and erythrocytes. The LDH-5 fractions from LDH isoenzymes considerably decreased in terms of their distribution in the lung and plasma, while fraction LDH-4 was increased. No change of isoenzyme activity in the lungs, plasma and erythrocytes was observed for ozone concentration of 0.5 ppm [263].

Saymor [264] has studied the behavior of mice upon prolonged exposure (120 days) to small doses of ozone. The DNA and RNA levels in the lung were decreased whereas the protein level was increased.

In vivo and *in vitro* high ozone concentrations are prerequisite for suppressed mitochondrial oxygen consumption and oxidative phosphorization and for increased membrane permeability. This can be associated with thiol groups oxidation and lipid oxidation [265].

Ozone exhibits a 100% bacteriological effect at concentrations of 0.023 mg/l without producing spores. Spores are destroyed at 0.61 mg/l [247]. At the same concentration, 87.5% of the hepatic virus B is inactivated. Large amounts of blood can be conserved without clotting by bubbling with oxygen-ozone mixture. Parallel to this, all the microbial materials in blood are oxidized, thus sterilizing it [266].

A detailed evaluation of the short-term influence of ozone on human health has been made in Refs. [260–270]. The transitory effect of ozone on the respiratory function was investigated by various respiratory measurements.

Miller et al. [271] have made comparative models of ozone operation regions in the respiratory tract, and Criebel and Smith suggested a new pharmacological model of ozone acute influence on human lung and have compared the effect of ozone on the lung (ozone 120 ppb, 2.5 h) in children and adults. They observed a considerable lowering of PFR and FEV values with adults compared to children. They succeeded in measuring the PFR baseline: 3.9 and 2.7% with adults and children, respectively. Adults made more symptomatic complaints than children. Stock and Kotchmar [272] reported that there is almost no risk of asthmatic attack in patients suffering from asthma at ozone concentrations over 120 ppb.

Many authors have investigated the influence of ozone on pulmonary lysosomes and mitochondria. Lysosomes contain quite many enzymes, mainly hydrolases, such as protease, lipases, nucleases, phosphatases, phosphodiesterases, etc. Dillar et al. [273] have studied the changes in lysosome levels upon exposure of mice to ozone (0.7 ppm). The activity of pulmonary lysosomal cathepsin A, cathepsin D, acidic phosphatase, β-N-acetylglucose aminidase, and benzyl arginine β-naphthylamide amidohydrolase was increased. This is due to inflammatory processes caused by ozone. The specific activity of protease and peptidase in the lung is enhanced when related to chronic obstructive genetic lack of α-1-antitrypsin factor.

Mitochondria are contained in every cell. A highly organized and specialized transportation chain, which transfers protons and electrons to molecular oxygen, is located in the internal mitochondrial membrane. It is composed by enzymes, redox systems, which organize proteins and lipids, and is called respiratory chain. The external membrane contains enzymes from the tricarboxylic acid cycle, fatty acid β-oxidation, etc. It has been established that at high ozone levels (2 ppm) the oxygen consumption of pulmonary mitochondria in rodents is decreased whereas this consumption is increased at 0.8 ppm ozone after inhalation for 20 days [274]. Mice, exposed to 0.8 ppm ozone, obtained an acute ozone intoxication due to blocking of the alveolar capillary function, followed by hemolysis and lethal hyloxia [275, 276]. The authors in Ref. [277] have studied in detail the cellular, biochemical, and functional effects of ozone on the lung.

Clearance of foreign materials (particles) from the lung and airways depends on the function of macrophages, ciliated cells and secretory cells, and on the physical and chemical properties of the alveolar cells. All these are affected by ozone exposure. Single acute exposure of animals and humans to ozone concentrations less than 0.6 ppm have been shown to accelerate clearance of particles from the tracheobronchial tree whereas acute exposures to ozone levels greater than 0.6 ppm caused a delay in particles clearance [79, 261]. Repeated exposures (2 h daily for 14 days) to 0.1 ppm ozone gave rise to acceleration in alveolar clearance of latex particles but had no effect on tracheobronchial clearance. These results, related to morphological studies, are consistent with certain adaptation [278–283].

The alveolar macrophage (AM) takes part in defending the alveoli and airways against inhaled particles and pathogens. To carry out this important function, AM maintain active mobility, phagocytic activity, membrane integrity and enzymatic capacity. Ozone exposures of AM *in vivo* have been shown to alter the number and morphology of AM, depending on dose and duration of exposure. A transient decrease in the absolute number of AM has been observed following exposures to high ozone concentrations [284, 285]. AM recovered from rats exposed for 16 h to 0.05, 0.10, 0.20 or 0.40 ppm ozone showed no differences in cell yield or cell viability. An increase in AM numbers has been also observed after *in vitro* ozone exposure [286]. Recruitment and persistence of AM result from chemotaxis and migration of cells into the lung alveoli and from local proliferation. A primary function of AM is the maintenance of sterility in the lung by phagocytosis of foreign particles. Changes in the phagocytic activity of AM have been observed following exposure to ozone. A marked depression of this activity was observed after exposure to 1.2 ppm ozone for 5 h [287] but enhanced phagocytic function has been observed after prolonged exposures for 20 days (3 times daily) to 0.8 ppm ozone. This difference in the phagocytic function of AM for various ozone concentrations and exposure conditions can be explained by changes in the cell populations. Ozone exposures also affect the secretory activity of AM and their ability to release factors which stimulate the migration of neutrophils and other cell types into the lung. A culture monolayer of rabbit

AM was exposed for 2 and 6 h to 0.1, 0.3 or 1.2 ppm ozone. Biochemical functions of alveolar macrophage can also be affected. Ozone exposure caused a reduction in intracellular concentrations of acid phosphatase and β-glucoronidase, as well as in lyposomal enzymes [289, 290].

Exposure of the upper airways of mice to ozone at 0.5–0.8 ppm for 1–24 days has induced a two-fold increase in IgA^+B-lymphocytes. In the deep lung, exposure to 0.7 ppm ozone for 14 days causes an accumulation of T-lymphocytes [291, 292]. The number of leucocyte cells in controls, $3.8.10^5$, was increased to $7.5.10^5$ cells in exposed mice. Ozone exposure of lymph nodes of euthymic mice indicated that the histopathologic canges were more extensive in the athymic animals [293].

Changes in the immune function depend on the organ, dose and duration of ozone exposure. Concanavalin A (Con A) stimulates the cell-mediated arm of the immune system. Con A reactivity was unaffected at days 4 or 7 of 0.7 ppm ozone exposure (20h/day) but increased by 1.5 and 3 fold after 14 and 28 days, respectively. In contrast, splenic cells of mice manifested decreased reactivity to T-cell mitogens *in vitro* (phytohemagglutinin, PHA, 38% decrease, Con A, 75 5 decrease) but not to a B-cell mitogen (lipopolysaccharide, LPS0 or to alloantigens [294].

In rats, 7 days of exposure to 1 ppm ozone enhanced the reactivity of splenic cells to PHA, Con A and LPS [295]. Ozone inhalation gives also rise to nonspecific changes. Exposure to ozone at concentrations of 0.4 to 0.8 ppm can induce weight losses and/or thymic and splenic atrophy. A similar effect has been observed if steroids are injected into mice, i.e., ozone can produce steroid-mediated effects on lymphocytes.

The immune response of the human organism to a foreign agent involves the following cell quartet: macrophage, B lymphocytes, T helper, and T suppresser. Amoruso et al. [296] have determined the anion peroxide radical (O_2^-), produced by alveolar macrophage from mice exposed to ozone levels of 3.2 to 10.5 ppm. They found decreasing concentration levels. The authors proposed that this decrease in anion peroxide radical amount cannot be due to a direct effect of ozone on cell viability but rather to an infection caused by ozone, i.e. the immune system of experimentally tested animals enhances its activity. Alveolar macrophage from test animals, if stimulated by phorbal myristat acetate, also decreased the concentration of O_2^-. This explanation seems to some extent confusing since the bactericide action of ozone which is very strong and in case of infection it would rather prevent than stimulate it.

Lung lesions observed after one week of exposure to 0.2, 0.5, 0.8 ppm ozone were more expressed than those observed after 3 months of continuous exposure. Some researchers [297, 298] suggested that the enzyme levels in the lung are increased which may account for the development of adaptation and tolerance to ozone.

Epithelial cels type I and squamous cells appear to be most sensitive to ozone [299]. An exposure to 0.5 ppm ozone increases the volume but reduces the area of epithelial cell types I [300]. These cells are regenerated to the normal condition after a week in fresh air and develop certain tolerance to ozone reexposure [301, 302]. The authors have studied tracheal cells (volume, density) by electron microscopy. They have found that all changes due to chronic ozone exposure were reversible, the normal values being regained if test animals are exposed to fresh air.

Pulmonary fibroses, defined as abnormal accumulation of collagen, are developed after prolonged exposure to ozone levels greater than 0.5 ppm [303, 304]. Chronic ozone effects at concentrations close to that in atmospheric air are hardly noticeable and cannot be measured. Other scientists [305] have exposed mice to 1 ppm ozone for 40 days and found increased levels of total lung collagen but this effect rapidly faded away after 10 days of exposure to fresh air. Chronic exposure to 0.06 ppm ozone (13h/day, 5 days weekly, 18 months) caused changes in the lung function including a decrease in the general respiratory effect.

7.4.3   OZONE SOLUTIONS

Ozone disinfectant and sterilizing properties were used to freshen water in Germany at Wiesbaden yet in 1902. On this basis, in 1934 the dentists started utilizing ozonized water for inflammatory processes. This was reported at the Sixth World Congress on Ozone held in Washington [306]. The following properties of ozone are used in this case: (a) disinfectant and sterilizing effect; (b) haemostatic effect, especially in cases of running hemorrhages; (c) accelerated wound healing, improved oxygen local supply and support to metabolic processes [307, 308]. Ozonized water, in the form of spray or stream, is used for the antrum in the following cases: gingivitis, paradontosis, white-gum, stomatitis. Preference is given to the spray mode, particularly in cases of caries for cleaning and disinfection, as well as for injuries and bleeding in the antrum. It is applied in cleaning nerve canals and pulpa tissue, in case of painful gingivitis and stomatitis. Ozonized water has also found wide application in orthopedy: artificial denture and crown cleaning prior to fixing, stopping of hemorrhages during fixing.

The ozone solutions in olive oils (ozone is present as ozonides) have found wide application in dermatology for treating: fistulae, decubitus and ulcer cruries. It is also successfully applied in mycosis and fungus finger and fingernail infections. Simply, patients have to put their fingers into a plastic sleeve covered by ozone-olive oil. Salzer et al. [309] have investigated the effect of ozonides olive oil on patients suffering from soorkolpitis. An especially positive effect was observed in its application in the vagina. The numbers of examined bacteria decreased very rapidly. These authors suggest that the positive effect of treatment is due to ozonides formed, as well as to tissue enrichment with oxygen. The same authors used autohemotherapy (100 ml blood enriched with oxygen-ozone; 1 ml contains 33 gammas ozone) in cases of chronic placenta insufficiency in pregnant women (20 patients, about thirtieth week). HPLC analysis indicated increased values. This therapy is also useful for weak bleedings during pregnancy.

Wermeister [310] described a technique, utilizing ozone gas of 80–100 μg/ml concentration, for wound cleansing and disinfection. Such an effect has not been observed during enzyme treatment. Often, the bacterially infected spot is washed out by ozone-oxygen mixture, which gives certain positive effect.

Ozonides that are prepared by means of the reaction between ozone and olefins do not cause inflammatory processes in skin and are not decomposed by catalase. Ozonides participate as an oxidizing agent without liberating peroxide groups, which might oxidize cells. About 1000 subcutaneous injections of castor-oil ozonides have been made without tissue response [311]. A 4% solution of the ozonide of sorbic acid in ethyl diglycol (ozonide SV) is used in cosmetics to preserve face creams and as disinfectant. The lethal dose is 10 ml/kg. In wheat-germ oil, containing vitamin E, the ozonide keeps the vitamin E levels unchanged.

7.4.4 STERILIZATION

A number of methods have been developed, according to which ozone is utilized for sterilizing bottles, cans, containers and glass vessels that find wide application in the pharmaceutical industry and cosmetics [312–314].

Usually sterilization is performed at sufficiently high temperatures that may cause damage on same materials used in food processing industry, pharmaceutical industry and cosmetics. The effect of ozone on bacteria is well known, especially its toxicity and reactivity characteristics that is why attention should be given to dosage and exposure time.

In [315] is discussed the application of ozone as sterilizing agent in the pharmaceutical industry. Physiological solution (0.9% NaCl), pH = 7, is treated with oxygen-ozone mixture containing 15% ozone for 15 min. The solution is allocated to glass or polyethylene banks, which are sealed up. Analysis showed that after 2 h the microbe number was 0 and after 6 h the ozone was completely destroyed. Sterilization for 2 h with 0.5 ppm ozone caused a 100% lost of *Legionella pneumoniae* [278].

An apparatus has been designed to produce oxygen-ozone mixture for sterilization of milk, diary products, wines, butter and liquid medicines. The experimental set-up is described in detail in a patent [279]. The quality of wine, beer and liquor is improved after exposure to ozone. Ozone levels of 0.01 ppm and rates of 1 l/min are employed in the pharmaceutical industry and food processing industry. Ozone has been employed fro cleaning dialyzers [316] and contact lens [317].

7.4.5 OZONATORS

Portable ozonators have been designed for medicine purposes covering the whole range of terapheutical doses from 0 to 100 µg/l. These units are manufactured in "Shumen" plant Ltd., Shumen, Bulgaria as P-1 trade mark.

7.5 CONCLUSION

This review is including information about cleaning of waste gases and water, purification of drinking water. Some chapters of review devoted of manufacture of organic and inorganic compounds by reactions with ozone. The information about application of ozone in medicine was included as well (therapeutic application, toxic action, ozone solutions and steriliza-tion). Design of ozonators is discussing.

KEYWORDS

- Aldehydes
- Autohemotherapy
- Fistulae
- Mucous colitis
- Ozone-oxygen
- Waste water

REFERENCES

1. Razumovskii, S. D.; Rakovski, S. K.; Shopov, D. M.; Zaikov. G. E. *Ozone and Its Reactions with Organic Compounds (in Russian)*. Publications of House of Bulgar-ian Academy of Sciences: Sofia, 1983.
2. Proskuryakov, V. A.; Shmidt, L. I. *Purification of the Sewage Waters from the Chemi-cal Industry*. Khimya: Leningrad, 1977.
3. Bailey, P.S. *Ozonattion in Organic Chemistry*, Academic Press: New York, v.I, 1978, v.II, 1982, *Adv. Chem. Ser. A*, (Ed. HH Wasserman), 39 I, 39 II, 1978, 1982.
4. Kuznetzov, I. E.; Tritzkaya, T. M. *Protection of the Air Space Pollute by Dangerous Compounds*. Khimya: Moscow, 1979.
5. *Itogi Nauki i Tekhniki* – Technology of Organic Compounds: Moscow, VINITI, Vol. 7, 1983.
6. *Itogi Nauki i Tekhniki*, -Protection and Reproduction of Natural Resources: Moscow, VINITI, Vol. 13, 1983.
7. Golubovskya, E. K. *Biological Bases of Water Purification*. Visshaya Shkola: Moscow, 1978.
8. Bokris, O. M., Ed.; *Environmental Chemistry*. Khimya: Moscow, 1982.

9. Pergut, E. A.; Gorelik, D. O. *Instrumental Methods to Control Air Pollution.* Khimiya: Leningrad, 1981.

10. Chanlett, E. T. *Environmental Protection;* McGraw-Hill, 1979.

11. Flaekt, A. B. *Res. Discl.,* **1991**, *326,* 453.

12. Ruck, W. *Vom Wasser,* **1993**, *80,* 253–272.

13. Mccoustra, M. R. S.; Horn, A. B. *Chem. Soc. Rev.,* **1994**, *23*(3), 195–204.

14. Tabata, K.; Matsumoto, I.; Fukuda, Y. *Jpn. Kokai Tokkyo Koho JP* 01,270,928 [89,270,928] (Cl. B01D52), 30Oct 1989, Appl. 88/98,632, 21 Apt 1988.

15. Pollo, I.; Jarosszynska-Wolinska, J.; Malicki, J.; Ozonek, J.; Wojcik, W. *Pol. PL,* **1987**, *137,*426 (Cl. B01D53/14); *Appl.,* **1983**, *245,* 514, 2pp.

16. De Guardia, A.; Bouzaza, A.; Martin, G.; Laplanche, A. *Pollut. Atmos.,* **1996**, *152,* 82–92.

17. Takeyama, K.; Nitta, K. *Jpn. Kokai Tokkyo Koho JP,* **1989**, *01,236,925* [89,236,925] (Cl. B01D53/34); *Appl.,* **1988**, *88/61,*038, 3.

18. Moortgat, G.K.; Horie, B. C. *Z. Pollut. Atmos.,* **1991**, 29–44.

19. Tomita, K. *Jpn. Kokai Tokkyo Koho JP* **1990**, 02,152,547 [90,152,547] (Cl. B01J20/18); *Appl.,* **1988**, 88/304,*185,* 4.

20. Vicard, J.F.; Vicard, G. *PCT Int. Appl. WO* **1992**, *92* 19,364 (Cl. B01D53/34), Fr *Appl.,* **1991**, 91/5, *853,* 15.

21. Iannicelli, J. *U.S. US,* **1990**, *4,* 923,688 (Cl. 423-224; B01D53/54); *Appl.* **1985**, *799,*494, 8.

22. Buettner, F.; Koch, P. *Ger. Offen DE,* **1991**, 9,931,*891* (Cl. B01D53/36); *Appl.,* **1989**, 3.

23. Wada, H.; Naoi, T.; Kuroda, Y. *Nippon Kagaku Kaishi,* **1994**, *9,* 834–840

24. Matt, K.; Guttenberger, H.G. *Eur. Pat. Appl. EP,* **1989**, *330,*028 (Cl. C02F1/78); *DE Appl.,* **1988**, *3,*805,906,, 8.

25. Waga, H.; Naoi, T.; Kuroda, Y. *Nippon Kagaku Kaishi,* **1995**, *4,* 306–313.

26. Tanaka, K. *PP. M.,* **1996**, *27*(12), 10–15.

27. Welch, J. F.; Siehwarth, J. D. *U.S. US,* **1989**, *4,* 861, 497 (Cl. 210-759; C02F1/74); *Appl.,* **1988**, *169,*851, 5.

28. Oeller, H. J.; Daniel, I.; Weinberger, G. *Water Sci. Technol.,* **1997**, *35*(2–3 Forest Industry Waste waters V), 269–76 Elsevier.

29. Moebius, C. H.; Cordes-Tolle, M. *Water Sci. Technol.,* **1997**, *35*(2–3 Forest Industry Waste waters V), 245–250, Elsevier.

30. Nakamura, Y.; Sawada, T.; Kobayashi, F.; Odliving, M. *Water Sci. Technol.,* **1997**, *35*(2-3 Forest Industry Waste waters V), 277–282, Elsevier.

31. Zhou, H.; Smith, D. W. *Water Sci. Technol.,* **1997**, *35* (2–3 Forest Industry Waste waters V), 251–259, Elsevier.

32. Huang, C. M.; Banks, M. K. *J. Environ. Sci. Health, Part B,* **1996**, *B31*(6), 1253–1266.

33. Sumimoto, H.; Yamazaki, T. *Jpn. Kokai Tokkyo Koho JP* **1991**, *03,111,670 [91,111,670]* (Cl. F03G7/05); *Appl.,* **1989**, 89/247,*809,* 3.

34. Kawamoto, K. *Kogaito Taisaku,* **1991**, *27*(7), 617–627.

35. Ukita, S. *Jpn. Kokai Tokkyo Koho JP,* **1991**, *03 80,996 [91 80,996]* (Cl. C02F1/50); *Appl.,* **1989**, *89*(218),*115,* 4.

36. Miriguchi, Y.; Hayashi, H.; Umetani, T. *Osaka-shi Suidokyoku Suishtsu Shikensho Chosa Kenkyu narabini Shiken Seiseki,* **1994,** *46,* 10–16, Pub. 1995.
37. Munz, C.; Galli, R.; Egli, R. *Chem. Oxid.,* **1992,** *2,* 247–63, Pub. 1994.
38. Kojima, H.; Katsura, E.; Ogawa, H.; Kaneshima, H. *Hokkaidoritsu Eisei Kenkyushoho,* **1991,** *41,* 71–73.
39. Ohashi, N.; Tsuchiya, Y.; Sasano, H.; Hamada, A. *Jpn. J. Toxicol. Environ. Health,* **1994,** *40*(2), 185–192.
40. Hapeman-Somish, C. J. *ASC Symp. Ser.,* **1991,** *459* (Pestic. Transform. Prod.: Fate Signif. Environ.), 133–147.
41. Terui, S.; Sano, K.; Nishikawa, K.; Inone, A. *Jpn. Kokai Tokkyo Koho JP,* **1990,** *02 139,017 [90,139,017]* (Cl. Bo1D53/36); *Appl.* , **1988,** 88(293),*929,* 3.
42. Terui, S.; Sano, K.; Kanazaki, T.; Mitsui, K.; Inoue, A. *Jpn. Kokai Tokkyo Koho JP 01,* **1989,** *56,124 [89 56,124]* (Cl. B01D53/36); *Appl.* **1987,** 87(209), *305,* 4.
43. Ogawa, K.; Seki, N. *Jpn. Kokai Tokkyo Koho JP,* **1992,** *04,243,597 [92,243,597]* (Cl. C02F1/78); *Appl.,* **1991,** 91/7,*486,* 4.
44. Hoshima, Y. *Jpn. Kokai Tokkyo Koho JP* **1990,** *02 40,289 [90 40,289]* (Cl. C02F1/78); *Appl.* **1988,** 88/190,*318,* 4.
45. Filippi, A.; Tilkes, F.; Beck, E. G.; Kirschner, H. *Dtsch. Zahnaerztl. Z.,* **1991,** *46*(7), 485–487.
46. Gehringer, P. *Oesterr. Forschungszent. Zeitbersdorf, [Ber.] OEFZS* **1995,** (OEFZS – 4738), 21.
47. Charles, L.; Pepin, D.; Puig, P. H. *J. Eur. Hydrol.,* **1996,** *27*(2), 175–191.
48. Thornhill, R. W. *UK Pat. Appl. GB,* **1989,** *2,219,790* (Cl. C02F1/50); *GB Appl.* **1988,** *88/14,222.*
49. Nitsuta, Y.; Nitta, Y. *Jpn. Kokai Tokkyo Koho JP,* **1992,** *04,367,799 [92,367,799]* (Cl. C02F11/00); *Appl.* **1991,** *91/165,204,* 3.
50. Gordon, G.; Pacey, G.E. *Chem.Oxid.,* **1992,** *2,* 230–246, Pub. 1994.
51. Bruckner, F. *Int. Tag. Ausstellung Umweltinf. Umweltkommun.,* **1992,** *2,* 211–222, Pub. 1993.
52. Preusser, M.; Ruholl, H.; Schneidler, H.; Wessling, E.; Wortmann, C. *Wiss. Umwelt,* **1990,** *3,* 135–140.
53. Rimpler, M. *Ger. Offen. DE 19,512,448,* (Cl. A62D3/00); *Appl.,* **1995,** *19/512,448,* 9.
54. Wessling, E. *Erzmetall,* **1991,** *44*(4), 196–200.
55. Mori, M; Yamakoshi, H; Nojima, M; Kusabayashi, S; Mccullough, K. J.; Griesbaum, K.; Kriegerbeck, P; Jung, I. C. *J. Chem. Soc. Perkin Trans.* I, **1993,** *12,* 1335–1343.
56. Mccullough, K. J.; Teshima, K.; Nojima, M. *J. Chem. Soc. Chem. Commun.,* **1993,** *11,* 931–933.
57. Sugiyama, T.; Yamakoshi, H.; Nojima, M. *J. Org. Chem.,* **1993,** 58, *16,* 4212–4218.
58. Ishiguro, K.; Nojima, T.; Sawaki, Y. *J. Phys. Org. Chem.,* **1997,** *10,* 11, 787–796.
59. Nojima, T.; Ishiguro, K.; Sawaki, Y. *J. Org. Chem.,* **1997,** *62*(20), 6911–6917.
60. Ishiguro, K.; Nojima, T.; Sawaki, Y. *J. Phys. Org. Chem.,* **1997,** *10*(11), 787–796.
61. Fukagawa, R.; Nojima, M. *J. Chem. Soc. Perkin Trans. 1,* **1994,** *17,* 2449–2454.
62. Mori, M.; Yamakoshi, H.; Nojima, M.; Kusabayashi, S.; Mccullough, K. J.; Griesbaum, K.; Kriegerbeck, P.; Jung, I. C. *J. Chem. Soc. – Perkin Trans. I,* **1993,** *12,* 1335–1343.

63. Mccullough, K. J.; Teshima, K.; Nojima, M. *J. Chem. Soc. – Chem. Commun.*,**1993**, *11*, 931–933.
64. Sugiyama, T.; Yamakoshi, H.; Nojima, M. *J. Org. Chem*, **1993**, *58*(16), 4212–4218.
65. Mccullough, K. J.; Sugimoto, T.; Tanaka, S.; Kusabayashi, S.; Nojima, M. *J. Chem. Soc. – Perkin Trans.*, **1994**, *1*(6), 643–651.
66. Satake, S.; Ushigoe, Y.; Nojima, M.; Mccullough, K. J. *J. Chem. Soc. – Chem. Comm.*, **1995**, 14, 1469–1470.
67. Teshima, K.; Kawamura, S. I.; Ushigoe, Y.; Nojima, M.; Mccullough, K. J. *J. Org. Chem.*, **1995**, *60*(*15*), 4755–4763.
68. Mccullough, K. J.; Tanaka, S.; Teshima, K.; Nojima, M. *Tetrahedron*, **1994**, *50*(25), 7625–7634.
69. Schobert, B. D. *Chim. Oggi*, **1995**, *13*(6), 21–24, */Pub.1995/*.
70. Appell, R. B.; Tomlinson, I. A.; Hill, I. *Syn. Comm.*, **1995**, *25*(22), 3589–3595.
71. (71) Yang, Y. T.; Li, T. S.; Li. Y. L. *Syn. Comm.*, **1993**, *23*(8), 1121–1124.
72. Jachak, M.; Mittelbach, M.; Kriessmann, U.; Junek. H. *Syn.- Stuthgart*, **1992**, *3*, 275–276.
73. Jellen, W.; Mittelbach, M.; Junek, H. *Monatsh. Chem.*, **1996**, *127*(2), 167–72.
74. Matsumoto, H.; Kanda, H.; Obada, Y.; Ikeda, H.; Murakami, T. *Jpn. Kokai Tokkyo Koho JP*, **1996**, *08 03,138 [96 03,138]* (Cl. C07D215/14); *JP Appl.*, **1994**, *94/28,596*, 10.
75. Allen, D. E.; Ticker, Ch. E.; Hobbs, Ch. C.; Chidambaram, R. Process for preparing 6-chloro-2-hexanone from 1-methylcyclopentane. *U.S. US 5,491,265 (Cl. 568-347; C07C45/51), 13 Feb 1996, Appl. 395,266*, 28 Feb 1995, 7pp.
76. Lee, H. W.; Ji, S. K.; Lee, I.-Y. Ch.; Lee, J. H. *J. Org. Chem.*, **1996**, *61*(7), 2542–2543.
77. Mikaev, F. Z.; Galin, F. Z. *Izv. Akad. Nauk SSSR, ser. khim.*, **1995**, *10*, 1984–1987.
78. Mccullough, K. J.; Mori, M.; Tabuchi, T.; Yamakoshi, H.; Kusabayashi, S.; Nojima, M. *J. Chem. Soc.-Perkin Trans.*, **1995**, *1*(1), 41–48.
79. Mori, M.; Tabuchi, T.; Nojima, M.; Kusabayashi, S. *J. Org. Chem.*, **1992**, *57*(6), 1649–1652.
80. Casey, M.; Ulshaw, A. J. *SYNLETT*, **1992**, *3*, 214–216.
81. Kuczkowski, R. L. Ozonolysis of Vinyl Ethers. *Advances in Oxygenated Process, 3(Series: Advances in Oxygenated Process: A Research Annual 3(1991)).*
82. Reiser, R.; Seeboth, R.; Suling, C.; Wagner, G.; Wang, J.; Schroder, G. *Chem. Ber.*, **1992**, *125*(1), 191–195.
83. Hon, Y.-S.; Yan, J.-L. *Tetrahedron*, **1997**, *53*(14), 5217–5232.
84. Marshall, J. A.; Garofalo, A. W.; Sedrani, R. C. *SYNLETT*, **1992**, 643–645.
85. Odinokov, V. N. *Bashk. Khim. Zhurn.*, **1994**, *1*(3), 29–33.
86. Odinokov, V. N. *Bashk. Khim. Zhurn.*, **1994**, *1*, (1), 11–14.
87. Ishmuratov, G. Yu.; Kkharisov, R. Ya.; Odinokov, V. N..; Tolstikov, G. A. *Uspehi Khimii*, **1995**, *64*(6), 580–608.
88. Malhotra, R.; Kumar, S.; Satyam, A. *J. Chem. Soc., Chem. Commun.*, **1994**, *11*, 1339–1340.
89. Zwilichovsky, G.; Gurvvich, V. *J. Chem. Soc., Perkin Trans. 1.*, **1995**, *19*, 2509–2515.
90. Mueller, T.; Schmidt, R. R. *Angew. Chem., Int. Ed. Engl.*, **1995**, *34*(12), 1328–1329.

91. Pitzer, K.; Hudlicky, T. *Synlet.,* **1995**, *8*, 803–805.
92. Kashima, Ch.; Maruyama, T.; Arao, H. *Yuki Gosei Kagaku Kyokaishi,* **1996**, *54*(2), 132–138.
93. Kashima, C.; Harada, K.; Hosomi, A. *Heterocycles,* **1992**, *33*(1), 385–390.
94. Wu, H.-J.; Lin, Ch.-Ch. *J. Org. Chem.,* **1995**, *60*(23), 7558–7566.
95. McVay, K. R.; Gaige, D. G.; Kain, W. S. *U.S. US,* **1996**, *5,543,565* (Cl. 562–523; C07C51/16); *Appl.* **1995**, *376,173,* 7.
96. Gesser, H. D.; Zhu, G.; Hunter, N.R. *Catalysis Today,* **1995**, *24*, 321–325.
97. Gesser, H. D.; Hunter, N. R.; Das, P. A. *Ñatalysis Lett.,* **1992**, *16*(1–2), 217–221.
98. Hitomi, S.; Ishibashi, T.; Takashi, M.; Kenkichi, T. *Tetrahedron Lett.,* **1991**, *32*(45), 6591–6594.
99. Suzuki, H.; Murashima, T.; Shimizu, K.; Tsukamoto, K. *Chem. Lett.,* **1991**, 817–818.
100. Suzuki, H.; Mori, T.; Maeda, K. *SYNTHESIS- Stutgart,* **1994**, *8*, 841–845.
101. Suzuki, H.; Yonezawa, S.; Mori, T.; Maeda, K. *J. Chem. Soc.-Perkin Trans.,* **1994**, *1*(11), 1367–1369.
102. Suzuki, H.; Murashima, T.; Tsukamoto, K. *Jpn. Kokai Tokkyo Koho JP,* **1992**, *04,282,368 [92,282,368]* (Cl. C07D213/61); *Appl.* **1991**, *91/67,642,* 3.
103. Suzuki, H.; Murashima, T. *Kagaku to Kagyo (Tokyo),* **1993**, *46*(1), 43–45.
104. Lyakumovich, A. G.; Samuilov, Y. D.; Goldberg, Y.; Kondrashova, M. N. *Russ. RU,* **1995**, *2,036,895* (Cl. C07C51/34); *Appl.,* **1993**, *93,015,890.*
105. Ebdon, J. R.; Flint, N. J. *Macromolecules,* **1994**, *27*, 6704.
106. Sawaki, T.; Watanabe, S.; Shimada, K. *Jpn. Kokai Tokkyo Koho JP,* **1990**, *02 26,868 [90,26,868]* (Cl. C04B35/56); *Appl.,* **1988**, *88/176,938,* 7.
107. Wasinger, E.; Hall, D. *U.S. US,* **1993**, *5 261,925* (Cl. 8-111; D06L3/14); *US Appl.,* **1990**, *560,357,* 7.
108. Bradley, R. H.; Clackson, I. L.; Sykes, D. E. *Appl. Surface,* **1993**, *72*(2), 143–147.
109. Connor, H. G.; Budgell, D. *PCT Int. Appl. WO,* **1996**, *96 18,761* (Cl. D01F2/00); *DE Appl. ,* **1994**, *4,444,700,* 21.
110. Miura, M.; Naito, T.; Hino, T.; Kuroda, H. *Jpn.Kokai Tokkyo Koho JP,* **1989**, *01,250,415 [89,250,415]* (Cl. D01F9/14)*;* **1988**, *Appl. 88/74,471,* 9.
111. Queslati, R.; Roudesli, S. *Eur. Polym. J.,* **1991**, *27*(12), 1383–1390.
112. Catoire, B.; Verney, V.; Hagege, R.; Michel, A. *Polymer,* **1992**, *33*(11), 2307–2311.
113. Kulik, E. A.; Ivanchenko, M.; Kato, K.; Sano, S.; Ikada, Y. *J. Polym. Sci., Part A- Polym. Chem.,* **1995**, *33*(2), 323–330.
114. Fujimoto, K.; Takebayashi, Y.; Inoue, H.; Ikada, Y. *J. Polym. Sci., Part A- Polym. Chem.,* **1993**, *31*(4), 1035–1043.
115. Wang, Ch.; Wang, A.; Che, B.; Zhou, C.; Su, I.; Lin, S.; Wang, B. *Gaofenzi Xuebao,* **1997**, *1*, 114–18.
116. Jian, Y.; Shiraishi, Sh. *Hyomen Gijutsu,* **1989**, *40*(11), 1251–1255.
117. Jian, Y.; Shiraishi, S. H. *Hyomen Gijutsu,* **1989**, *40*(11), 1256–1260.
118. Jian, Y.; Shiraishi, S. H. *Hyomen Gijutsu,* **1990**, *41*(3), 273–277.
119. Yamamoto, H. *Jpn. Kokai Tokkyo Koho JP,* **1990**, *02,198,819 [90,198,819]* (Cl. B29C47/04)*; Appl.,* **1989**, *89/16,184,* 6.
120. Yoshito, M. *Jpn. Kokai Tokkyo Koho JP,* **1991**, *03,45,604 [91,145,604]* (Cl. C08F8/06); *Appl.,* **1989**, *89/181,558,* 2.

121. Shiraishi, S. H.; Ken, U.; Suchiro, K.; Nitsuta, K. *Jpn. Kokai Tokkyo Koho JP*, **1991**, *03,103,448 [91,103,448]* (Cl. C08J7/12); *Appl.*, **1989**, *89/239/357*, 5.
122. Marshall, R. A.; Parker, D. K.; Hershberger, J. W. *U.S. US*, **1996**, *5,498,693* (Cl. 528-483; C08F6/24); *Appl.* **1994**, *313,504*, 5.
123. Iwasaki, T. *Kobunshi Ronbunshu*, **1993**, *50*(2), 115–120.
124. Hosoda, Y.; Ikishima, K.; Yanai, A. *Jpn. Kokai Tokkyo Koho JP*, **1992**, *04,197,472 [92,197,472]* (Cl. B05D3/10); *Appl.*, **1990**, *90/331,521*, 4.
125. Ebdon, J. R.; Flint, N. J. *J. Polym. Sci., Part A- Polym. Chem.*, **1995**, *33*(3), 593–597.
126. Rl, L.; Ebdon, J. R.; Hodge, P. *Polymer*, **1993**, *34*(2), 406.
127. Eftushenko, A. M.; Chikhachova, I. P.; Timofeeva, G. V.; Stavrova, S. T.; Zubov, V. P. *Visoko molek. Soed., Ser. A i B*, **1993**, *35*(6), Â312–315.
128. Tashibana, S. H.; Yokoyama, R. *Jpn. Kokai Tokkyo Koho JP*, **1989**, *01,248,339 [89,248,339]* (Cl. G11B7/26); *Appl.*, **1988**, *88/75,079*, 4.
129. Rimmer, S.; Ebdon, J. R.; Sherpherd, J. M. *React. Funct. Polym.*, **1995**, *26*, 145–155.
130. Chtourou, H.; Riedl, B.; Kokta, B. V.; Adnot, A.; Kaliguine, S. *J. Appl. Polym. Sci.*, **1993**, *49*, 361–373.
131. Vallet, A.; Delmas, M.; Fargere, T. H.; Sacher, G. *PCT Int. Appl. WO*, **1996**, *96 18,484* (Cl. B29B17/02); *FR Appl.*, **1994**, *94/15,174*, 16.
132. Kawabe, K.; Hatakeyama, Y. *Jpn. Kokai Tokkyo Koho JP*, **1997**, *09 01,665 [91 01,665]* (Cl. B29C65/02); *Appl.*, **1995**, *95/172,699*, 8.
133. Czerwiec, W.; Kwaitek, A. *Eur. Pat. Appl. EP*, **1990**, *383,969* (Cl. C08F8/50), *Appl.*, **1989**, *89/103,098*, 3.
134. Shimizu, S. H.; Iwata, N. *Jpn. Kokai Tokkyo Koho JP*, **1996**, *08 337,867 [96 337,867]* (Cl. C23C8/18); *Appl.*, **1995**, *95/141,448*, 6.
135. Lagana, V. *EUR Pat. Appl. EP*, **1992**, *504,621* (Cl. C23C8/12); *IT Appl.*, **1991**, *91/ MI715*, 7.
136. Berneron, R.; de Gelis, P. *EUR Pat. Appl. EP*, **1989**, *340,077* (Cl. C23C8/36); *FR Appl.* **1988**, *88/5,091*, 10.
137. Lu, H. H.; Diguette, D. J. *Report 1989, Order No.AD-A213804*, 37pp.
138. Havlik, T.; Skrobian, M. *Rudy*, **1989**, *37*(10), 295–301.
139. Havlik, T.; Skrobian, M. *Can. Metall. Q.*, **1990**, *29*(2), 133–139.
140. Clough, T.; Siebert, J. W.; Riese, A. C. *Pat. Specif. (Aust.) AU*, **1991**, *607,523* (Cl. C22B11/04), *US Appl.*, **1988**, *213,884*, 40.
141. Ye, Y.; Jang, W. H.; Yalamanchili, M. R.; Miller, J. D. *Trans. Soc. Min., Metall., Explor.*, **1990**, *288*, 173–9, /Pub. 1991/.
142. Chkoniya, T. K. *U.S.S.R. SU*, **1990**, *1,541,190* (Cl. C01G45/12); *Appl.*, **1987**, *4,246,700*.
143. Saski, T.; Sasaki, H.; Iida, K. *Jpn. Kokai Tokkyo Koho JP*, **1996**, *08 169,769 [96 169,769]* (Cl. C04B35/64); *Appl.*, **1994**, *94/316,464*, 4.
144. Saski, T.; .Sasaki, H. *Jpn. Kokai Tokkyo Koho JP*, **1995**, *07 277,821 [95 277,821]* (Cl. C04B35/495); , *Appl.*, **1994**, *94/66,030*, 4.
145. Caropreso, F. E.; Castrantas, H. M.; Byne, J. M. *U.S. US*, **1990**, *4,935,391* (Cl. 501-146; C04B20/06); *Appl.*, **1989**, *331,423*, 6.
146. Emel'yanova, O. V.; Emel'yanov, B. V. *U.S.S.R. SU*, **1990**, *1,567,520* (Cl. C01G13/04), *Appl.*, **1987**, *4,342,566*.

147. Ruhlandtsende, A. D.; Bacher, A. D.; Muller, U. *Zeitschrift fur Naturforschung, Section B*, *47*(12), 1677–1680.

148. Belfadhel, H.; Ratel, A.; Ouederni, A.; Bes, R. S.; Mora, J. C. *Ozone: Sci. Eng.*, **1995**, *17*(6), 637–645.

149. Achutharaman, V. S.; Beauchamp, K. M.; Chandrasekhar, N.; Spalding, G. C.; Johnson, B. R.; Goldman, A. M. *Thin Solid Films*, **1992**, *216*(1), 14–20.

150. Locquet, J. P. *J. Vacuum Sci. Tech.*, *A. Vacuum Surfaces and Films*, **1992**, *10*(5), 3100–3103.

151. Berkley, D. D.; Goldman, A. M.; Johnson, B. R.; Morton, J.; Wang, T. *Rev. Sci. Instrum.*, **1989**, *60*(12), 3769–74.

152. Usui, T.; Osanai, Y. *Jpn. Kokai Tokkyo Koho JP*, **1989**, *01 122,921 [89 122,921]* (Cl. C01G3/00); *Appl.*, **1987**, *87/278,772*, 5.

153. Odan, K,; Miura, H.; Bando. Y. *Jpn. Kokai Tokkyo Koho JP 01 224,258 [89 224,258] (Cl. C04B25/00), 07 Sep 1989, Appl. 88/45,774, 01 Mar 1988, 3pp*

154. Goldman, A. M.; Berkley, D. D.; Johnson, B. R. *PCT Int. Appl. WO 90 02,215 (Cl. C23C14/26), 08 Mar 1990, US Appl. 234,421*, 19 Aug 1988, 29pp.

155. Nakamori, T. *Jpn. Kokai Tokkyo Koho JP 01 179,752 [89 179,752] (Cl. C04B35/00), 17 Jul 1989, Appl. 88/648*, 07 Jan 1988, 3pp.

156. Johnson, B. R.; Beauchamp, K. M.; Berkley, D. D.; Liu, J. X.; Wang, T.; Goldman. A. M. *Proc. SPIE-Int. Soc. Opt. Eng.*, *1187 (Process. Films High Tc Supercond. Electron)*, **1990**, 27–36.

157. Vig, J. R. UV/ozone cleaning of surfaces. *Proc. Electrochem. Soc., 90–99 (Semicond. Clean. Tachnol. 1989)*, **1990**, 105–113.

158. Suemitsu, M.; Kaneko, T.; Nobuo. N. *Jpn. J. Appl. Phys., Part 1*, **1989,** *28*(12), 2421–2424.

159. Kaneko, T.; Suemitsu, M.; Miyamoto. N. *Jpn. J. Appl. Phys., Part 1*, **1989**, *28*(12), 2425–2429.

160. Inada, A.; Kawasumi. K. *Jpn. Kokai Tokkyo Koho JP 01 135,023 [89 135,023] (Cl. H01L21/30), 26 May 1989, Appl. 87/292, 016 20 JNov 1987, 2pp.*

161. Ohmi, T. *Int.*, **1996**, *19*, (8), 323-4, 326, 328, 330, 332, 334, 336, 338.

162. Kamikawa, Y.; Matsumura. K. *Jpn. Kokai Tokkyo Koho JP 02 106,040 [90 106,040] (Cl. H01L21/304), 18 Apr 1990, Appl. 88/259,204, 20 Oct 1988, 3pp.*

163. Maeda, K. *Kurin Tekunoroji*, **1994**, *4*(2), 62–9.

164. Choi, H.; Kyoo, K.; Jung, S.D.; Jeon. H. *Han'guk Chaehyo Hakhoechi*, **1996**, *6*(4), 395–400.

165. Hansen, R. W. C.; Welske, J.; Wallace, D.; Bissen. M. *Nucl. Instrum. Methods Phys. Res., Sect. A*, **1994**, *347*(1–3), 249–253.

166. Ohmi, T.; Sagawa, T.; Kogure, M.; Imaoka. T. *J. Electrochem. Soc.*, **1993**, *140*(3), 804–810.

167. Takaaki, K.; Akimasa, Y.; Yasuji. M. *Jpn. J. Appl. Phys.*, **1992**, *31*, 2925–30, part 1, No.9A.

168. Hishamune, Y. *Jpn. Kokai Tokkyo Koho JP 01 82,634 [89 82,634] (Cl. H01L21/316), 28 Mar 1989, Appl. 87/242,113, 25 Sep 1989, 6pp.*

169. Kurihara, H.; Saito, M.; Kozuki. T. *Jpn. Kokai Tokkyo Koho JP 09 08,026 [97 08,026] (Cl. H01L21/316), 10 Jan 1997, JP Appl. 95/118,005, 19 Apr 1995, 8pp.*

170. Homma, T. *Mater. Chem. Phys.*, **1995**, *41*(4), 234–9.

171. Toris, K.; Saiton, S.; Ohji. Y. *Jpn. J. Appl. Phys.*, *Part 1- Regular Papers Short Notes & Review Papers;* **1994**, *33*(9B), 5287–5290.
172. Haluska I. A. Fr. *Demande FR 2,697,852 (Cl. C23C16/22), 13 May 1994, Appl. 91/565, 18 Jan 1991, 23pp.*
173. Nishida, S. H. Ger. *offen DE 3,916,983 (Cl. C23C16/40), 30 Nov 1989, JP Appl. 88/125,792, 25 May 1988, 14pp.*
174. Hollinger, G.; Gallet, D.; Gendry, M.; Besland, M. P.; Joseph. J. *Appl. Phys. Lett.*, **1991**, *59*(13) 1617–1619.
175. Tawada, H.; Saitoh, M.; Isobe. C. H. Eur Pat. Appl. *EP 407,945 (Cl. C30B29/28), 16 Jan 1991, JP Appl. 89/178,586, 11 Jul 1989, 23pp.*
176. Tamada, H.; Yamada, A.; Saitoh. M.; *J. Appl. Phys.*, **1991**, *70*(5), 2536–2541.
177. Sadamoto, M.; Ashida, Y.; Fukuda. N. Jpn. *Kokai Tokkyo Koho JP 08 148,707 [96 148,707] (Cl. H01L31/04), 07 Jun 1996, Appl. 94/280,407, 15 Nov 1994, 14pp.*
178. Bianchetti, M. F.; Canepa, H. R.; Larrondo, S.; Walsel de Reca. N. E. *An Asoc. Quim. Argent.*, **1996**, *84*,(1), 73–77.
179. Enomoto, R.; Asai. M. Ger. *offen DE 3,913,966 (Cl. C09J3/00), 09 Nov 1989, JP Appl. 88/104,044, 28 Apr 1988, 15pp.*
180. Binz. *Real Encyclopedi der gesamten Helkunder* Bd. 18, S. 189, Berlin und Wien, 1898.
181. Lambert, B. K. *Nature*, **1958**, *181*, 1202.
182. Shepetinovskii, Z.; Mikashinovich. I. Second All-Union Conference "Ozone obtaining and application". Theses of reports. p. 157, Moscow, 1991.
183. Eur. Pat. Appl. EP 261 032 (1988).
184. Carpendiae, J.; Freeberg. *J. Clin. Gastroenterol.* **1993**, *17*, 2, 142.
185. Belianin, E,; Shmelev. I. Ter. *Arkh.*, **1994**, *66*, 2, 29.
186. Sato. *Exp. Anim.*, **1990**, *39*, 2, 223.
187. Grits, A Formichev. *Microbiol.*, **1990**, *59*, 5, 831.
188. Heindl. *Zentralbl. Hyg. Unwelmed.*, **1993**, *194*, 5/6, 464.
189. Ancowski. *Complement Inflammation*, **1990**, *7*, 2, 57.
190. Kontorshtikova, KN,; Andreeva. N. N. Second All-Union Conference "Ozone obtaining and application. Theses reports, p. 157, Moscow, 1991.
191. Kitanova, S.; Koynova, G.; Rakovsky, S.; Kagan. V. *Xenobiotic Metabollism and Toxicity Workshop of Balkan Countries*, p 35, March 10-21st 1991, Novi Sad, Yugoslavia.
192. Rakovsky, S. Pross. I-st Bulg.-Russ. Sympos. "Free Radicals and Biostabilzers" I-st Bulg. – Russ. Symp. p 113, 23-26 November 1987, Sofia.
193. Rakovsky, S. *International Conference on "Regulation of Free Radical Reactions"* – (Biomedical Aspects), Abstracts, No.140, 13-16 September, 1989.
194. Rakovsky, S.; Fotty, R.; Ershov. V. *International Conference on "Regulation of Free Radical Reactions"* – (Biomedical Aspects), Abstracts, No.141, 13-16 September, 1989.
195. Rakovsky, S.; Todorova, O.; Fotty, R.; Terebenina, A.; Borisov. G. *International Conference on "Regulation of Free Radical Reactions" – (Biomedical Aspects)*, Abstracts, No.142, 13-16 September, 1989.

196. Haruf, M.; Serbinova, E.; Erin, A.; Ulhin, L.; Savov, V.; Rakovsky, S.; Kagan. V. *International Conference on "Regulation of Free Radical Reactions" – (Biomedical Aspects)*, Abstracts, No.198, 13-16 September, 1989.

197. Aliev, M. A.; Ioffe. L. *Second All-Union Conference "Ozone obtaining and application"*. Theses reports, p. 161 Moscow, 1991.

198. Krivatkina, S. D. *Second All-Union Conference "Ozone obtaining and application"*. Theses reports, p. 156, Moscow, 1991.

199. Gloor, M.; Lipphardt. B. *Z. Hautkr.*, **1976**, *51*, 3, 97.

200. Sweet, F. *Science (Washington-D.C.)*, **1980**, *209* (4459), 931.

201. Richter, A. *J. Toxicol. Environ. Health*, **1988**, *25*, 3, 383.

202. Shinriki, N. *Eur. Pat.* Appl. EP **1994**, *601*,891, Cl A61K9/08.

203. Burleson, G. R.; Murray. T. M. *Appl. Microbiol.*, **1975**, *29*, 340.

204. Karzenelson, E.; Koerner, G.; Biederman. N. *Appl. Environ. Microbiol.*, **1979**, *37*, 715.

205. J.Vaughn, J.; Chen. Y. S. *Appl. Environ. Microbiol.*, **1987**, *53*, 9, 2218.

206. Lee, R. E. *Semin. Oncol.*, **1974**, *1*, 254.

207. Rokitansky. *Hospitalis*, **1982**, *52*, 643.

208. Stroczynski, Z Antoszewski. *Pol. Tyd. Lek.*, **1992**, *47*(42–43), 964.

209. Turzynski, J Stroczynski. *Pol. Tyd. Lek.*, **1991**, *46*(37–39), 700–710.

210. Rilling, S. *"Ozone" Sci. Eng. Int.*, **1985**, *7*(4), 257.

211. Schulzs. *Lab. Anim.*, **1986**, *20*(1), 41–48.

212. Pribuda, S. *Semana Med.*, **1963**, *123*(26), 10–269.

213. Wolff. *Das medizinische Ozon. vfm Publications*, Heilderberg, 1979.

214. Payr. *Munchner Medizinische Wochenschrift*, **1935**, *82*, 220.

215. Dorstewitz, H. *Kongres-bericht der artztlichen geselltschafen fur ozontherapie*, Baden;-Baden, 1981.

216. Stadlaender, S.H.; S. H. *Erfahrungsheilkunde*, **1981**, *30*, 4, 277.

217. Rokitansky. *Hospitalis*, **1982**, *52*, 711.

218. Rotter, M.; Mittermayer. H. *Wien Klin. Wochenschr.*, **1974**, *86*, 24, 776.

219. Washuttl, J. *Erfahrungsheilkunde*, **1979**, *28*, 811; **1983**, *32*, 815.

220. Pryor, W. *Chem. Biol. Interact.*, **1991**, *79*, 41.

221. Washuttl, J.; Viebahn. R. *"Ozone" Sci. Eng. Int.*, **1989**, *11*(4), 411.

222. Sweet, F. *Science*, **1980**, *202*, 931.

223. Viebahn, R, R. Steiner. I. *"Ozone" Sci. Eng. Int.*, **1989**, *12*, 65.

224. Brinkman, R.; Lamberts. H. *Nature*, **1958**, *181*, 1202.

225. Chow, C.; Mustafa. M. *Environ. Physiol. Biochem.*, **1975**, *5*, 142.

226. Brickley, R.; Hackney, J.; Clark. K. *Arch. Environ. Health*, **1975**, *30*, 40.

227. Goldstein, B. D. *Arch. Environ. Health*, **1975**, *26*, 279.

228. Vender, R. *Toxicol. Ind. Health*, **1994**, *10*, 1/2, 53.

229. Lamberts, H.; Veninga. T. S. *Lancet*, **1964**, *1*, 133.

230. Schulz, E.; Moore, G.; Calabrese, E. *Bull. Environ. Contam. Toxicol.*, **1981**, *26*, 273.

231. Chow, C.; Tappel. A. L. *Arh. Environ. Health*, **1973**, *26*,205.

232. Sato, S.; Kawakami. M. *Am. Rev. Resp. Dis.*, **1976**, *113*, 809.

233. Chow, C.; Kanenko. J. *Environ. Res.*, **1979**, *19*, 49.

234. Chan, P. C.; Kindya. R. J. *J. Biol. Chem.*, **1977**, *252*, 23, 8537.

235. Savino, A, A.; Peterson. ML .*Environ. Res.*, **1978**, *15*, 1, 65.

236. Canada, A, A.; Airriess. G. *Nutrit Res.*, **1987**, *7*, 797.
237. Murphy, S.; Davis. H. *Toxicol. Appl. Pharmacol.*, **1964**, *6*, 528.
238. Giri, S.; Hollinger. M. *Environ. Res.*, **1980**, *21*, 467.
239. Veninga, T.; Wagenaar. J. *Environ. Health Perspect.*, **1981**, *39*, 153.
240. Shimasaki, H.; Takatori. T. *Biochem. Biophys. Res. Comm.*, **1976**, *68*, 1256.
241. Jankovski, S. ; Doroszkiewicz. W. *Complement Inflammation*, **1990**, *7*, 2, 57.
242. Jegier, Z. *J. Am. Ing. Hyg. Assoc.*, **1974**, *137*, 6, 329.
243. M.Kleinfeld, M.; Gielrnynb. P. *Am. J. Med. Sci.*, **1956**, *231*, 638.
244. Laperwerff, M. *Aerospace Med.*, **1963**, *34*, 479.
245. Trams, E.; Lauter. C. *Arch. Environ. Health.*, **1972**, *24*, 153.
246. Berney, W. B.; Dyer. R. S. *U.S. Environ. Prot. Agency Off Res. Dev.*, *EPA*, 1, PB-**1976**, *264*,233,555.
247. Hackney, J. D. From *Gov. Rep. Announce Index (U.S.)*, **1977**, *77*, 14, 98.
248. Takanashi, Y.; Miura. T. *J. Toxicol. Environ. Health*, **1985**, *15*, 855.
249. Kitanova, S..; Koynova, G.; Rakovsky, S. ; Kagan. V. *Xenobiotic Metabollism and Toxicity Workshop of Balkan Countries*, P34, March 10-21st 1991, Novi Sad, Yugoslavia
250. Kitanova, S.; Koynova, G.; Rakovsky, S.; Kagan. V. Pross. II-nd USSR Conference *"Ozone – Obtainment and Application"*. Moscow, p. 168, 30.01-1.02.1991.
251. Zelac, R.; Crompoy. H. *Environ. Res.*, **1971**, *4*, 325; **1971**, *4*, 262.
252. Vaughan, T.; Moorman. W. *Toxicol. Appl. Pharmacol.*, **1971**, *20*, 404.
253. Iwao, U. *Environ. Res.*, **1989**, *48*, 1, 78.
254. Fairchild, E.; Grahm. S. *Toxicol. Appl. Pharmacol.*, **1964**, *6*, 607.
255. Gordon, T.; Taylor, B.; Amdur. M. *Arch. Environ. Health*, **1981**, *36*,6, 284.
256. Eskerv, M.; Scheuchenzuber. M. *Environ. Res.*, **1976**, *40*, 274.
257. Criegee, R. *Rec. Chem. Prog.*, **1957**, *18*, 111.
258. Menzel, D. B. *Adv. Mod. Environ. Toxicol.*, **1982**, *5* , 183.
259. Chow, C. K.; Hussain, M. Z. *Exprl. Path.*, **1976**, *25*, 182.
260. Last, J. A.; Greenberg. D. B. *Toxicol. Appl. Pharm.*, **1979**, *51*, 247.
261. Hussain, Z. M.; Goss. C. E. *Life Sci.*, **1976**, *18*, 897.
262. Veninga, T. S. *Environ. Health Perspect.*, 1981, *39*, 153.
263. Chow, C. K.; Goss. C. E. *J. Toxicol. Environ. Health*, **1977**, *3*, 5/6, 877.
264. Saymor. *Arch. Environ. Health*, **1974**, *29*, 3, 164.
265. Mohammad, D Anthany. *J. Chest*, **1974**, *66*, 1, 165, suppl. pt 21.
266. Summer, W. *Process Biochem.*, **1976**, *11*, 7, 26.
267. Plopper, C. G. *Toxicol. Appl. Pharmacol.*, **1994**, *121*, 1, 124.
268. Derbal, E Kimberly. *Environ. Health Perspect. Suppl.*, **1994**, *102*, 61.
269. Kelly, F. J. *Hum. Exp. Toxicol.*, 1994, *13*, 6, 407.
270. Drechsler, P.; Deborah, M. *Exp. Geront.*, **1995**, *30*, 1, 65.
271. Miller, F. J.; Oventon, J. H.; Kaskot, R. H. *Tox. and Appl. Pharm.*, **1985**, *79*, 11.
272. Stock, T. H.; Kotchmar, D. J. *J. Air Pollut. Contr. Assoc.*, **1985**, *35*, 1266.
273. Dillar, C.; Urribarri, N. *Arch. Environ. Health*, **1972**, *25*, 6, 426.
274. Mohammad, D Anthony. *J. Lab. Clin. Med.*, **1973**, *82*, 3, 357.
275. Peister, A. *Presse Term. Clim.*, **1977**, *1*, 58.
276. Godstein, B.; Belchum, O. *Toxicol. Appl. Pharmacol.*, **1974**, *27*, 2, 330.
277. Elaine, S.; Dziedzic, D. *Toxicol. Lett.*, **1990**, *51*, 125.

278. Schesinger, R. B.; Driscoll, K. E. *J. Toxicol. Environ. Health*, **1987**, *20*,125.
279. Abraham, W. M.; Sielezak, W. *Eur. J. Respir. Dis.*, **1986**, *68*, 114.
280. Foster, WM, W.M.; Costa, D. L.; Langenbak, E. G. *J. Appl. Physiol.*, **1987**, *63*, 996.
281. Boorman, G. A.; Schwartz, LW .; Dungworth, D. L. *Lab. Invest.*, **1980**, *43*, 108.
282. K.Nikua, K.; Wilson, D. W.; Giri, S. N. *Am. J. Pathol.*, **1988**, *131*, 373.
283. Harkema, J. R.; Plopper, C. G.; Hyde, D. M. *Am. J. Pathol.* **1987**, *128*, 29, 44.
284. Kirkpatrik, J. R.; Henderson R. F. *Exp. Lung. Res.* **1989**, *15*, 1.
285. Schlesinger, R. B. *J. Toxicol. Environ. Health*, **1987**, *21*, 27.
286. Lum, H.; Plopper, D. G. *Exp. Lung Res.*, **1983**, *5*, 61.
287. Michael, W. F. *J. Appl. Physiol.*, **1993**, *75*, 5, 1938.
288. Driscoll, K. E.; Schleinzenger, R. B. *Toxicol. Appl. Pharmacol.*, **1988**, *93*, 312.
289. Kimura, I.; Godstein, E. *J. Infect. Dic.*, **1981**, *143*, 247.
290. Sherwood, R. L.; Goldstein, E. *Environ. Res.*, **1986**, *41*, 378.
291. Bleavins, M. R.; Whith, H. J. *Toxicologist*, **1987**, *7*, 129.
292. Blevins, M. R.; Dziedzic, D. *Toxicologist*, **1988**, *8*, 198.
293. Sell, S. *Immunology, Immunopathology and Immunity*; Elsevier: New York, 1987.
294. Aranyi, C.; Vana, S. C.; Thomas, P. T. *J. Toxicol. Environ. Health*, **1983**, *12*, 55.
295. Eskew, M. L.; Scheuchenzuber, W. J.; Scholz, R. W. *Environ. Res.*, **1986**, *40*, 274.
296. Amoruso, M. A.; With, G. *Life Sci.*, **1981**, *28*, 20, 2215.
297. Nikola, K. J.; Wilson, D. W.; Dungworth, D. L.; Plopper, C. G. *Toxicol. Appl. Pharmacol.*, **1989** *93*, 394.
298. Clark, T. *Toxicol. Letter*, **1994**, *72*,1–3, 279.
299. Stephens, R. J.; Sloan, M. F.; Reeman, G. *Am. J. Pathol.*, **1973**, *74*, 31.
300. Evans, M. J.; Dekker, N.P. *Exp. Mol. Phatol.*, **1985**, *42*, 366.
301. Plopper, C. G.; Chow, C. K.; Dunworth, D. L. *Exp. Mol. Pathol.*, **1978**, *29*, 400.
302. Nikula, K. J.; Wilson, D. W.; Dungworth, D. L.; Plopper, C. G. *Am. J. Pathol.*, **1988**, *131*, 373.
303. Hussain, M. Z.; Cross, C. E. *Life Sci.*, **1976**, *18*, 897, 904.
304. Last, J. A.; Greenberg, D. B. *Toxicol. Appl. Pharmacol.*, **1979**, *51*, 257.
305. Fukase, O.; Hashimoto, K. *Nippon Eiseigaku Zassi.*, **1982**, *37*, 694.
306. Kramer F. *Erfahrungsheilkunde*, **1975**, *24*, 120.
307. Kramer F. *Ozonachrichten*, **1983**, *2*, 65.
308. Turk, R. *Erfahrungsheilkunde*, **1976**, *25*, 177.
309. Salzer, N.; Metka, M.; Shonbauer, M. *Erhfahrungsheilkunde*, **1983**, *32*, 9731.
310. Werkmeister, H. *Erfahrungsheilkunde*, **1976**, *25*, 180.
311. Geabelin, K. *Seifen, Oile, Fette, Wasche*, **1986**, *112*, 1, 1718.
312. Eur. Pat. 222 309 (1987).
313. Ger.Offen 3,806,203 Cl A 61 L9/01.
314. US Pat. 5,266,275 Cl 422-116 A 61 L2/20.
315. Pat. G. **1969**, 66 489, Cl A611.
316. Gal, G . *Intern. J. Arti. Organs*, **1992**, *15*, 8, 461.
317. Hiromichi. JP 05329, **1993**, *196*(93,329,196) Cl A 61 L 2/18.

CHAPTER 8

FUNCTIONAL MODELS OF FE(NI) DIOXYGENASES: SUPRAMOLECULAR NANOSTRUCTURES BASED ON CATALYTIC ACTIVE NICKEL AND IRON HETEROLIGAND COMPLEXES

LUDMILA I. MATIENKO, VLADIMIR I. BINYUKOV,
LARISA A. MOSOLOVA, ELENA M. MIL,
and GENNADY E. ZAIKOV

CONTENTS

8.1 Introduction 186
8.2 Background 187
8.3 Experimental 188
8.4 Results and Discussion 188
    8.4.1 Role of H-Bonding in Stabilization of Catalytic
    Complexes Nix(acac)y(OAc)z(L2)n(H2O)m 191
    8.4.2 Role of H-Bonding in Formation of Catalytic Complexes
    Fex(acac)yL2m(H2O)n (L2 = 18C6, CTAB) 194
8.5 Conclusion 199
Keywords 201
References 201

8.1   INTRODUCTION

Research of structure and catalytic activity of complexes of nickel in different reactions of oxidation with molecular O_2 causes heightened interest of researchers last years in connection with that nickel as an essential catalytic cofactor of enzymes is found in eubacteria, archaebacteria, fungi, and plants. These enzymes catalyze a diverse array of reactions that include both redox and nonredox chemistries [1, 6].

The methionine salvage pathway (MSP) plays a critical role in regulating a number of important metabolites in prokaryotes and eukaryotes. Acireductone dioxygenases (ARDs) Ni(Fe)ARD are enzymes involved in the methionine recycle pathway, which regulates aspects of the cell cycle. The relatively subtle differences between the two metal protein complexes are amplified by the surrounding protein structure, giving two enzymes of different structures and activities from a single polypeptide [2–5].

The products of the two enzymatic Acireductone oxidations differ in that one enzyme $Fe^{II}ARD$ or ARD') converts Acireductone to the R-ketoacid precursor of methionine, and formate, while the other enzyme $Ni^{II}ARD$ (or ARD) produces methylthiopropionate, CO, and formate. The precise function of the $Ni^{II}ARD$ reaction in the K. pneumoniae is unclear. Its products are not precursors for methionine. The $Ni^{II}ARD$-catalyzed reaction may thus be considered a shunt in the methionine salvage pathway, aiding in the regulation of methionine [2–5].

We have offered the new approach to research of mechanism of catalysis with heteroligand complexes of nickel (iron) which can be considered as model $Ni^{II}(Fe^{II})ARD$.

We developed the method for enhancing the catalytic activity of transition metal complexes in the processes of alkylarens (ethyl benzene, cumene) oxidation with dioxygen to afford the corresponding hydro peroxides [6, 7], intermediates in the large-scale production of important monomers [8]. This method consists of introducing additional mono- or multidentate modifying ligands into catalytic metal complexes. The activity of systems $\{ML^1_n + L^2\}$ (M=Ni, Fe, L^1 = acac⁻, L^2 = N-methylpirrolidon-2 (MP), HMPA, MSt (M=Na, Li, K), crown ethers or quaternary ammonium salts) is associated with the fact that during the ethyl benzene oxidation, the primary

$(M^{II}L^1_2)_x(L^2)_y$ complexes and the real active catalytic heteroligand $M^{II}_xL^1_y(L^1_{ox})_z(L^2)_n(H_2O)_m$ complexes are formed to be involved in the oxidation process [6,7,9]. We established that the mechanism of formation of active heteroligand $M^{II}_xL^1_y(L^1_{ox})_z(L^2)_n(H_2O)_m$ complexes is analogous to mechanism of action of NiII(FeII)ARD (and Fe-Acetylacetone Dkel[20]) dioxygenases. Therefore $M^{II}_xL^1_y(L^1_{ox})_z(L^2)_n(H_2O)_m$ complexes seem to be useful as models of $Ni^{II}(Fe^{II})$ARD (and Dkel). The conclusions, received in the present work, can be use at discussion of mechanisms of $Ni^{II}(Fe^{II})$ARD dioxygenases operation.

8.2   BACKGROUND

Nanostructure science and supramolecular chemistry are fast evolving fields that are concerned with manipulation of materials that have important structural features of nanometer size (1 nm to 1 μm) [10, 11]. Nature has been exploiting no covalent interactions for the construction of various cell components. For instance, microtubules, ribosomes, mitochondria, and chromosomes use mostly hydrogen bonding in conjunction with covalently formed peptide bonds to form specific structures. The self-assembled systems and self-organized structures mediated by transition metals are considered in connection with increasing research interest in chemical transformations with use of these systems [12].

H-bonding can be a remarkably diverse driving force for the self-assembly and self-organization of materials. H-bonds are commonly used for the fabrication of supramolecular assemblies because they are directional and have a wide range of interactions energies that are tunable by adjusting the number of H-bonds, their relative orientation, and their position in the overall structure. H-bonds in the center of protein helices can be 20 kcal/mol due to cooperative dipolar interactions [13, 14].

As one of the reasons of stability of heteroligand complexes $M^{II}_xL^1_y(L^1_{ox})_z(L^2)_n(H_2O)_m$ ($L^1_{ox} = CH_3COO^-$) in the conditions of oxidation there can be a formation of intermolecular H-bonds [6, 9]. At the other hand the different activity of nickel and iron complexes in the oxidation processes as well as in biological systems can be connected with different structures characteristics of formed complexes. At the present article AFM method was used in analytics aims, for research of possibility of the different supramolecular structures formation on basis of nickel and iron heteroligand complexes: $Fe_x(acac)_y18C6_m(H_2O)_n$, and $Fe_x(acac)_y(CTAB)_p(H_2O)_q$, or $Ni_2(OAc)_3(acac)L^2\cdot2H_2O$ ($L^2 = MP$) – with the assistance of H-bonding.

8.3 EXPERIMENTAL

AFM SOLVER P47/SMENA/ with Silicon Cantilevers NSG11S (NT MDT) with curvature radius 10 nm, tip height: 10–15 μm and cone angle ≤ 22° in taping mode on resonant frequency 150 KHz was used. As substrate the polished silicone surface special chemically modified was used. Hydrophobic modified silicone surface was exploit for the self-assembly-driven growth due to H-bonding of heteroligand nickel and iron complexes $Ni_2(OAc)_3(acac)MP\cdot2H_2O$ and $Fe_x(acac)_y18C6_m(H_2O)_n$ with Silicone surface. The solutions of complexes in water (or $CHCl_3$) was put on a surface, maintained some time, and then water deleted from a surface by means of special method – spin-coating process.

8.4 RESULTS AND DISCUSSION

It is well known that transition metal β-diketonates are involved in various substitution reactions. Methine protons of chelate rings in β-diketonates complexes can be substituted by different electrophiles. Formally, these reactions are analogous to the Michael addition reactions [15, 16]. This is a metal-controlled process of the C–C bond formation [17]. The complex $Ni^{II}(acac)_2$ is the most efficient catalyst of such reactions.

In our works we have established that the electron-donating ligand L^2, axially coordinated to $M^{II}L^1_2$ (M=Ni, Fe, L^1 =acac⁻), controls the formation of primary active complexes $M^{II}L^1_2\cdot L^2$ and the subsequent reactions in the outer coordination sphere of these complexes. The coordination of an electron-donating extra-ligand L^2

with an $M^{II}L^1_2$ complex favorable for stabilization of the transient zwitter-ion $L^2[L^1M(L^1)^+O_2^-]$ enhances the probability of regiose-lective O_2 addition to the methine C–H bond of an acetylacetonate ligand, activated by its coordination with metal ions. The outer-sphere reaction of O_2 incorporation into the chelate ring depends on the nature of the metal. Transformation routes of a ligand (acac)$^-$ for Ni and Fe are various, but lead to formation of similar heteroligand complexes $M^{II}_xL^1_y(L^1_{ox})_z(L^2)_n(H_2O)_m(L^1_{ox}=CH_3COO^-)$ [6, 7]. Thus for nickel complexes, the reaction of acac-ligand oxygenation follows a mechanism analogous to those of Ni^{II}-containing Dioxygenase ARD [2–5] or Cu- and Fe-containing Quercetin 2,3-Dioxygenases [18,19]. **Namely,** incorporation of O_2 into the chelate acac-ring was accompanied by the proton transfer and the redistribution of bonds in the transition complex leading to the scission of the cyclic system (**A**) to form a chelate ligand OAc$^-$, acetaldehyde and CO (in the Criegee rearrangement) (Scheme 1).

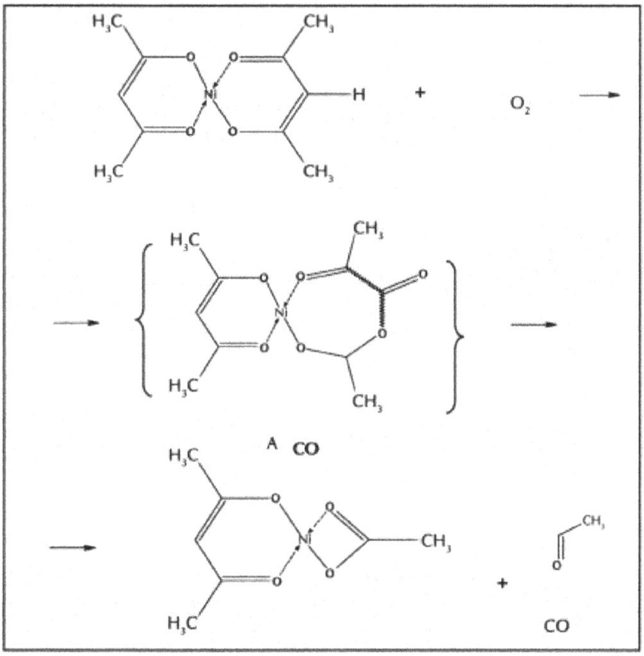

SCHEME 1

In the effect of Fe^{II}-acetylacetonate complexes, we have found [6, 7] the analogy with the action of Fe^{II}ARD (ARD´) [2] and with the action of Fe(II) containing Acetylacetone Dioxygenase (Dke1) [20]. For iron complexes oxygen adds to C–C bond (rather than inserts into the C=C bond as in the case of catalysis with nickel(II) complexes) to afford intermediate, i.e., a Fe complex with a chelate ligand containing 1,2-dioxetane fragment **(B)** (Scheme 2). The process is completed with the formation of the (OAc)¯ chelate ligand and methylglyoxal as the second decomposition product of a modified acac-ring (as it has been shown in [20]) (see Scheme 2).

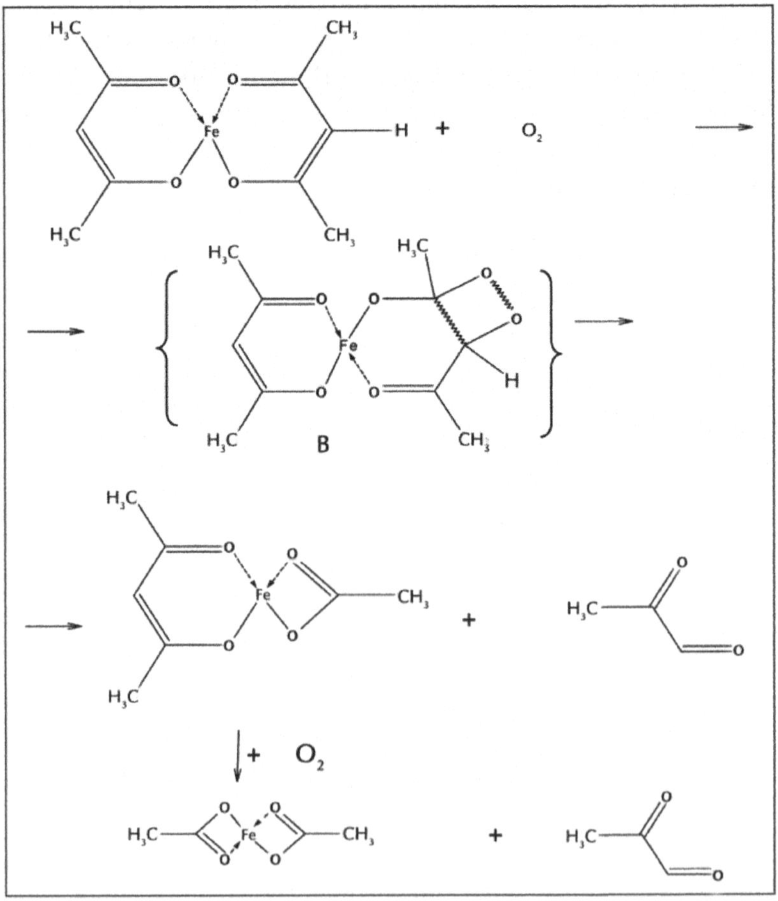

SCHEME 2

8.4.1   ROLE OF H-BONDING IN STABILIZATION OF CATALYTIC COMPLEXES $NI_x(ACAC)_y(OAC)_z(L^2)_N(H_2O)_M$

It is known that heteroligand complexes are more active in relation to reactions with electrophiles in comparison with homoligand complexes [6]. Thus the stability of heteroligand complexes $Ni_x(acac)$ $_y(OAc)_z(L^2)_n(H_2O)_m$ with respect to conversion into inactive form seems to be due to the formation of intramolecular H-bonds.

The complex, formed in the course of ethyl benzene oxidation, catalyzed with system $\{Ni^{II}(acac)_2 + MP\}$, has been synthesized by us and its structure has been defined with mass spectrometry, electron and IR spectroscopy and element analysis [6, 7]. The certain structure of a complex $Ni_2(OAc)_3(acac)\cdot MP\cdot 2H_2O$ corresponds to structure that is predicted on the basis of the kinetic data [6, 7]. Prospective structure of the complex $Ni_{2(OAc)_3}(acac)\cdot MP\cdot 2H2O$ is presented with Scheme 3.

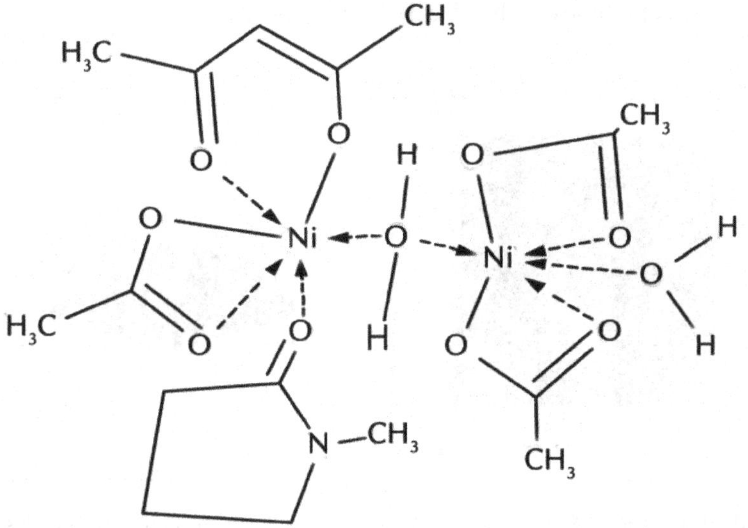

SCHEME 3

On the basis of the known from the literature facts it was possible to assume that binuclear heteroligand complexes $Ni_2(OAc)_3(acac)$ $MP \cdot 2H_2O$ are capable to form macro structures with the assistance of inter- and intramolecular H-bonds (H_2O – MP, H_2O – acetate (or acac$^-$) group) [21, 22].

The association of $Ni_2(AcO)_3(acac) \cdot MP \cdot 2H2O$ to supramolecular structures as result of H-bonding is demonstrated on the next Figs. (1–3).

On Figs. 1–3 three-dimensional and two-dimensional AFM image of the structures formed at drawing of a uterine solution on a hydrophobic surface of modified silicone are presented. It is visible that the majority of the generated structures have rather similar form of three almost merged spheres.

As it is possible to see in Figs.1a and 3, except particles with the form reminding three almost merged spheres (Fig.1b), there are also structures of more simple form (with the height approximately equal 3–4 nm). A profile of one of the particles of the 3–4 nm on height is presented on Fig. 3c.

The distribution histogram (Fig. 2) shows that the greatest number of particles is particles of the size 3–4 nm on height.

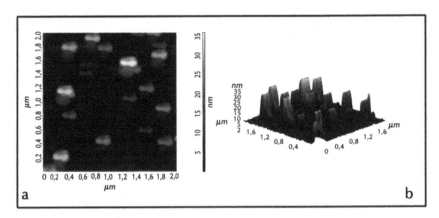

FIGURE 1   The AFM two- (a) and three-dimensional (b) image of nanoparticles on the basis $Ni_2(AcO)_3(acac) \cdot L^2 \cdot 2H_2O$ formed on the hydrophobic surface of modified silicone.

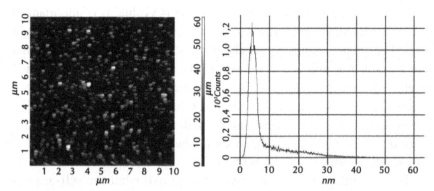

FIGURE 2   AFM image of structures on the hydrophobic modified silicone surface 10×10 μm (at the left). The distribution histogram on height of nanoparticles (to the right).

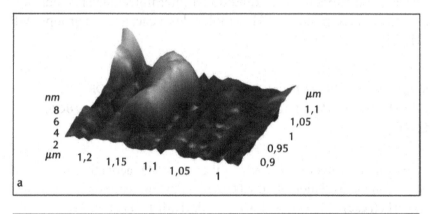

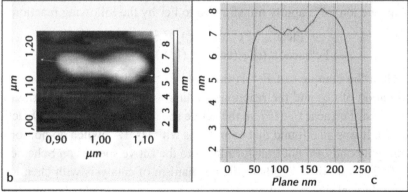

FIGURE 3   The AFM three- (a) and two-dimensional (b) image and profile of the structure (c) with minimum height along the greatest size in plane XY.

From the data in Figs. 1–2 the following is visible. It is important to notice that for all structures the sizes in plane XY do not depend on height on Z. They make about 200 nm along a shaft, which are passing through two big spheres, and about 150 nm along a shaft crossing the big and smaller spheres (particles with the form reminding three almost merged spheres). But all structures are various on heights from the minimal 3–4 nm to ~20–25 nm for maximal values. In distribution on height there is a small quantity of particles with maximum height 20–25 nm and considerably smaller quantity with height to 35 nm (Fig. 1b).

Thus, in the present article we have shown what the self-assembly-driven growth seems to be due to H-bonding of binuclear heteroligand complex $Ni_2(OAc)_3(acac) \cdot MP \cdot 2H_2O$ with a surface of modified silicone, and further due to directional intermolecular H-bonds, apparently at precipitation of H_2O molecules, acac, acetate groups, MP [21, 22].

8.4.2 ROLE OF H-BONDING IN FORMATION OF CATALYTIC COMPLEXES $FE_x(ACAC)_Y L^2{}_M (H_2O)_N$ (L^2 = 18C6, CTAB)

Participation of hydrogen bonds in the step of O_2 activation was found with us in the oxidation of ethylbenzene in the presence of catalytic systems $\{Fe(acac)_2 + L^2\}$ (L^2 = 18C6, CTAB, DMF). In the oxidation of ethylbenzene ion Fe^{III} rapidly transformed to Fe^{II} by the following reaction:

$$Fe^{III}(acac)_3 \, ((Fe^{III}(acac)_2)_m \cdot (L^2)_n) + RH \rightarrow$$
$$\rightarrow Fe^{II}(acac)_2 \, ((Fe^{II}(acac)_2)_x \cdot (L^2)_y) \ldots Hacac + R^{\cdot}$$

Then the complex of Fe^{II} is involved in the chain initiation reaction (activation of O_2) and the reactions leading to the conversion of primary complexes $Fe^{II}(acac)_2)_x \cdot (L^2)_y$ in the active species (through step of O_2 activation). It was also found that complexes with HMPA, which do not form hydrogen bonds are not transforming into the active species on Scheme 2 [6, 7]. The role of H-bonds in the mechanism of catalysis with chemical and biological systems follows from the AFM data presented below.

The surrounding protein structure, give two enzymes $Ni^{II}(Fe^{II})ARD$ of different structures and activities. Conversion from monomeric to multimeric forms could be one of ways of regulating $Ni^{II}(Fe^{II})ARD$ activity [4]. Association of the catalyst in macrostructures with the assistance of the intermolecular H-bonds may be one of reasons of reducing $Ni^{II}ARD$ activity in mechanisms of $Ni^{II}(Fe^{II})ARD$ operation. On the other hand the $Fe^{II}ARD$ operation seems to comprise the step of oxygen activation ($Fe^{II}+O_2 \rightarrow Fe^{III}-O_2^{-\cdot}$) (by analogy with Dke1 action [20]). Specific structural organization of iron complexes may facilitate following regioselective addition of activated oxygen to Acireductone ligand and following reactions leading to formation of methionine. Here for the first time we demonstrate the different structures organization of complexes of nickel and iron in aqueous and hydrocarbon medium.

At first we received UV-spectrum data, testified in the favor of the complex formation between $Fe(acac)_3$ and 18C6. On the next Fig. 4 the spectrums of solutions of $Fe(acac)_3$ (red) and mixture $\{Fe(acac)_3+18C6\}$ (blue) are presented.

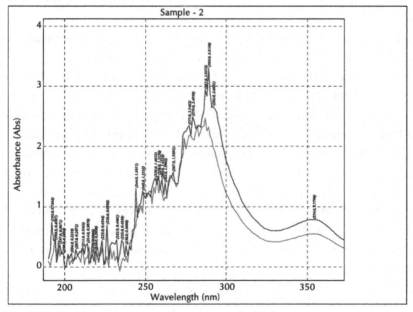

FIGURE 4   Absorption spectra of $CHCl_3$ solutions: of $Fe(acac)_3$ (red), mixture $\{Fe(acac)_3 + 18C6\}(1:1)$ (blue), 20°C.

One can see that at the addition of a solution of 18C6 to the $Fe(acac)_3$ solution (1:1) an increase in absorption intensity of acetylacetonate ion $(acac)^-$, broadening of the spectrum and a bathochromic shift of the absorption maximum from $\lambda \sim 285$ nm to $\lambda = 289$ nm take place. The similar changes in the intensity of the absorption band and shift of the absorption band are characteristic for narrow, crown-unseparated ion-pairs [6]. Earlier similar changes in the UV- absorption band of Co(II) $(acac)_2$ solution we observed in the case of the coordination of macrocyclic polyether 18C6 with $Co(II)(acac)_2$ [6, 7]. The formation of a complex between $Fe(acac)_3$ and 18C6 occurs at preservation of acac ligand in internal coordination sphere of Fe^{III} ion because at the another case the short-wave shift of the absorption band should be accompanied by a significant increase in the absorption of the solution at $\lambda = 275$ nm, which correspond to the absorption maximum of acetylacetone [6, 7]. It is known that Fe(II) and Fe(III) halogens form complexes with crown-ethers of variable composition (1:1, 1:2, 2:1) and structure dependent on type of crown-ether and solvent [23]. It is known that $Fe(acac)_3$ forms labile OSCs (Outer Sphere Complexes) with $CHCl_3$ due to H-bonds [24].

In an aqueous medium the formation of supramolecular structures of generalized formula $Fe^{III}_x(acac)_y 18C6_m(H_2O)_n$ is quite probable.

On the Fig. 5 three-dimensional (a) and two-dimensional (b) AFM image of the structures on the basis of iron complex with 18C6 $Fe^{III}_x(acac)_y 18C6_m(H_2O)_n$, formed at drawing of a uterine solution on a hydrophobic surface of modified silicone are presented. It is visible that the generated structures are organized in certain way forming structures resembling the shape of tubule micro fiber cavity (Fig. 6c). The heights of particles are about 3–4 nm. In control experiments it was shown that for similar complexes of nickel $Ni^{II}(acac)_2 \cdot 18C6 \cdot (H_2O)_n$ (as well as complexes $Ni_2(OAc)_3(acac) \cdot MP \cdot 2H_2O$) this structures organization is not observed. In addition, in the absence of the aqueous environment as these iron constructions are not formed.

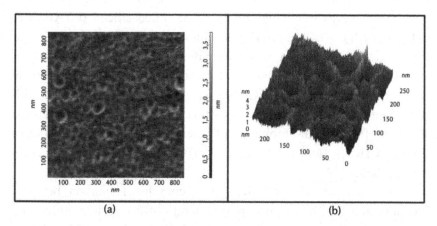

FIGURE 5   The AFM two- (a) and three-dimensional (b) image of nanoparticles on the basis $Fe_x(acac)_y 18C6_m(H_2O)_n$ formed on the surface of modified silicone.

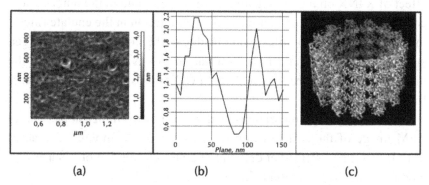

FIGURE 6   The AFM two-dimensional image (a) of nanoparticles on the basis $Fe_x(acac)_y 18C6_m(H_2O)_n$ formed on the hydrophobic surface of modified silicone.

The section of a circular shape with fixed length and orientation is about 50–80 nm (b), (c) The structure of the cell microtubules.

As the other example we researched the possibility of the supramolecular nano structures formation on the base of $Fe_x(acac)_y CTAB_m$

at putting of solutions of $Fe_x(acac)_y CTAB_m$ in $CHCl_3$ (Fig. 7) or H_2O (Fig. 8) on the hydrophobic surface of modified silicon ($CTAB=Me_3$ (n-$C_{16}H_{33}$)NBr). We used CTAB concentration in the 5–10 times less on comparison with $Fe(acac)_3$ with aim to decrease the possibility micelles formation in water. But formation of sphere micelles at these conditions cannot be excluded. It is known that salts QX can form complexes with metal compounds of variable composition, which depends on the nature of solvent [6,7]. So the formation of heteroligand complexes $Fe_x(acac)$ $_y CTAB_m(CHCl_3)_p$ (and $Fe^{II}_x(acac)_y(OAc)_z(CTAB)_n(CHCl_3)_q$ also) seems to be probable [24].

Earlier we established outer-sphere complex formation between $Fe(acac)_3$ and R_4NBr with different structure of R – cation. Unlike the action of 18C6 in the presence of salts Me4NBr, Me3(n-C16H33)NBr (CTAB), Et4NBr, Et3PhNCl, $_{Bu4}$NI and Bu_4NBr, a decrease in the maximum absorption in this band and its bathochromic shift ($\Delta\lambda \approx 10$ nm) were observed (in $CHCl_{3)}$. Such changes in the UV spectra reflect the effect of R4NX on the conjugation in the acac ligand in the outer-sphere coordination of R_4NX. A change in the conjugation in the chelate ring of the acetylacetonate complex could be due to the involvement of oxygen atoms of the acac ligand in the formation of covalent bonds with the nitrogen atom or hydrogen bonds with CH groups of alkyl substituents [6].

On the Fig. 7 three-dimensional (a) and two-dimensional (b, c) AFM image of the structures on the basis of iron complex – $Fe_x(acac)$ $_y CTAB_m(CHCl_3)_p$, formed at drawing of a uterine solution on a surface of modified silicone are presented.

As one can see in this case also the generated supramolecular nano structures are organized like the generated structures on the base of $Fe_x(acac)_y 18C6_m(H_2O)_n$, but with less clear form than presented on Figs. 5, and 6, resembling the shape of tubule micro fiber cavity (Fig. 6c). The heights of particles on the base of $Fe_x(acac)_y CTAB_m(CHCl_3)_p$ are about 7-8 nm (Fig. 7c), that are higher than in the case of $Fe_x(acac)_y 18C6_m(H_2O)_n$. We showed that complexes $Ni^{II}(acac)_2 \cdot CTAB$ (1:1) (in $CHCl_3$) does not form similar structures.

The section of a circular shape with fixed length and orientation is about 50–70 nm (c).

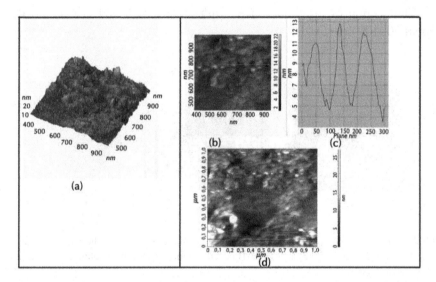

FIGURE 7 The AFM three- (a) two-dimensional image (b, d) of nanoparticles on the basis $Fe_x(acac)_y CTAB_m(CHCl_3)_p$ formed on the hydrophobic surface of modified silicone.

On the Fig. 8 two-dimensional (a–c) AFM image of the structures on the basis of iron complex with CTAB: $Fe_x(acac)_y CTAB_p(H_2O)_q$, – formed at drawing of a uterine water solution on a hydrophobic surface of modified silicone are presented. In this case, we observed apparently phenomenon of the particles, which are the remnants of the micelles formed. Obviously, the spherical micelles, based on $Fe(acac)_3$ and CTAB, in a hydrophobic surface are extremely unstable and decompose rapidly. The remains of the micelles have a circular shaped structures, different from that observed in

Figs. 6 and 7. The height of particles on the base of $Fe_x(acac)_yCTAB_p(H_2O)_q$ is greater and is about 12–13 nm.

The particles resemble micelles in sizes (100–120 nm in-plane XY (Fig. 8c)) and also like micelles, probably due to the rapid evaporation of water needed for their existence, are very unstable and destroyed rapidly (even during measurements).

The section of a circular shape with fixed length and orientation is about 70–100 nm (c).

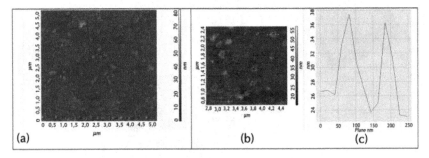

FIGURE 8   The AFM two-dimensional image (a, b) of nanoparticles (the remains of the micelles) on the base of $Fe_x(acac)_yCTAB_p(H_2O)_q$ formed on the hydrophobic surface of modified silicone.

8.5   CONCLUSION

AFM method was used by us in the analytical purposes to research the possibility of the formation of supramolecular structures on basis of hetero-ligand complex $Ni_2(OAc)_3(acac)\cdot MP\cdot 2H_2O$ (L^2 =), $Fe_x(acac)_y 18C6_m(H_2O)_n$ and $Fe_x(acac)_yCTAB_m(CHCl_3)_p$ with the assistance of intermolecular H-bonding.

Here for the first time we show the different structures organization of nickel and iron complexes with the assistance of intermolecular H-bonds in aqueous and hydrocarbon medium.

The experimental data, presented on Figs. 1–3, specifies in quite probable possibility of formation supramolecular structures on the basis of complex $Ni_2(OAc)_3(acac) \cdot MP \cdot 2H_2O$ with the assistance of intermolecular H-bonds in the course of alkylarens oxidation. H-bonding seems to be one of the factors, responsible for the stability of real catalysts – heteroligand complexes $Ni_2(OAc)_3(acac) \cdot MP \cdot 2H_2O$, in the course of alkylarens (ethyl benzene, cumene) oxidation by dioxygen into hydro peroxide (intermediates in the large-scale production of important monomers) in the presence of catalytic systems $\{Ni^{II}(acac)_2 + L^2\}$.

The received in present work data can be useful in treatment of biological effects in $Ni^{II}(Fe^{II})$ARD enzymes operation. Conversion of monomeric to multimeric forms could be one of ways of regulating $Ni^{II}(Fe^{II})$ ARD activity [4]. Association of catalyst in different macrostructures with the assistance of the intermolecular H-bonds may be one of reasons of different actions of Ni^{II}- and Fe^{II}ARD. The different supramolecular nanostructures organization based on nickel and iron heteroligand complexes may facilitate the understanding mechanisms of $Ni^{II}(Fe^{II})$ARD operation. Specific structural organization of iron complexes may facilitate the first step in Fe^{II}ARD operation: O_2 activation and following regioselective addition of activated oxygen to Acireductone ligand (unlike mechanism of regioselective addition of nonactivated O_2 to Acireductone ligand in the case of Ni^{II}ARD), and reactions leading to formation of methionine.

At the same time it is necessary to mean that important function of Ni^{II}ARD in cells is established now. Namely, carbon monoxide, CO, is formed as a result of nickel – containing dioxygenase Ni^{II}ARD. It is established, that CO is a representative of the new class of neural messengers, and like NO, seems to be a signal transducer [2–5].

KEYWORDS

- **H-bonding**
- Heteroligand nickel
- **β-diketonates**
- **Acireductone ligand**

REFERENCES

1. Li, Y.; Zamble, D. B. *Chem. Rev.*, **2009**, *109*, 4617.

2. Dai, Y.; Pochapsky, Th. C.; Abeles, R. H. *Biochemistry*, **2001**, *40*, 6379.

3. Al-Mjeni, F.; Ju. T.; Pochapsky, Th. C.; Maroney, M. J. *Biochemistry*, **2002**, *41*, 6761.

4. Sauter, M.; Lorbiecke, R.; OuYang, Bo, Pochapsky, Th.C.; Rzewuski, G. *Plant J.*, **2005**, *44*, 718.

5. Chai, S. C.; Ju, T.; Dang, M.; Goldsmith, R. B.; Maroney, M. J.; Pochapsky, Th. C. Biochemistry, **2008**, *47*, 2428.

6. Matienko, L. I.; Mosolova, L. A.; Zaikov, G.E. *Selective Catalytic Hydrocarbons Oxidation*. New Perspectives, Nova Science Publications, Inc.: New York, N. Y., p 150.

7. Matienko, L. I. In: *Reactions and Properties of Monomers and Polymers*, D'Amore, A. Zaikov, G., Eds.; Chapter 2, Nova Sience Publications, Inc.: New York, NY, 2007, p 21.

8. Weissermel, K.; Arpe, H.-J. *Industrial Organic Chemistry*, 3nd ed., transl. by Lindley, C. R. VCH: New York, N. Y., 1997.

9. Matienko, L.I.; Mosolova, L.A. *Communications*, **2010**, *33*, 830.

10. Leninger, S. T.; Olenyuk, B.; Stang, P. J. *Chem. Rev.*, **2000**, *100*, 853.

11. Stang, P. J.; Olenyuk, B. *Chem. Res.*, **1997**, *30*, 502.

12. Beletskaya, I.; Tyurin, V. S.; Tsivadze, A. Yu.; Guilard, R.; Stem, Ch. *Chem. Rev.*, **2009**, *109*, 1659.

13. Drain, C. M.; Varotto, A.; Radivojevic, I. *Chem. Rev.*, **2009**, *109*, 1630.

14. Cheng-Che, C.; Raffy, G.; Ray, D.; Del Guerzo, A.; Kauffmann, B.; Wantz, G.; Hirsch, L.; Bassani, D.M. J. Am. *Chem. Soc.*, **2010**, *132*, 12717.

15. Uehara, K.; Ohashi, Y.; Tanaka, M. *Bull. Chem. Soc. Jpn.*, 1976, **49**, 1447.

16. Nelson, J. H.; Howels, P. N.; Landen, G. L.; De Lullo, G. S.; Henry, R. A. In: *Fundamental Research in Homogeneous Catalysis*, Vol. 3, Plenum: New York, London, 1979, p 921.

17. Daolio, S.; Traldi, P.; Pelli, B.; Basato, M.; Corain, B.; Kreissl, F. *Inorg. Chem.*, **1984**, *23*, 4750.

18. Gopal, B.; Madan, L. L.; Betz, S. F.; Kossiakoff, A. A. *Biochemistry*, **2005**, *44*, 193.

19. Balogh-Hergovich, É.; Kaizer, J.; Speier, G. J. Mol. Catal. A: *Chem.*, **2000**, *159*, 215.

20. Straganz, G. D.; Nidetzky, B. *J. Am. Chem. Soc.*, 2005, *127*, 12306.

21. Basiuk, E. V.; Basiuk, V. V.; Gomez-Lara, J.; Toscano, R. A. *J. Incl. Phenom. Macrocycl. Chem.*, **2000**, *38*, 45.

22. Mukherjee, P.; Drew, M. G. B.; Gómez-Garcia, C. J.; Ghosh, A. *Inorg. Chem.*, **2009**, *48*, 4817.

23. Belsky, V. K.; Bulychev, B. M. *Usp. Khim.*, 1999, *68*, 136. *Russ. Chem. Rev.*, 1999, *68*, 119].

24. Nekipelov V. M.; Zamaraev K. I. *Coord. Chem. Rev.*, **1985**, *61*, 185.

CHAPTER 9

ASSOCIATION BETWEEN CALCIUM CHLORIDE AND CAFFEINE AS SEEN BY TRANSPORT TECHNIQUES AND THEORETICAL CALCULATIONS

MARISA C. F. BARROS, ABILIO J. F. N. SOBRAL,
LUIS M. P. VERISSIMO, VICTOR M. M. LOBO,
ARTUR J. M. VALENTE, MIGUEL A. ESTESO,
and ANA C. F. RIBEIRO

CONTENTS

9.1   Introduction.. 206
9.2   Experimental.. 207
    9.2.1   Diffusion Measurements: Equipment and Procedure 207
    9.2.2   Density Measurements 209
    9.2.3   Viscosity Measurements... 209
    9.2.4   Ab initio Studies 209
9.3   Experimental Results and Discussion... 210
    9.3.1   Diffusion Coefficients, D11, D12, D21 and D22 of
            $CaCl_2$ in Aqueous Solutions of Caffeine.......................... 210
    9.3.2   Densities of $CaCl_2$ in Aqueous Solutions of Caffeine 213
    9.3.3   Viscosities of $CaCl_2$ in Aqueous Solutions of Caffeine.... 215
    9.3.3   Ab Initio Calculations and the Ca^{2+}/Caffeine
            Interactions 216
9.4   Conclusion 217
Keywords.. 218
References.. 218

9.1 INTRODUCTION

Caffeine (3,7-dihydro-1,3,7-trimethyl-1H-purine-2,6-dione) is naturally occurring in some beverages and is also used as a pharmacological agent being a central nervous system stimulant [1–3]. The interactions of metal ions with caffeine are of major biological interest, due to the fact that these complexes play a dominant role in many biochemical interactions [1–3]. Among them, we are interested in systems containing calcium ions and caffeine due to their influence in health sciences.

While some studies have been carried out on the structure analysis of the systems involving calcium ions with caffeine using spectroscopy techniques (for example, Fourier Transform Infrared (FTIR) and UV-visible spectroscopies used to determine the cation binding mode and the association constants, K(caffeine–Ca) = 29.8 M^{-1}) [1], few have taken into account the transport behavior of those complex chemical systems. Therefore, the characterization of the transport and thermo-dynamic properties (diffusion, density and viscosity) in solutions of calcium chloride in the presence of caffeine is important both for fundamental reasons, for understanding the nature of the structure of those aqueous electrolytes, and for practical application in fields such medicine and pharmacy.

The main scope of our work shown here is to present the experimental studies concerning the transport properties of systems containing calcium ions and caffeine, and some *ab initio* calculations. On the other words, we present ternary diffusion coefficients D_{11}, D_{22}, D_{12} and D_{21} for the system containing calcium chloride and caffeine, for concentrations of each component between 0.0025 M and 0.05 M and at 25°C and 37°C, using a specially designed apparatus built for measuring diffusion coefficients based on the Taylor technique (studies already done and indicated in the literature [4]). In addition, for the same systems and for the same temperatures, we report new viscosity and densities measurements. To obtain a better understanding of the structure of the chemical species formed, we have complemented these studies with *ab initio* calculations.

9.2 EXPERIMENTAL

9.2.1 DIFFUSION MEASUREMENTS: EQUIPMENT AND PROCEDURE

The theory of the Taylor dispersion technique is well described in literature [5–10] (Fig. 1), and we only indicate some relevant points concerning this method on the experimental determination of binary diffusion coefficients, D, and ternary diffusion coefficients, respectively.

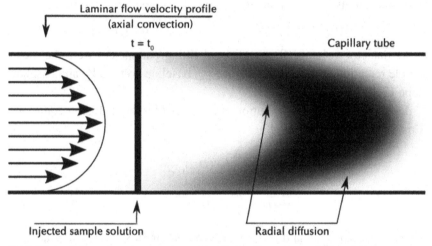

FIGURE 1   Schematic representation of the Taylor dispersion technique [6].

The above method is based on the dispersion of small amounts of solution injected into laminar carrier streams of solvent or solution of different composition, flowing through a long capillary tube [5–10]. The length of the Teflon dispersion tube used in the present study was measured directly by stretching the tube in a large hall and using two high quality theodolytes and appropriate mirrors to accurately focus on the tube ends. This technique gave a tube length of 3.2799 ($\pm$0.0001) × 10^4 mm, in agreement with less-precise control measurements using a good-quality measuring tape. The radius of the tube, 0.5570 ($\pm$0.00003) mm, was calculated from the tube volume obtained by accurately weighing (resolution 0.1 mg) the tube when empty and when filled with distilled water of known density.

At the start of each run, a 6-port Teflon injection valve (Rheodyne, model 5020) was used to introduce 0.063 ml of solution into the laminar carrier stream of slightly different composition. A flow rate of 0.17 ml min^{-1} was maintained by a metering pump (Gilson model Minipuls 3) to give retention times of about 1.1×10^4 s. The dispersion tube and the injection valve were kept at 298.15 K and 303.15 K (± 0.01 K) in an air thermostat.

Dispersion of the injected samples was monitored using a differential refractometer (Waters model 2410) at the outlet of the dispersion tube. Detector voltages, $V(t)$, were measured at accurately 5 s intervals with a digital voltmeter (Agilent 34401 A) with an IEEE interface. Binary diffusion coefficients were evaluated by fitting the dispersion equation:

$$V(t) = V_0 + V_1 t + V_{max} \, (t_R/t)^{1/2} \exp[-12D(t - t_R)^2/r^2 t] \tag{1}$$

to the detector voltages. The additional fitting parameters were the mean sample retention time t_R, peak height V_{max}, baseline voltage V_0, and baseline slope V_1.

Diffusion in a ternary solution is described by the diffusion equations [Eqs. (2) and (3)],

$$-(J_1) = (D_{11})_v \frac{\partial c_1}{\partial x} + (D_{12})_v \frac{\partial c_2}{\partial x} \tag{2}$$

$$-(J_2) = (D_{21})_v \frac{\partial c_1}{\partial x} + (D_{22})_v \frac{\partial c_2}{\partial x} \tag{3}$$

where J_1, J_2, $(\partial c_1/\partial x)$ and $(\partial c_2/\partial x)$ are the molar fluxes and the gradients in the concentrations of solute 1 and 2, respectively. Main diffusion coefficients give the flux of each solute produced by its own concentration gradient. Cross diffusion coefficients D_{12} and D_{21} give the coupled flux of each solute driven by a concentration gradient in the other solute. A positive D_{ik} cross-coefficient ($i \neq k$) indicates co-current coupled transport of solute i from regions of higher concentration of solute k to regions of lower concentration of solute k. However, a negative D_{ik} coefficient indicates counter-current coupled transport of solute i from regions of lower to higher concentration of solute k.

Extensions of the Taylor technique have been used to measure ternary mutual diffusion coefficients (D_{ik}) for multicomponent solutions. Theses D_{ik} coefficients, defined by Eqs. (2) and (3), were evaluated by fitting the ternary dispersion equation [Eq. (4)] to two or more replicate pairs of peaks for each carrier-stream.

$$V(t) = V_0 + V_1 t + V_{max} (t_R / t)^{1/2} \left[W_1 \exp\left(-\frac{12D_1(t-t_R)^2}{r^2 t} \right) + (1-W_1) \exp\left(-\frac{12D_2(t-t_R)^2}{r^2 t} \right) \right] \quad (4)$$

Two pairs of refractive-index profiles, D_1 and D_2, are the eigenvalues of the matrix of the ternary D_{ik} coefficients.

In these experiments, small volumes of ΔV of solution, of composition $\bar{c}_1 + \overline{\Delta c_1}$, $\bar{c}_2 + \overline{\Delta c_2}$ are injected into carrier solutions of composition, $\bar{c}_1$, $\bar{c}_2$ at time $t = 0$.

9.2.2 DENSITY MEASUREMENTS

Densities of these solutions were obtained using an Anton Paar DMA5000M densimeter with a sensibility of 1×10^{-6} g cm^{-3} and a reproducibility of $\pm 5 \times 10^{-6}$ g cm^{-3} in the ranges of 0-90°C of temperature and 0–10 bars of pressure. This apparatus is provided with a Peltier system, which allows keeping constant the temperature of the samples into the vibrating-tube within $\pm 0.005°$. The measurements were carried out at 25°C and 37°C. The density value for each solution studied was the mean of at least four sets of measurements. Density values were reproducible within $\pm 0.001\%$.

9.2.3 VISCOSITY MEASUREMENTS

Viscosity measurements at 25°C and 37°C have been performed with a Ubbelohde automatic type viscometer, previously calibrated with water [10]. Viscosities in Table 4 are the mean value of four sets of flow times for each solution. The efflux time was measured with a stopwatch with a resolution of 0.2 s. Viscosity values have been reproducible within $\pm 0.1\%$.

9.2.4 AB INITIO STUDIES

The potential surfaces were calculated after an *ab initio* geometry minimization using a RHF algorithm with a small (3-21G) basis set. Calculations were performed in a HP Evo dc7700 workstation using the HyperChem v7.5 software package from Hypercube Inc., 2000, USA. The geometry optimization used a Polak-Ribiere conjugated gradient algorithm for energy minimization in vacuum, with a final gradient of 0.1 kcal/(A.mol).

9.3   EXPERIMENTAL RESULTS AND DISCUSSION

9.3.1   DIFFUSION COEFFICIENTS, D_{11}, D_{12}, D_{21} AND D_{22} OF CACL$_2$ IN AQUEOUS SOLUTIONS OF CAFFEINE

Tables 1 and 2 show the experimental diffusion coefficients for the ternary system CaCl$_2$+caffeine+water at 25°C and 37°C, D_{11}, D_{12}, D_{21} and D_{22} [4]. These results are, in general, the average of 3 experiments performed on consecutive days. The main diffusion coefficients D_{11} and D_{22}, giving the molar fluxes of the CaCl$_2$ (1) and caffeine (2) components driven by their own concentration gradient, were generally reproducible within $\pm0.010\times10^{-9}$ m^2 s^{-1}. The cross diffusion coefficients, D_{12} and D_{21}, were reproducible within $\pm0.020\times10^{-9}$ m^2 s^{-1} and $\pm0.010\times10^{-9}$ m^2 s^{-1}, respectively, being, however, almost zero, within the imprecision of this method. Previous papers reporting data obtained with this technique have shown that the error limits of our results should be close to the imprecision, therefore giving an experimental uncertainty of 1 to 3% (e.g., [4, 5]).

The main coefficients D_{11} and D_{22}, were compared with those obtained for binary systems at the same temperatures and with the same technique (Tables 1 and 2), that is, for CaCl$_2$ [11,12] and caffeine [13] in aqueous solutions, respectively, being, in general, these coefficients lower than the binary diffusion coefficients of aqueous CaCl$_2$ and caffeine for both temperatures (deviations between 1% and 6%). In addition, we verified that the gradient in the concentration of CaCl$_2$ produces counter-current coupled flows of caffeine, while the gradient in the concentration of caffeine produces co-current coupled flows of CaCl$_2$. On the other words, we may say that, at the concentrations used, a mole of diffusing caffeine co-transports at most 0.02 mol of CaCl$_2$, increasing the co-transport with the increase of its concentration.

TABLE 1 Ternary diffusion coefficients, D_{11}, D_{12}, D_{21} and D_{22}, for aqueous calcium chloride (1) + caffeine (2) solutions and the respective standard deviations, S_D, at 25°C [4].

c_1[a]	c_2[a]	R[b]	$D_{11}\pm S_D$ / $(10^{-9}\ m^2\ s^{-1})$	$D_{12}\pm S_D$ / $(10^{-9}\ m^2\ s^{-1})$	$D_{21}\pm S_D$ / $(10^{-9}\ m^2\ s^{-1})$	$D_{22}\pm S_D$ / $(10^{-9}\ m^2\ s^{-1})$	D_{12}/D_{22}[c]	D_{21}/D_{11}[d]
0.000	0.0025					0.749[f]		
0.000	0.0050					0.738[f]		
0.000	0.0100					0.703[f]		
0.000	0.0500					0.594[f]		
0.0025	0.000		1.270[e]					
0.0050	0.000		1.260[e]					
0.0100	0.000		1.215[e]					
0.0500	0.000		1.115[e]					
0.0025	0.0050	0.5	1.258±0.012 (−0.9%)[g]	0.062±0.011	−0.042±0.004	0.701±0.010 (−5.0%)[h]	0.088	−0.033
0.0050	0.0025	2	1.212±0.004 (−3.8%)[g]	0.056±0.007	−0.041±0.006	0.711±0.012 (−5.1%)[h]	0.079	−0.034
0.0050	0.0050	1	1.191±0.006 (−5.5%)[g]	0.116±0.003	−0.158±0.008	0.635±0.009 (−13.9%)[h]	0.183	−0.133
0.0100	0.0100	1	1.215±0.006 (0.0%)[g]	0.016±0.002	−0.071±0.011	0.698±0.006 (0.7%)[h]	0.023	−0.058
0.050	0.050	1	1.162±0.013 (+4.2%)[g]	0.082±0.028	−0.197±0.028	0.582±0.002 (−2.0%)[h]	0.141	−0.170

[a] c_1 and c_2 in units of mol.dm⁻³. [b] $R = c_1/c_2$ for ternary systems. [c] D_{12}/D_{22} gives the number of moles of CaCl$_2$ co-transported per mole of caffeine. [d] D_{21}/D_{11} gives the number of moles of caffeine counter-transported per mole of CaCl$_2$. [e] Our experimental binary D values for aqueous CaCl$_2$ [11, 12]. [f] Taylor binary D values for aqueous caffeine [13]. [g] These values indicated in parenthesis represent the relative deviations between the experimental values of D_{11} and the binary values for the same concentration, D (see note e). [h] These values indicated in parenthesis represent the relative deviations between the experimental values of D_{22} and the binary values for the same concentration, D (see note f).

TABLE 2   Ternary diffusion coefficients, D_{11}, D_{12}, D_{21} and D_{22}, for aqueous calcium chloride (1) + caffeine (2) solutions and the respective standard deviations, S_D, at 37°C [4].

c_1[a]	c_2[a]	R[b]	$D_{11} \pm S_D$ / (10^{-9} m^2 s^{-1})	$D_{12} \pm S_D$ / (10^{-9} m^2 s^{-1})	$D_{21} \pm S_D$ / (10^{-9} m^2 s^{-1})	$D_{22} \pm S_D$ / (10^{-9} m^2 s^{-1})	$\dfrac{D_{12}}{D_{22}}$[c]	D_{21}/D_{11}[d]
0.000	0.0025					1.021[f]		
0.000	0.0050					0.980[f]		
0.000	0.0100					0.944[f]		
0.000	0.0500					0.822[f]		
0.0025	0.000		1.602[e]					
0.0050	0.000		1.601[e]					
0.0100	0.000		1.590[e]					
0.0500	0.000		1.530[e]					
0.0025	0.0050	0.5	1.581±0.005 (−1.3%)[g]	0.016±0.007	−0.030±0.008	0.987±0.001 (0.71%)[h]	0.016	−0.019
0.0050	0.0025	2	1.558±0.001 (−2.7%)[g]	0.018±0.005	−0.091±0.007	0.943±0.012 (−7.6%)[h]	0.019	−0.058
0.0050	0.0050	1	1.605±0.001 (0.25%)[g]	0.051±0.009	−0.070±0.004	0.959±0.006 (−2.1%)[h]	0.053	−0.044
0.0100	0.0100	1	1.565±0.008 (−1.6%)[g]	0.040±0.004	−0.055±0.016	0.947±0.005 (0.32%)[h]	0.042	−0.035
0.050	0.050	1	1.482±0.003 (−3.1%)[g]	0.029±0.005	−0.015±0.007	0.788±0.009 (−4.1%)[h]	0.037	−0.010

[a]c_1 and c_2 in units of mol.dm^{-3}. [b]$R = c_1/c_2$ for ternary systems. [c]D_{12}/D_{22} gives the number of moles of CaCl$_2$ co-transported per mole of caffeine. [d]D_{21}/D_{11} gives the number of moles of caffeine counter-transported per mole of CaCl$_2$. [e]Our experimental binary D values for aqueous CaCl$_2$ [11,12]. [f]Taylor binary D values for aqueous caffeine [13]. [g] These values indicated in parenthesis represent the relative deviations between the experimental values of D_{11} and the binary values for the same concentration, D (see note e). [h] These values indicated in parenthesis represent the relative deviations between the experimental values of D_{22} and the binary values for the same concentration, D (see note f).

Through D_{21}/D_{11} values, at the same concentrations, we can expect that a mole of diffusing $CaCl_2$ counter-transports at most 0.03 mol of caffeine. Similarly, at 310.15 K, the number of moles of $CaCl_2$ co-transported per mole of caffeine, D_{12}/D_{22} is at most 0.02 mol while the number of moles of caffeine counter-transported per mole of calcium chloride, D_{21}/D_{11}, is at most 0.02 mol. From these results, it is evident that $CaCl_2$ species exert an influence on the diffusion of caffeine in aqueous solution (that is, in general, for [$CaCl_2$]/[Caffeine] ratio values ≥ 0.5), increasing this effect with the increase of these ratios. Some calcium ions can be present in solution as aggregates. Consequently, they will have less mobility and, they can be responsible for relatively large decreases in D_{22}. This effect is less relevant when we consider the effect of caffeine on transport of $CaCl_2$, probably due to the similarity of the mobilities of caffeine free species and eventual aggregates of $CaCl_2$ and caffeine.

9.3.2 DENSITIES OF CACL₂ IN AQUEOUS SOLUTIONS OF CAFFEINE

Experimental density values, ρ, are reported in Table 3 for the different concentrations and two temperatures studied.

TABLE 3 Density values, ρ, for ternary aqueous solutions of $CaCl_2$ (1) and caffeine (2) at 25°C and 37°C.

c_1 / (mol.dm⁻³)	c_2 / (mol.dm⁻³)	/(g.cm⁻³) ($T = 25°C$)	$10^4\sigma$/(g.cm⁻³) [a] ($T = 25°C$)	ρ/(g.cm⁻³) ($T = 37°C$)	$10^4\sigma$/(g.cm⁻³) [a] ($T = 37°C$)
0.00249	0.00498	0.9994	2.8	0.9942	2.9
0.00501	0.00250	0.9986	2.3	0.9949	1.9
0.00499	0.00502	0.9980	1.2	0.9943	0.8
0.01010	0.01020	0.9992	2.1	0.9953	0.2
0.05009	0.05005	1.0033	0.8	1.0021	0.7

[a] σ represent the standard deviation of these measurements.

These values were adequately fitted using a least-squares regression method, using equation $\rho/(g\ cm^{-3})) = \rho_0\{1 + b_1[c/(mol\ dm^{-3})] + b_2[c^2/(mol^2\ dm^{-6})]\}$, where ρ_0 and c stand for the pure water density value and the molarity, respectively, and b_1 and b_2 are adjustable coefficients [(Eqs. (5) and (6)].

$$\rho = 0.9988 + 0.0205\ c - 2.1953\ c^2 \tag{5}$$

$$\rho = 0.9939 + 0.1363\ c - 0.5530\ c^2 \tag{6}$$

Figure 2 shows the plot of the experimental density values against molarity. As can be seen, density values for ternary aqueous solutions of caffeine and $CaCl_2$ increase with the rising of the solution concentration for both temperatures, showing a different behavior, depending on the temperature. That is, the lines are not parallel between them, implying that temperature influences the behavior of this physicochemical property and that the changes in density should not be only due to the breaking links derived from the thermal effect, but could also be a consequence of the breaking of more complex structures present in the solution like caffeine-caffeine, caffeine-water, Ca^{2+}-water, Cl^--water, or even caffeine-$CaCl_2$.

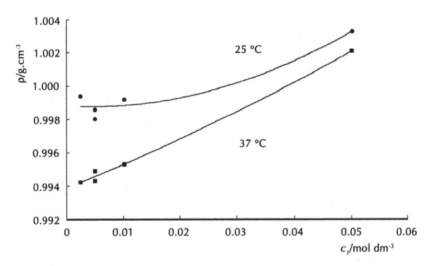

FIGURE 2 Density values for ternary aqueous solutions of $CaCl_2$ (1) and Caffeine (2) at 25°C and 37°C.

In addition, the density values for ternary aqueous solutions of caffeine and $CaCl_2$ were compared with those obtained for binary systems at the same temperatures and with the same technique, that is, for $CaCl_2$ [12] and caffeine [14] in aqueous solutions, respectively. Because significant deviations appear (<1.0%), we may conclude that solute-solute and solute-water interactions have been enhanced.

9.3.3 VISCOSITIES OF CACL$_2$ IN AQUEOUS SOLUTIONS OF CAFFEINE

Measured viscosity values, η, are collected in Table 4 for both the concentration and temperature ranges studied. These experimental values were fitted by using a least-squares regression method [Eqs. (7) and (8)]. The viscosity values show a similar parabolic behavior in the array of concentrations studied for both temperatures (Fig. 3). The viscosity values increase linearly in the array of concentrations studied.

$$\eta = 0.8953 + 0.2891\ c + 7.1653\ c^2 \tag{6}$$

$$\eta = 0.6908 + 1.7914\ c - 20.7380\ c^2 \tag{7}$$

As expected the temperature has influence in the behavior of this transport property. The viscosity values show a similar parabolic behavior in the array of concentrations studied for both temperatures.

TABLE 4 Viscosity values for ternary aqueous solutions of $CaCl_2$ (1) and caffeine (2) at 25°C and 37°C.

c_1 /(mol.dm^{-3})	c_2 /(mol.dm^{-3})	η/cp (25°C)	$10^4\sigma$/cp [a]	η/cp (37°C)	$10^4\sigma$/cp [a]
0.0050	0.0025	0.8972	0.8	0.6972	0.6
0.0025	0.0050	0.8954	2.1	0.6973	1.1
0.0050	0.0050	0.8965	4.9	0.6980	3.2
0.0102	0.0101	0.8996	3.8	0.7080	6.5
0.0500	0.0501	0.9277	0.6	0.7285	8.9

[a]σ represent the standard deviation of these measurements.

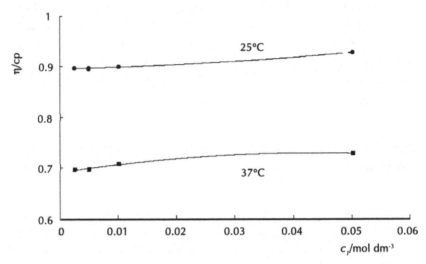

FIGURE 3 Viscosity values for ternary aqueous solutions of $CaCl_2$ (1) and Caffeine (2) at 25°C and 37°C.

Moreover, from the no negligible deviations between the viscosities values for ternary aqueous solutions of caffeine and $CaCl_2$, and those obtained for binary systems at the same temperatures and with the same technique (<1%), we may also conclude that caffeine affects the behavior of the viscosity of solutions of $CaCl_2$.

9.3.3 *AB INITIO* CALCULATIONS AND THE CA^{2+}/CAFFEINE INTERACTIONS

The strong interaction observed between Ca^{++} and caffeine leads us to look for some theoretical evidence that could support those results. A simple *ab initio* calculation of the electrostatic potential map of caffeine (Fig. 4) using a small 3-21G basis set showed as expected two important electronegative zones (A and B) on the carbonyl oxygen's. Those two regions show that a single molecule of caffeine can coordinate with two Ca^{2+} ions with equal probability.

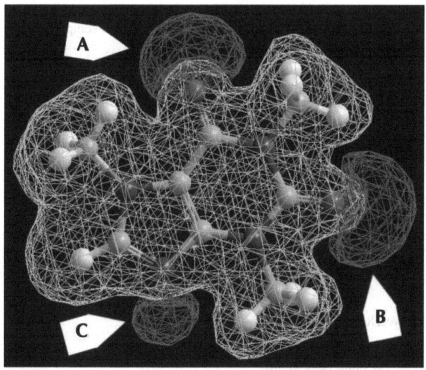

FIGURE 4 Calculated (*ab initio*/3-21G) electrostatic potential map of caffeine (represented in red are the electronegative regions and in green the electropositive regions).

However, the calculated electrostatic potential map of caffeine shows that, besides those expected carbonyl regions, another possibility for Ca^{2+} coordination in the exposed nitrogen (C). That extra coordination that may be understood as a supplemental point for the coordination of Ca^{2+} with caffeine and in some way contribute to the stronger than usual caffeine/Ca^{2+} coordination.

9.4 CONCLUSION

From our measured transport properties of the aqueous systems containing calcium chloride and caffeine (i.e., ternary diffusion measurements and viscosities), we can conclude that, for infinitesimal concentration, the solutes are not interacting. However, for finite concentrations, there is strong evidence of possible interactions (eventually as complexes 1:1), confirmed by our molecular mechanics calculations.

For example, from our cross coefficients D_{21} and D_{21} values different from zero at finite concentrations, and having in mind that main coefficients D_{11} and D_{22} are not identical to the binary diffusion coefficients of aqueous $CaCl_2$ and caffeine, we can conclude that the diffusion of $CaCl_2$ in aqueous solutions at both temperatures, 25°C and 37°C, may be affected by the eventual presence of new different species resulting from various equilibria.

Thus, our measured data for aqueous solutions of $CaCl_2$ and caffeine is important for fundamental reasons (helping to understand the nature of aqueous electrolyte structure) and, also, for practical applications, once they provide transport data necessary to model the diffusion for various chemical and pharmaceutical applications.

KEYWORDS

- *Ab initio* calculations
- Caffeine free species
- Ternary diffusion
- Theodolytes

REFERENCES

1. Nafisi, S.; Shamloo, D. S.; Mohajerani, N.; Omid, A. *J. Mol. Struct.*, **2002**, *608*, 1–7.
2. Nafisi , S.; Monajemi; M.; Ebrahimi. S.; *J. Mol. Struct.*, **2004**, *705*, 35–39.
3. Kolayli, S.; Ocak, M.; Küçük, M.; Abbasoğlu, R. *Food Chem.* **2004**, *84*, 383–388.
4. Ribeiro, A. C. F.; Barros, M. C. F.; Lobo, V. M. M.; *J. Chem. Eng Data* 2010, *55*(2), 897–900.
5. Ribeiro, A. C. F.; Gomes, J. C. S.; Santos, C. I. A. V., Lobo, V. M. M.; Esteso, M. A.; Leaist, D. G. *J Chem Eng Data*, **2011**, *56,* **4696–4699.**
6. Callendar, R.; Leaist, D.G. *J. Sol. Chem.*, **2006**, *35*, 353–379.
7. Barthel, J.; Feuerlein, F.; Neuder, R.; Wachter, R. *J. Solution Chem.* **1980**, *9*, 209–212.

8. Tyrrell, H. J. V; Harris, K. R. *Diffusion in Liquids* (2nd ed.), Butterworths: London, 1984.

9. Robinson, R.A.; Stokes. R. H. *Electrolyte Solutions* (2nd ed.), Butterworths: London, 1959.

10. Ribeiro, A. C. F.; Esteso, M. A.; Lobo, V. M. M.; Burrows, H. D.; Valente, A. J. M.; Santos, C. I. A. V.; Ascenso, O. S.; Cabral, A. C. F.; Veiga, F. J. B. *J. Chem. Eng. Data* **2008**, *53*, 755–759.

11. Lobo, V. M. M. *Handbook of Electrolyte Solutions*. Elsevier: Amsterdam, 1990.

12. Ribeiro, A. C. F.; Barros, M. C. F.; Teles, A. S. N.; Valente, A. J. M.; Lobo, V. M. M.; Sobral, A. J. F. N.; Esteso, M. A. *Electrochim. Acta.* **2008**, *54*, 192–196.

13. Leaist, D. G.; Hui, L. *J. Phys. Chem.* **1990**, *94*, 8741–8744.

14. Esteso, M. A.; Santos, C. I. A. V.; Teijeiro, C.; Ribeiro, A. C. F.; Lobo, V. M. M.; Burrows, H. D. Determination of Density and Viscosity Values for β-Cyclodextrin-Caffeine Aqueous Solutions at 298.15 K and 310.15 K. *J. Chem. Eng. Data*, 2008.

CHAPTER 10

SYNTHESIS OF BIOLOGICALLY AWAKE ANTIOXIDANTS IN REACTIONS OF ESTERIFICATION 2-(N-ACETYLAMID)-3-(3′,5′-DI-TERT. BUTYL-4′-HYDROXYPHENYL)-PROPIONIC ACID

G. E. ZAIKOV, A. A. VOLODKIN, L. N. KURKOVSKAJA,
N. M. EVTEEVA, S. M. LOMAKIN, L. J. GENDEL,
and E. J. PARSHINA

CONTENTS

10.1 Introduction 222
10.2 Experimental 222
10.3 Discussion of Results 224
10.4 Conclusion 228
Keywords 229
References 229

10.1   INTRODUCTION

Publications of last years testify to synthesis urgency biologically active materials with properties of antioxidants [1–4]. In development of the yielded direction perspective are synthesis and researches of properties esters   2-(N-acetylamid)-3-(3',5'-di-tert.butyl-4'-hydroxyphenyl)-propionic acid. Properties of esters 2-(N-acetylamid)-3-(3',5'-di-tert.butyl-4'-hydroxyphenyl)-propionic acid which antioxidant properties on a series of parameters are close to properties 4-methyl-2,6-di-tert.butylphenol was earlier investigated (an inhibition constant $\kappa_7 \cong 2.10^4$ l.mol^{-1}s^{-1}). Presence in a molecule of esters 2-(N-acetylamid)-3-(3',5'-di-tert.butyl-4'-hydroxyphenyl)-propionic acid acetylamid group leads to change of antioxidative properties, and on value of a constant of inhibition come nearer to efficacy connatural antioxidant – tocopherol ($\kappa_7 \cong 2.10^6$ l.mol–1.s^{-1}) [5].

On the basis of quantum-chemical calculations of frames methyl ester 3-(3', 5'-di-tert.butyl-4'-hydroxyphenyl)-propionic acid esters 2-(N-acetylamid)-3-(3',5'-di-tert.butyl-4'-hydroxyphenyl)-propionic acid it is positioned that distinctions in antioxidative parameters can be bound to geometry of communications between kernels of atoms and influence of an electronic field on transferring of one of electrons to the top occupied orbital.

10.2   EXPERIMENTAL

NMR spectrums ^{1}H wrote down on the device "Avance-500 Bruker" rather TMS. IR-spectra wrote down on a spectrometer PERKIN-ELMER 1725-X in crystals a method of diffusive reflectance. For LC used the device «Bruker LC-31». Biological researches made with erythrocytes of blood of white rats on a method [7]. Morphological changes erythrocytes under the influence of antioxidants in concentration 10^{-3}–10^{-5} mol·l^{-1} fixed a method of optical microscopy.

Quantum-chemical calculations are used with program «Mopac2009» [8]. Optimisation of molecule frames made with use of procedure EF on a method of Hartrii-Foka.

2-(N-acetylamid)-3-(3', 5'-di-tert.butyl-4'-hydroxyphenyl)-propi-onic acid (1) received on a method [6], m.p. 203–204°C. $SOCl_2$ marks «Serva» (Germany), refined fractionation, m.p.79°C.

Methyl ester 2-(N-acetylamid)-3-(3', 5'-di-tert.butyl-4'-hydroxyphenyl)-propionic acid (2) To solution compound **1** in 30 ml of MeOH have added of 3.35 g (0.01 mol) 5 ml $SOCl_2$, maintained at 20°C during 15 mines, solvent have evaporated, to residue have added 20 ml of water, have heat up to boil and filtrated. The residual on the filter was obtained . A yield of compound **2** 3.2 g (~92 %); m.p.148–149°C. Spectrum NMR ^{1}H (CDCl$_3$, δ, J/Hz) 1.44 (s, 18 H, ^{t}Bu); 2.01 (s 3 H, COCH$_3$); 3.06 (s, 1H, CH$_2$); 3.08 (s, 1H, CH$_2$) 3.78 (s, 3H. O-CH$_3$); 4.86 (m. 1H CH-CH$_2$);; 5.14 (s, 1 H, OH); 5.91 (br, 1 H, NH); 6.87 (s, 2H, Ar); IR-spectrum, ν / cm^{-1}:3367 (NHCO); br. 3187); 2960 (CH); 1739 (COO-); 1649 (CONH); 1213 (C-O-C). It is found %: C 68.66; H 8.74; N 3.77. $C_{20}H_{31}NO_4$. It is calculated %: C 68.74; H 8.94; N 3.85.

Ethyl ester 2-(N-acetylamid)-3-(3',5'-di-tert.butyl-4'-hydroxyphenyl)-propionic acid (3) are received similarly from 3.35 g (0.01 mol) compound **1**, 10 ml EtOH and 5 ml $SOCl_2$ Yield of **3** 3.5 g; m.p.135–136°C. Spectrum NMR ^{1}H (acetone d$_6$ δ, J/Hz):1.68 (t, 3H, CH$_3$CH$_2$, J = 7,1 Hz) 1.42 (s, 18 H, ^{t}Bu); 1.91 (s, 3 H, COCH$_3$); 2.02 (d.d., 1H$_a$, J=6.2 Hz); 3.11 (d.d. 1H$_b$, J=6.3 Hz); 4,09 (2H, CH$_3$CH$_2$, J=7,1 Hz) 4.60-4.64 (m, 1 H$_c$); 5.98 (s, 1 H, OH); 6.95 (s, 2H, Ar); 7.34 (d, 1 H, NH, J = 7.15 Hz). IR-spectrum, ν / cm^{-1}: 3354 (NHCOCH$_3$); 3189 (br., OH); 1732 (COOC$_2$H$_5$); 1647 (HNCO). It is found (%): C, 69.59; H, 9.13; N 3.94; $C_{21}H_{33}NO_4$. It is calculated (%): C, 69.40; H, 9.15; N, 3.85.

Propyl ester 2-(N-acetylamid)-3-(3',5'-di-tert.butyl-4'-hydroxyphenyl)-propionic acid (4) are received similarly from 3.35 g (0.01 mol) of compound **1**, 10 ml iso-PrOH and 5 ml $SOCl_2$.

Yield of **4** 3.4 g (90%), m.p.127–128°C. Spectrum NMR ^{1}H (DMSO d$_6$, δ, J/Hz): 0.99 (s. 3H, (CH$_3$)$_2$): 1.14 (s. 3H, (CH$_3$)$_2$); 1.36 (s, 18 H, ^{t}Bu); 1.83 (s, 3 H, COCH$_3$); 2.81 (s. 1H, CH-HCH); 2.83 (s.1 H, HC-HCH); 4.32 (m. 1H, CH-CH2); 4.76 (m, 1 H, CH- (CH$_3$)$_2$); 6.77 (s, 1 H, OH); 6.91 (s, 2H, Ar); 8.27 (d, 1 H, NH, J=7.45 Hz); IR-spectrum, ν / cm^{-1}: 3356 NHCO) br.3197; 2949 (CH); 1719 (COO-); 1640 (CONH); 1202 (C-O-C). It is found %: C 69.88; H 9.32; N 3.77. $C_{22}H_{35}NO_4$. It is calculated %: C 69.99; H 9.34; N 3.71.

n-Butyl   ester   2-(N-acetylamid)-3-(3',5'-di-tert.butyl-4'-hydroxyphenyl)-propionic acid (5) are received similarly from 3.35 g (0.01 mol) of compound **1**, 10 ml n-BuOH and 5 ml $SOCl_2$. Yield of **5** 3.6 g (~92 %); m.p. 115°C. Spectrum NMR ^1H ($CDCl_3$, δ, J/Hz): 0.92 (t. 3H, $CH_2CH_2CH_2CH_3$ J=7.35 Hz.); 1.37 (m. 2H, $CH_2CH_2CH_2CH_3$); 1.43 (s, 18 H, ^tBu); 2.0 (s, 3 H, $COCH_3$); 3.04 (d.d 1H, CH-HCH, J=5.2 Hz; $Jgem$ = 6.1 Hz),); 3.06 (d.d. 1H, CH-HCH J = 6.6 Hz; $Jgem$ = 6.1Hz); 4.07 (m. 2H. OCH_2); 4.85 (m, 1 H, CH); 5.13 (s, 1 H, OH); 5.91 (d, 1 H, NH J=7.6 Hz); 6.80 (s, 2H, Ar); IR-spectrum, ν / cm^{-1}: 3367 (NHCO); br.3187; 2960 (CH); 1739 (COO-); 1649 (CONH); 1213 (C-O-C). It is found %: C 70.68; H 9.42; N 3.47. $C_{22}H_{35}NO_4$. It is calculated %: C 70.55; H 9.52; N 3.58.

n-Nonyl   ester   2-(N-acetylamid)-3-(3',5'-di-tert.butyl-4'-hydroxyphenyl)-propionic acid (6) are received by gaining to solution of 3.35 g (0.01 mol) compound **1**, 0.1 ml n-nonyl alcohol in 15 ml Bu^tOH within of 5 ml $SOCl_2$ 30 minutes at 20°C After 6 h solvent have separated, the residual crystalizatied from acetone. Yield of **6** 4.1 g (88 %), m.p.104-106 °C.

Spectrum NMR ^1H ($CDCl_3$, δ, J/Hz): 0.90 (t. 3H, $CH_2CH_2CH_2CH_3$ J = 6.71 Hz.); 1.28 (m. 14 H, $CH_2CH_2CH_2CH_3$); 1.43 (s, 18 H, ^tBu); 2.01 (s, 3 H, $COCH_3$); 306 (d.d., 1H, HCHAr J = 4.75 Hz, $Jgem$ = 4.85 Hz); 3.08 (d.d., 1H, HCHAr, J = 3.95 Hz, $Jgem$ = 4.85 Hz); 4.11 (m, 2H, O-CH_2-).; 4.85 (m, 1 H, CH); 5.13 (s, 1 H, OH); 5.93 (br., 1 H, NH); 6.89 (s, 2H, Ar); IR-spectrum, ν / cm^{-1}: 3351 (NHCO); br. 3169 (OH); 2958 (CH); 1720 (COO-); 1645 (CONH); 1252 (C-O-C). It is found %: C 78.78; H 10.34; N 3.07. $C_{28}H_{47}NO_4$. It is calculated %: C 78.84; H 10.26; N 3.03.

Methyl ester 3-(3',5'-di-tert.butyl-4'-hydroxyphenyl)-propionoic acid (7). To solution of 4.88 g (0.01 mol) 2,6-di-tert.butylphenolate potassium in 4 ml DMSO at 115°C are added 2.5 ml (0.03 mol) methyl acrylate. At 3 h after refrigerating have neutralized 10 % HC1 and after crystalizated from hexane have received 5.16 g (88 %) compound **1**, m.p. 66°C (compare [9] 66°C).

10.3 DISCUSSION OF RESULTS

The most perspective is the esterification method in solution of alcohols (methanol, ethanol, butanol) in the conditions of catalysis $SOCl_2$ Yield of esters to 98%. Esterification reaction proceeds with *iso*-propanol. In solution Bu^{t}OH compound 1 with SOCl$_2$ did not interaction. This property is used for prepare synthesis of esters from higher alcohols (n-nonyl alcohol) (the schema 1). In inorganic acids (HCl, H$_2$SO$_4$) products of desalkylation are formed that corresponds to the data [10].

R = CH$_3$ (2), C$_2$H$_5$ (3)   C$_3$H$_7$, (4), C$_4$H$_9$ (5), (CH$_2$)$_8$-CH$_3$ (6)

The constitution of the received compounds 2–6 is confirmed by NMR ^{1}H spectrums. Identification IR the spectrums received by a method of reflectance in a solid phase, frequency of phenolic hydroxyl is in area of 3100 cm^{-1} that can be a consequence of intermolecular interaction of a proton group OH bunches with the functional group The similar effect is considered in work [11].

Results of inhibition of oxidation with antioxidant participation (τ) and coefficient of stopping of chain (f) at constant speed initiation (W_i=1.5.10^{-8} mol · l^{-1}s^{-1}) defined a method [12]. The specified parameters are bound by expression: f=τ · W_i · [InH]$^{-1}$, where InH – compounds 2–7. Specific reaction rate value (k_7) defined from dependence Δ [O$_2$] / [RH] =-k_2/k_7·ln (1–t/ τ), where k_2 – kinetic constant ROO · c isopropyl toluene (k_2 = 1.75 l·mol^{-1} · s^{-1}), [RH] – concentration of isopropyl toluene (7.18 mol·l^{-1}), Δ [O$_2$] – quantity absorbed oxygen (Fig. 1).

Results of measurements of kinetics of oxygen uptake by nature an antioxidant specify in dependence on frame that is expressed as in character of change of oxygen uptake, and value of coefficient of stopping of chain. From dates comparisons of dependences of oxygen uptake in the presence of antioxidants 2 and 7 follows that compound 2 (k_7 = 2·10^6 l·mol-1·s^{-1}) possesses higher antioxidantive properties in comparison with 7 (k$_7$=2·10^4 l·mol-1·s^{-1}). Results of calculations of constants of reactions peroxy radical from isopropyl toluene with compounds 2–6 and values of coefficient of stopping of chain are presented in Table 1.

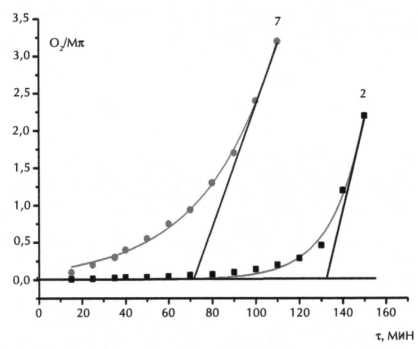

FIGURE 1 Kinetic dependences of oxygen uptake in reaction of the initiated oxidation of isopropyl toluene at 50 °C in the presence of compounds 2, 7 and azodiisobutyronitrile; Wi = $1.5 \cdot 10^{-8}$ mol.l^{-1}.c^{-1}. $[2]_o$ =$3.1 \cdot 10^{-5}$; $[7]_o$ =$3.6 \cdot 10^{-5}$ mol.l^{-1}.

TABLE 1 Constant of reactions peroxy radical from isopropyl toluene with antioxidants 2-6 and values of coefficient of stopping of chain.

Compound	f	k$_7$, 10^6 l·mol^{-1}·s^{-1}
2	2	2±0.2
3	2	2±0.2
4	1	1 ±0.1
5	2	2±0.2
6	2	1±0.1

Alkyl group in frames of compounds 2–6 affect on values к$_7$ and parameters f, however, this influence is insignificant in comparison with similar values of efficacy of compound 2. In received natural to laws in comparison with earlier known of the spatially-complicated phenols it is impossible to explain the found dependence of change of antioxidantive properties within the limits of the theory of a cleavage O–H communication.

Results of calculations of frames of compounds by **2** and **7** quantum-chemical method specify that in these compounds of value of energy of a homolytic cleavage (D $_{(OH)}$) are close enough (Table 2) whereas geometry of their molecules are various.

TABLE 2 Power parameters of compounds 2 and 7 according to calculation of frames by a method of Hartrii-Foka (RHF).

	$-E°_{InH}/E°_{In.}$ kJ·mol^{-1}	H° kJ·mol^{-1}	S° J·K·mol^{-1}	E_{total} kJ·mol^{-1}	D $_{(OH)}$ кJ ·mol^{-1}
2a	871/741 846/712	81.6	837	70.3	349.4
26	667/536	80.3	807	84.5	354.0
7		67.7	711	87.8	348.5

Calculation D $_{(OH)}$ is based on energy of phenols formations ($-E°_{InH}$), corresponding a radical

$-E°_{(In.)}$ and atom of hydrogen also it is expressed by an interrelation [13]:
D $_{(OH)}$ = $-E°_{(In.)}$ + E°$_{(H)}$ – ($-E°_{InH}$); E°$_{(H)}$ = 218.0 кJ·mol^{-1} (~52.1 kcal·mol^{-1})

In approach PM6 two frames 2a and 2b with minima energies of formations of E°$_{(InH)2a}$ =871 and E°$_{(InH)2b}$= 846 кJ·mol^{-1} (Fig. 2) have been calculated.

To geometry of these frames correspond to two structure, different on entropy (S° = 807 and 837 unit.), a free energy (E_{total}) and energy of a homolytic cleavage O-H communication D $_{(OH)}$ in an molecule.

It is known [14] that geometrical parameters and entropy of compound are interconnected and, the inspector, entropy can influence antioxidative properties, through changes of a free energy of compound. Areas with the raised electronic density that conducts to the mutual pushing away of electrons (steric effect) occur a consequence of such change.

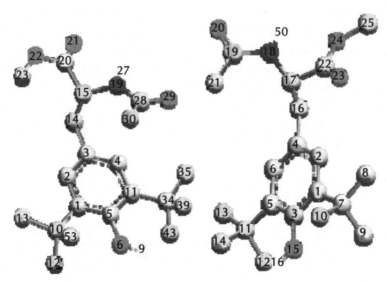

FIGURE 2   Rotomeres (2a and 2b) of compound 2.

In researches with use erythrocytes and antioxidants to made the confirmation that these enter interaction with a cellular membrane and incorporated in intra membranous space of a cage. Frame influence alkyl group on intercalation process is revealed. The presently data specifies in allocation of the most hydrophobic components of molecules of antioxidants in inside-it a monomolecular layer erythrocyte membranes where they display antioxidative actions.

10.4 CONCLUSION

Thus, it is possible to assume that properties esters 2-(N-acetylamid)-3-(3', 5'-di-tert.butyl-4'-hydroxyphenyl)-propionic acid are bound to geometry of communications between kernels of atoms in a molecule and possibility of electron transition with an antioxidant on the lowest vacant orbital of molekule peroxy radical.

KEYWORDS

- Desalkylation
- Isopropyl toluene
- Molekule peroxy radical
- Steric effect

REFERENCES

1. Dyubchenko, O. I.; Nikulina, V. V.; Terah, E. I.; Prosenko, A. E.; Grigoriev, I. A. *Izv. Akad. Nauk SSSR, Ser. Khim.*, Russian, **2007**, *14*, 1107.
2. Prosenko, A. E.; Skorobogatov, A. A.; Djubchenko, O. I.; Pinko, P. I.; Kandalintseva, N. V.; Shakirov, M. M. *Izv. Akad. Nauk SSSR, Ser. Khim.* **2007**, *12*, 1078.
3. Storozhok, N. M.; Perevozkina, M. G.; Nikiforov, G. A. *Izv. Akad. Nauk SSSR, Ser. Khim.* Russian, **2005**, *19*, 323.
4. Arefev, D. V.; Belostotskaja, I. S.; Voleva, V. B.; Domina, N. S.; Komissarova, N. L.; Sergeeva, O. J.; Hrustaleva, R. S. *Izv. Akad. Nauk SSSR, Ser. Khim.*, Russian, **2007**, *22*, 751.
5. Tsepalov, V. F. *Research of Synthetic and Connatural Antioxidants in vivo and in vitro.* Nauka: Moscow, Russian, 1992, p 16.
6. Volodkin, A. A.; Zaikov, G. E.; Evteeva, N. M. *Izv. Akad. Nauk SSSR, Ser. Khim.*, Russian, **2009**, *34*, 900.
7. Parshina, E. J.; Gendel, L.J.; Ruban, V. *Biophysics*, Russian, **2004**, *49*, 1094.
8. Stewart, J. J. P. *J. Mol. Mod.* **2007**, *13*, 1173
9. Volodkin, A. A.; Zaikov, G. E. *Izv. Akad. Nauk SSSR, Ser. Khim.*, Russian, **2002**, 2031.
10. Prosenko, A. E.; Skorobogatov, A. A.; Djubchenko, O. I.; Pinko, P. I.; Kandalintseva, N. V.; Shakirov, M. M. *Izv. Akad. Nauk SSSR, Ser. Khim.*, Russian, **2007**, 1078.
11. Huang, J.; Chen, Sh.; Guzel, A.; Yu, L. *J. Amer. Chem. Soc.*, **2006**, *128*, 11985.
12. Emanuel, N. M.; Denisov, E. T.; Majzus, Z. K. *Chain Reactions of Oxidation of Hydrocarbons in a Fluid Phase*; Nauka: Moscow, Russian, 1965.
13. Hursan, S. L. *Gomodesmichesky Method of Definition of Thermochemical Characteristics of Organic Compounds. Oxidation, Antioxidants*: Moscow, Russian 2010, p. 195.
14. Gribov, L. A. *Elements of the Quantum Theory of a Constitution and Properties of Moleculas;* Moscow, Russian 2010, p. 172.

CHAPTER 11

ATTEMPTING TO CONSIDER MECHANISM OF ORIGINATING OF DAMAGES OF CHROMOSOME IN THE DIFFERENT PHASES OF MITOTIC CYCLE

LARISSA I. WEISFELD

CONTENTS

11.1 Introduction 232
11.2 Materials and Methodology 237
11.3 Results and Discussion 238
11.4 Conclusion 247
Keywords 248
References 248

11.1   INTRODUCTION

Scientists attracted are drawn attention to the chemical compounds that cause heritable changes. There are remain unsolved a question of mechanism of their action on the chromosome. The action of many compounds is similar to ionizing radiation; it cause mutations of genes, disruptions of cell division and rearrangement of chromosomes.

It is known that ethylene imine and its derivatives alkylate DNA and proteins (see review [1]). They induce mutations as it was shown in various model objects (*Drosophila*, higher plants, fungi, bacteria, viruses and others) and breakage of the chromosome apparatus [2, 3]. Chemical compounds that cause mutations and breakage of chromosomes usually are called "chemical mutagens."

In the history of the study of chemical mutagens a large role belongs to works of J. A. Rapoport [4, 5]. He is the discoverer of the phenomenon of the chemical mutagenesis and presented theoretical substantiation of the phenomenon of chemical mutagenesis, its difference from the radiation [6]. He revealed out super-mutagens and discovered possibilities their application in the breeding of crops and in other areas of agricultural production. J.A. Rapoport organized synthesis of super-mutagens. Every year, he organized since 1959 All-Union conferences for scientists and breeders on chemical mutagenesis and its application in agriculture. In these meetings served in those years a good genetic school, especially for young breeders, who have been trained by the unscientific method of Lysenko. On the basis of the methodology developed by Rapoport and with the help of mutagens, which he distributed free of charge, breeders created the source material of crops and introduced new varieties. The number of investigations on the application of the chemical mutagen ethylene imine and on the mechanism of its action in winter wheat was conducted and is still conducting by N. S. Eiges [7, 8], which is the follower of J. A. Rapoport.

The most effective and available way to analyze the mutagenesis is a cytogenetic method, that is the study of rearrangements (aberrations) of chromosomes observed in dividing cells (mitosis, karyokinesis) and disorders in passing of the mitotic cycle.

Ionizing radiation damages the chromosomes immediately after irradiation at all stages of the mitotic cycle—"undelayed" effect. In the cells that came into mitosis from G_2 phase and S, arise aberrations of chromatid type arise and in cells that come in mitosis from phase G_1 aberrations of chromosome type (double bridges, paired fragments at anaphase) occur.

Under treatment by chemical mutagens of asynchronous cell cultures no rearrangements are detected during the first 2–4 h (depending on the duration of G_2 phase in different objects). This phenomenon is named "delayed" effect. Chromosomal rearrangements appear later, after the entry to mitosis of cells treated at the beginning or during DNA synthesis (phase S) or before phase S (named pre-synthetic phase G_1). Chromosome rearrangements are usually analyzed in the ana-telophases or meta-phases. A large number of chromosomes in many objects make the identification of chromosome aberrations in metaphase plates difficult. In this case fragments of chromosomes or broken bridges cannot be identified. In this case, scientists analyze the anaphase and early telophases (ana-telophase method).

For the estimation of environmental contamination the method of analyzes of ana-telophases is sufficient. The estimation of pollution on the meristem of birch in the industrial area of the city of Staryj Oskol is an example [9].

In Moscow in the 60–70-th years extensive cytogenetic studies of chemical and radiation mutagenesis were carried by N. P. Dubinin, his colleagues and followers. The work was begun in the laboratory of radiation genetics of Institute of biophysics of the Academy of Science of the USSR and was continued in the Institute of general genetics. A large number of articles and monographs about cytogenetic effects of ionizing radiation or chemical compounds were published [10]. N. P. Dubinin formulated the idea of the mechanism of action of chemical mutagens: mutagens cause potential changes in the chromosomes at all stages of the mitotic cycle that are realized in a number of cell generations. He called it "chain processes" in mutagenesis [11–13].

At the same time period the staff of the laboratory under the direction of B. N. Sidorov and N. N. Sokolov (Andreev V. S., Generalova M. V., Grinih L. I., Durymanova S. E., Kagramanyan R. G., Protopopova E. M., Shevchenko V. V., and others) were conducted extensive work on the induction of chromosomal damage, their localization in the chromosomes, their association with the phases of the mitotic cycle and DNA synthesis after treatment of seedlings by ethylene imine, tio-TEF, radiation and other mutagens on the model object *Crepis capillaris*. They studied chromosome aberrations in metaphase plates, which make it possible to take into account the polyploidy of metaphases.

In the early 60's, working with N.P. Dubinin, I investigated the cytogenetic effect of alkylating agent phosphemidum (lat. synonym phosphemid, phosphasin)—di-(ethylene imid)-pyrimidyl-2-amidophosphoric acid (Fig. 1).

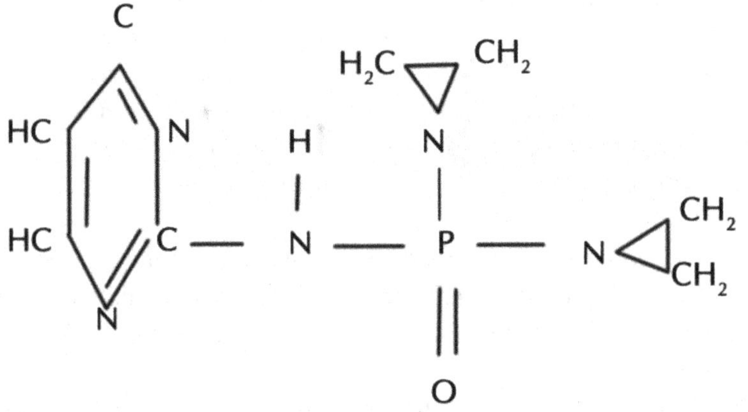

FIGURE 1   Phosphemidum (syn. phosphasin, phosphemid).

This compound is interesting because it consists of pyrimidine base and two molecules of ethylene imine. It was assumed that phosphemid will cause a lot of damages in DNA synthesis and thereby would inhibit tumor growth. The drug was synthesized in the laboratory of V. A. Chernov et al. [14], the All-Union Scientific Research Chemical-Pharmaceutical Institute, and was given to N. P. Dubinin for cytogenetic studies. The drug is a white crystalline powder, dissoluble in water and alcohol. Chernov et al. [14–15] has shown that phosphemid (then it was named phosphasin) inhibits the growth of tumors in laboratory animals, but at the same time it causes leukopenia, leukocytosis, and suppressed erythropoiesis. But currently the drug is used in medicine for the treatment of certain tumors such as leukemia, lymphoma. In the early work with phosphemid (1963–1964gg), it was important to approach the mechanism of its antineoplastic action. It was assumed that due to pyrimidine bases and ethylene imine groups of drug will directly affect DNA of the during a synthesis and thereby destroys actively dividable tumor cells.

The work was carried out on the primary culture of embryonic tissues — the mouse and human fibroblasts [16, 17]. We analyzed chromosomal aberrations in cells at the stages of anaphase or telophases (ana-telophases), after treatment culture by phosphemid in a concentration of $1 \times 10^{-4}M$.

It was shown that phosphemid delays entry of cells into mitosis and inhibits the mitotic activity of fibroblasts during the total period of the culture growth. The waves of fall and of rise of mitotic activity generally are repeating waves of mitotic activity in control group (without treatment), but at a lower level. The average frequency of mitoses made 54% from control. Lesions spindle was not observed, as a rule, but in individual cells were visible clumping of the chromosomes in the form of "stars." At the later stages of fixing the number of nuclei was decreased by 3–4 times, that indicates on cell death or the loss of their contact with surface of glass. At consideration of the types of chromosomal rearrangements mainly appears. At later stages of fixation after 26 h of the growth of culture double bridges (chromosomal type) were observed. They could have arisen as a result of doubling of the chromatid bridges in the second cycle (as a result of their passing to one of poles during the first division) or as a result of the breaking-fusion of chromosomes before synthesis DNA (phase G_1) with the subsequent doubling during the synthesis of DNA, in accordance with the thesis of N. P. Dubinin.

It was necessary to set connection with the phases of mitotic cycle. To do this, it would be desirable to find an object that, firstly, would be synchronized (or at least partially synchronized) with the terms of the phases of the mitotic cycle, and secondly, it would be convenient for the analysis of chromosomes in metaphase. To do this, it is desirable to find an object which, firstly, would be synchronized (or at least partially be synchronized) with phases of the mitotic cycle, and secondly, was suitable for the analysis of chromosomes in metaphase.

In this aspects *Crepis capillaris* (L.) Wallr. serves as ideal object. Mikhail S. Navashin studying chromosomes in the metaphase plates of seedlings published a number of works about taxonomy of genus *Crepis* [18]. He showed emergence of a large number of chromosome rearrangements when seeds became older.

At metaphase plate *Cr. capillaris* three pairs of homologous chromosomes are clearly identified (Fig. 2). Analysis of rearrangements here is certain.

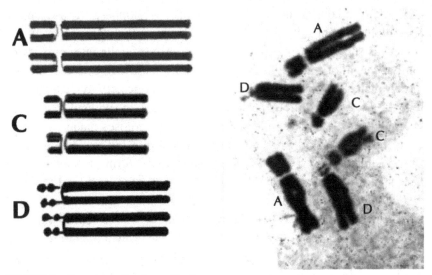

FIGURE 2   Karyotype *Crepis capillaris.*

According to the results of experiments with ionizing radiation it was assumed, that the cells of seeds (not seedlings!) *Cr. capillaris* are in phase G_1. The phase of G_1 in germinal cells of seeds of *Crepis* is heterogeneous, as it was shown in [19] the first marked cellular nuclei appear in seedlings through 10 h after the treatment of grain by thymidine H^3.

As far as the seeds germinate, the cells from the phase of S come in mitosis at first come and then from the phase of G_1. If the chemical mutagen interacts with the chromosome before the start of germination, i.e., before the start of DNA synthesis—in phase G_1, then in metaphases of seedling germinating rearrangements of chromosome type must appear. The data of our experiments, presented below, show that after the treatment of seeds in the $2n$-methaphases by phosphemid appear rearrangements of chromatid type.

11.2   MATERIALS AND METHODOLOGY

Air-dried seeds of *Crepis capillaris* of crop of 1967 were analyzed in 1968: April (8 months of storage after harvest), June (10 months of storage), and July (11 months of storage). In the control and in the experiments we used distilled water. The chromosome aberrations were analyzed in seedlings (meristem of root tip).

A certain amount of seeds (100 pieces) were treated in an aqueous solution phosphemid of concentrations: $1 \times 10^{-2}M$ (22.4 mg was dissolved in 10 ml water), $2 \times 10^{-2}M$ (22.4 mg was dissolved in 50 ml) or $2 \times 10^{-3}M$ (2.24 mg dissolved in 50 ml of water). The treatment was carried out at room temperature (19–21°C) for three h. Then the seeds were washed with running water for 45 min. The washed seeds were placed in Petri dishes on filter-paper moistened with a solution of colchicine (0.01%). Seeds were germinating in a thermostat at 25°C, but in July 1968 the temperature in the thermostat could reach 27°C because of hot weather. In parallel control experiments, the seeds were treated with aqua distillate.

In 24, 27 and 31 h after soaking of seeds we choose "arisings." The term *"arising"* refers to those seedlings that emerge after the beginning of soaking of dry seeds and have a size of less than 1 mm. These seedlings were selected in a Petri dish for further germination and subsequent fixation. In Russian the term "arising" is called "proklev."

For the analysis of chromosomes in metaphase plates, root tips (0.5–1 cm) were cut off with a razor and placed in the solution: 96% ethanol 3 parts + 1 part of glacial acetic acid. Solution was poured out through 3–4 h. Seedlings were washed 45 min in 70% alcohol. These seedlings were kept in 70% alcohol. We were preparing temporary pressure preparations: fixed root tips were stained acetous carmine and crushed in a solution of chloral hydrate between the slide and cover slip. We analyzed chromosome aberrations in metaphase plates of seedlings in the first division after treatment of seeds ($2n$-karyotype). In each seedling were all metaphases counted up. Intact seeds of harvest of 1966, 1067 and 1969 years served as control. The seedlings were fixed at different time intervals from 3 to 24 h.

Mitotic activity in the seedlings was determined by two criteria:
— the number of metaphase plates in relation to the number of nuclei in seedlings in the control and experiment (after seed treatment

phosphemid). We used the seedlings after 2, 4, 6, 8, 10 and 20 h after the "arising." In each seedling, we counted between 500 and 1,000 nuclei; in an each seedling counted from 500 to 1000 nuclei; — the number of seedlings with metaphase to all watched seedlings at all stages of fixation.

In all the experiments was estimated standard deviation from the mean.

11.3 RESULTS AND DISCUSSION

The natural level of mitotic activity-control was estimated by the criterion of the number of metaphases in seedlings: of the 18 000 counted nuclei metaphase plates were $2.83 \pm 0.12\%$. Table 1 shows that level of natural mitotic activity for 1966, 1967, and 1969, determined by criterion of number of metaphases in all studied roots, is in a range of 90–99%

On the average, over the years and months, were studied 217 seedlings. Of these, 94,9% contained metaphases. The frequency of metaphases with rearrangements in 1966 was in the range 0.24–0.44%. In 1967 there was a clear tendency to increase the frequency of rearrangements, as they are stored. Mitotic activity in 1967 was about 90%. The number of rearrangements of chromosomal type was various and small. For all the years we have discovered 50 such changes (Table 1). It should be noted that there is a tendency to increase their number during the aging of seeds (harvest 1967, 19 months of storage). The average level of metaphases with rearrangements in different years was less than 2%, rearrangements of chromosomal type in average of 0.55%; the largest number of them was in 1967.

As the count the nuclei in the seedlings after treatment by phosphemid at concentrations of 1×10^{-2} to 5×10^{-4} M of 68 500 counted nuclei were 743 metaphases, an average of $1.08 \pm 0.06\%$, thus phosphemid suppressed mitotic activity more than twice (see above 2.82% in control).

After exposure of seeds by phosphemid the average were analyzed 17 513 seedlings with metaphases for all concentrations and of fixations analyzed 17 513 seedlings with metaphases, among them were discover 2306 metaphases with rearrangements (13.17%). Alterations of chromosomal type made less than 1–0.113%. This magnitude is similar to the natural frequency. At a concentration of phosphemid of $2 \times 10^{-3}\,M$ mitotic activities was on the average 51.5% (Table 2), i.e., two times lower than in the controls (94.9%) (Table 1).

TABLE 1 The natural level of chromosome rearrangements in the $2n$-meristem cells of *Crepis capillaris* after soaking the seeds in water and germinating in 0.01% solution of colchicine. Harvest of 1966, 1967, 1969 years.

Year of harvest	Month, year studies	Number of investigating seedlings		Metaphases		Rearrangements of chromosomal type	
				Σ	With rearrangements, %±		
		Σ	With metaphases			Σ	%±
1966	XII, 1966	25	25	1125	0.44±0.198	0	–
	I, 1967	27	27	2120	0.24±0.105	2	–
	III, 1967	43	42	1692	0.24±0.118	0	–
Average:		95	94/99.0%	4937	0.31±0.008	2	0.04±0.029
1967	IV, 1968	49	48	1169	0.77±0.256	0	–
	VI, VII, 1968	20	14	350	1.14±0.569	3	0.86±0.492
	III, 1969	20	18	1316	2.96±0.468	24	1.73±0.369
Average:		89	80/89.9%	2835	1.62±0.214	27	0.81±0.155
1969	IV, 1970	33	32	1384	1.59±0.006	21	1.52±0.329
Total (1996–1969rr):		217	206(94.9 ±1.49%)	9156	1.05±0.134	50	0.55±0.080

Note: Some metaphase with numerous damages of the spindle and chromosomes are not taken into account.

Data of Table 2 show that at the weak concentration of preparation – 2 $\times 10^3$ with the increase of term of fixation in every "arising" mitotic activity increases, and frequency of alterations diminishes.

TABLE 2　Aberrations of chromosomes in 2n-meristem cells of *Crepis capillaris* after 24 and 27 h from the start of treatment of seeds in a solution of phosphemid $2 \cdot 10^{-3}M$ (2.24 mg, 50 ml of water).

Time, in h		Number of investigating seedlings			Metaphases		Rearrangements of chromosomal type
From soaking up to "arising"	From soaking up to fixation	Σ	With metaphases		Σ	With rearrangements, %±	
			Σ	%			
24	3	57	24	42.1	741	9.4±1,08	0
	6	56	32	57.1	1270	17.2±1.06	1
	8	54	36	66.7	1815	14.0±0.82	0
27	3	47	16	34.0	373	20.6±2.10	1
	5	19	12	63.2	313	15.3±2.04	0
Total		233	120	**51.5±3.28**	4512	**14.8±0.53**	**0.043±0.031%**

Fixation through 3–8 h after "arising". Seeds were germinating in 0.01% solution of colchicine. Harvest in 1967. The analysis in April 1968.

At max concentration – $1 \times 10^{-2} M$ (Table 3), we registered a large number of rearrangements – nearly 60% through 27 h after "arising" in term 24 h, of which 16.4% were mitosis with multiple rearrangements (it is marked with asterisk). On the average, considerably the number of strongly damaged metaphases is increasing up to 13.3±2.9%. Rearrangements of chromosomal type at this dilution also was not found.

Average frequency of mitotic activity of phosphemid in a concentration of $1 \times 10^{-2} M$ at the same quantity of seeds, as shown in Table 2, was lower almost twice—25% (Table 3). Average number of metaphases with rearrangements increased—nearly 22% versus 14.8%.

TABLE 3 Rearrangements of the chromosomes in the $2n$-meristem cells of *Crepis capillaris* seedlings at 24 and 27 h from the start of treatment of seeds in a solution of phosphemid $1 \cdot 10^{-2}M$ (22.4 mg, 10 ml water).

Time, in h		Number of investigated seedlings		Metaphases		
From soaking up to "arising"	From soaking up to fixation	Σ	With metaphases	Σ	With rearrangements	
					Σ	%±
24	3	21	6	148	25	16.9±3.09
	6	19	3	365	24(2*)	6.6±1.23
27	3	20	2	73	7(2*)	9.59±3.47
	24	12	7	185	110(18*)	59.46±3.62
Total		72	16/**25.0**±5.14%	771	166(22*)	**21.53**±2.85

Note. (*) the number of greatly damaged metaphases with multiple rearrangements. Fixation through 3–24 h after "arising". Seeds were germinating in 0.01% solution of colchicine. Harvest in 1967. The analysis in April 1968.

Thus the average level of metaphases with chromatid aberrations was increasing with increasing of dose of preparation.

Rearrangements of chromosomal type didn't arising in spite of increasing concentration of preparation.

During storage of untreated seeds and after subsequent treatment of seeds – namely in June and July of 1968 patterns of mitotic activity and frequencies of rearrangements were different, despite the use of the same concentration of phosphemid: $2 \times 10^{-2}M$ (Figs. 3a, b and 4a, b). Mitotic activity was higher than in April. With increasing time from "arising" to fixation after 12 h frequency of seedlings with mitoses close up to 90% and was slightly below control.

Through 24 and 27 h from the start of soaking of seeds the mitotic activity was steadily increasing to depending on time of fixation in each "arising". At a later "arising"—36 h (Fig. 3a) or 31 h (Fig. 4a) mitotic activity was fluctuating at the higher level as compared to the early stages of "arising" and terms of fixation from "arising." The data on Figs. 3 and 4 are statistically significant. The frequency of metaphases with rearrangements increased steadily in the fraction of the 24 h from "arising" from 3 to 12 h (Fig. 3b) and from 3 to 9 h (Fig. 4b).

Through 24 h after the soakage of seeds the frequency of metaphases with rearrangements was growing steadily with time of fixing after "arising," average frequency of rearrangements was higher through 27, 31 and 36 h (Figs. 3b and 4b).

At the 36 h "arising" frequency of metaphases with rearrangements increases with time from "arising" before fixation (Fig. 3b). Through 12 h after "arising" the level was almost 23% in 31 h "arising" (Fig. 4b), at 31 h "arising" (Fig. 4b), at fixations through 3 and 4 h frequency of metaphases with rearrangements was at the high level—17–18%.

Rearrangements of chromatid type were observed at all terms of fixation and only solitary rearrangements of the chromosomal type were found – an average of less than 1%, that is similar to the control level. This fact is important, because it shows, that the chemical preparation doesn't break chromosomes before synthesis DNA.

From the data of Figs. 3 and 4 at the first "arising" 24 h under any fixation and concentration the frequency of aberrations increased from 3 h and more. The level of rearrangements in the early fixations was lower than in the later ones, apparently due to the fact that the mutagen affected the chromosome during the shortest period of time.

It is possible to explain The facts reflected in Figs. 3 and 4 by (1) the preparation does not influence on chromosomes during phase G_1, but remains in seeds; (2) chromosomes are damaged by the preparation regardless of the stage of mitotic cycle, but this damage shows up only during synthesis of DNA (presence exclusively chromatid rearrangements). The number of rearrangements of chromosomes increases with the increase of terms of fixation; (3) existence of both factors is possible. Below the works of B. N. Sidorov and N. N. Sokolov will be described they convincingly explain this effect exceptionally by chromatid alterations.

Increased mitotic activity of the seedlings with increasing terms of fixation indicates the possibility of washout of the preparation from the cells during the growth of seedlings. Increasing number of rearrangements in "arisings" while increasing time before fixation and keeping of high level of metaphases with rearrangements in the seedlings shows to suggest that phosphemid penetrating in the seeds from the beginning of their treatment, is included in the metabolism of cells and stores there for a long time.

Phosphemid suppressed mitotic activity the stronger than concentration is higher (see Tables 2, and 3; Figs. 3, and 4). Phosphemid also interacts with proteins of spindle, as at higher concentrations of the preparation the number of heavily damaged mitoses significantly increased (see Table 3). Such metaphases often form a sort of a star at the center of the cell. A similar pattern sometimes was observed in the culture of fibroblasts. In addition to these disruptions, in some cells we have seen that all the chromosomes were fragmented.

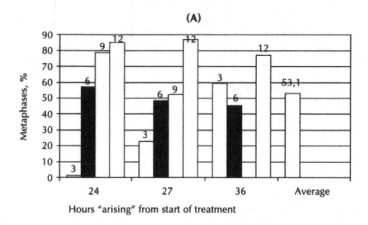

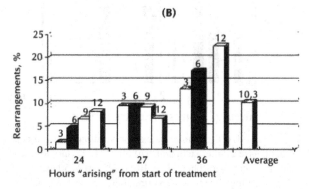

FIGURE 3 *Crepis capillaris*: the mitotic indexes in seedlings (*a*) and rearrangements of chromosomes in metaphase (*b*) after 24, 27 and 36 h from the start of seed soaking in the solution of phosphemid $2 \cdot 10^{-2}M$ (22,4*mg*, 50*ml* of water). Fixation: through 3–12 h after "arising." Seeds germinate in 0.01% colchicine. Seeds are of harvested in 1967. Analysis: June 1968.

(A)

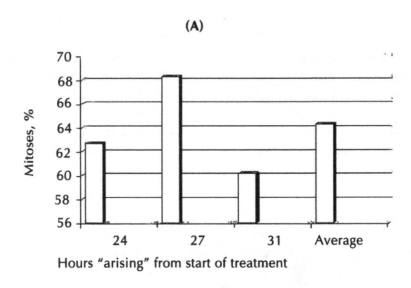

Hours "arising" from start of treatment

(B)

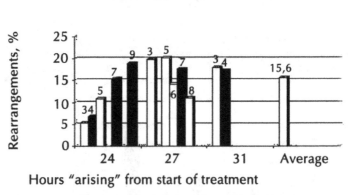

Hours "arising" from start of treatment

FIGURE 4 *Crepis capillaris*: the mitotic indexes in seedlings (*a*) and rearrangement of chromosomes in metaphase 2n-meristem cells (*b*) after 24, 27 and 36 h from the start of seed soaking in the solution of phosphemid $2 \cdot 10^{-2}M$ (22,4 mg, 50 ml of water). Fixation: through 3–9 h after seedling. Seeds were germinating in 0.01% colchicine. Seeds are harvest 1967. Analysis: July 1968.

Mitotic activity estimated as of the number of metaphases to the number of non-dividing nuclei through 2–10 h after "arising" in the control amounted to 2.11% at 16,000 nuclei. After treatment of the seeds the number of metaphases in the seedlings over the same period (2–10 h) was lower and depended on the concentration of phosphemid. At a high concentration of $1 \times 10^{-2}M$ at an average around 16,000 nuclei was 0.73% of metaphases, with preparation concentration $5 \times 10^{-3}M$ – 0.82% of metaphases around 26,500 nuclei, at a concentration of $5 \times 10^{-4}M$ observed 1.42% of metaphases around 21,000 nuclei. However, by 20 h after the start of treatment the frequency of mitoses increased both in control (3.40%) and at concentration of phosphemid $5 \times 10^{-3}M$ – 2.38%. These data also reflect the decline of mitotic activity with the increase of concentration of phosphemid and it's increasing with reduction of concentration of preparation.

Mutagen may be remains in the seeds in conjunction with other cellular proteins and that brakes transitions of phases of mitotic cycle, thus influencing on the chromosomes longer and therefore stronger and damaging the synthesis of larger number of loci of chromosomes.

In the 60th—70th year's outstanding scientists N. N. Sokolov, B. N. Sidorov [20–22] was conducting workings about effects of ethylene imine on seedlings *Cr. capillaris*. They cultivated seedlings in a solution of colchicine for five cell generations [20]. These researchers applied an original technique. They grew up the seedlings after effecting of ethylene imine, in solution colchicine and found new rearrangements of chromatid type in tetraploid and higher polyploidy cells, which were arising de novo, but not as result duplicating.

The authors explained this phenomenon by the fact that the mutagen is saved in the cells and there are new rearrangements. In a next article [21] the authors applied even more original technique. Seedlings were treating of ethylene imine. These seedlings were washed in running water within 2 h. From them after 48 h "thin gruel" were preparing. Intact seedlings of *Cr. capillaris* were treated by this thin gruel. In the seedlings treated by thin gruel rearrangements of chromatid type appeared.

The authors suggested that ethylene imine formed active secondary mutagens, connected with the components of the cell, including nucleic acids. In the third article of this series [21], the authors added ethylene imine to amino acids *in vitro*: glycine and histidine, to the hexamine (hexamethylenetetramine), to vitamins: thiamin (vitamin B_1), nicotinic acid.

Consequent treatment of seedlings by these preparations did not cause rearrangements of chromosomes. The frequency of them was at the level of control. Ethylene imine (concentration 0.05%) caused about 15% of rearrangements, while in mixture with thiamine caused significantly more rearrangements (+22.27%). Treatment by ethylene imine in mixture with nicotinic acid, glycine, and histidine showed even some protective effect. Treatment by ethylene imine together with adenine, guanine (derivative of purine), cytosine (derivative of pyrimidine) showed absence of effect or a small excess of rearrangements above the level of rearrangements of pure ethylene imine.

Significant excess of frequency of rearrangements was received under the influence of mixtures: uracil + ethylene imine gave an increase nearly 19%, thymine + ethylene imine caused nearly 61% of rearrangements (+37.25%). Guanine and cytosine in a mixture with ethylene imine showed an insignificant action. In the mixture with thio-TEPA (three ethylene imine groups), only thiamine gave significant excess of frequency of rearrangements (+25.97%) over control (ethylene imine). After treatment of seedlings by a mixture of ethylene imine with thymine, the excess of frequency of rearrangements was not found. All the rearrangements were chromatid type. The authors explain this phenomenon by the formation of secondary mutagens in cells. Thusly these experiments [20–22] showed that mutagens do not cause aberrations of chromosomes before or after phases of the mitotic cycle G_2, G_1 and causes only chromatid aberrations effecting on the chromosomes in the course of DNA synthesis. Modern microscopy suggests that mutagen penetrating into the cell is remaining in "space around a chromosome" [23] in bonding with proteins or DNA, but its effect is revealing when mutagen passes through phase of DNA synthesis and becomes visible in the form of chromosome rearrangements during mitosis. The same can be an evidence of the phenomenon of fragmentation of chromosomes, strong destruction of mitosis and the spindle during treatment by high doses of the alkylating agent (in our case by phosphemid) in culture of fibroblasts of human and mouse or in germinal cells of seeds *Cr. capillaris*. Perhaps the same principle is working in the course of "chain process" of Dubinin. Someday, 4D-microscopy will reveal the mechanism of interaction of chromosomes with the chemical mutagens. However, the question of mutagen preservation in cytoplasm appeared not such simple.

Sokolov and Sidorov's collaborators [24] were determinate that in seedlings of "old" seeds (3–7 years) ethylene imine caused rearrangements of chromatid and chromosomal types. Using a method of a thin gruel, in a combination with ethylene imine, on "old" seeds authors found in seedlings of rearrangements of chromosomal type. They also assumed that when aging seeds in cytoplasm having non-delayed effect is formed.

Understanding the mechanism of chemical mutagenesis in the present time is fundamentally important in view of the global contamination of the surrounding nature by different chemicals that damage the genetic structure of organisms.

11.4 CONCLUSION

1. We have shown above that in cultivated fibroblasts phosphemid is suppressing mitotic index and induces rearrangements of chromosomes.

2. In seedlings *Crepis capillaris* phosphemid also causes inhibition of the mitotic cycle. The average number of metaphases on the number of nuclei in seedlings after treatment phosphemid decreased twice.

3. Phosphemid after treatment of seeds *Cr. capillaris* causes rearrangements in the cells of seedlings, regardless of the age treated seeds, but depending on the concentrations of the drug. The greatest number of rearrangements occurs with the highest concentration of the preparation.

4. After treatment of the seeds *Cr. capillaris* by phosphemid only chromatid type rearrangements were found in the seedlings. The number of rearrangements of chromosomal type was at the level of controls or smaller. This means that the preparation works the same as other chemical mutagens, i.e., chromosomes break during DNA synthesis.

5. Phosphemid in treatment seeds *Cr. capillaris* showed heterogeneity of germinal cells in the seeds during the G_1 phase. The frequen-

cy of chromosomal rearrangements varies depending on the time between fixation and "arising."

6. At high concentrations $(1 \times 1^{-2}M)$ phosphemid caused the destruction of mitotic spindle and multiple fractures of the chromosomes.

The author is grateful to F. Tatarinov and A. Bukonin for their assistance in the preparation of the text.

KEYWORDS

- **Arisings**
- **Chemical mutagens**
- **Delayed/undelayed effect**
- *Drosophila*
- **Proklev**
- **Stars**

REFERENCES

1. Ross, W. *Biological Alkylating Agents/ Fundamental Chemistry and Design of Compounds for Selective Toxicity.* London. 1962. Translation edited by A. Ja. Berlin's: Moscow. – Medicine, 1964, p 259.

2. Loveless, A. *Genetic Allied Effects of Alkylating Agents*; London. 1966. Translation edited by N. P. Dubinin's: Nauka, Moscow, 1970, p 255.

3. Kihlman, B. A. *Aberrations Induced by Radiomimetic Compounds and their Relations to Radiation Induced Aberrations.* Radiation-Induced Chromosome Aberrations: NY, London, –1963, p 260.

4. Stroeva, O. G. *Josef Abramowitz Rapoport, 1912-1990*; Nauka: Moscow, p 215.

5. Rapoport, J. A. *Scientist, Warrior, Citizen. Essays, Memoirs, Materials.* Compiled by O.G. Stroeva; Nauka: Moscow, 2003, p 335.

6. Rapoport J. A. *Discovery of Chemical Mutagenesis. The Chosen Works.* Nauka: Moscow, 1993, p 304.

7. Eiges N. S. *Characteristic Features of Chemical Mutagenesis Method* I.A. Rapoport and its use in breeding of winter wheat. Penza, 2008, pp 14–17.

8. Eiges N. S.; Vaysfel'd L. I.; Volchenko G. A.; Volchenko S.G. *Some Aspects of the Securities Chemomutant Characters a Collection of Winter wheat and Characterization of These Characters.* International teleconference number 1: Basic Science and Practice. January 2010. Website: http://tele-conf.ru/nasledstvennyie-morfologicheskie-kle-

tochnyie-fakto/aspektyi-ispolzovaniya-tsennyih-hemomutantnyih-priznakov-kollekt-sii-ozimoy-pshenitsyi-i-ih-har-ka-chast-1.html
9. Kalaev, V. N.; Butorina, A.K.; Sheluhina, O. Yu. *Ecol. Genet.*, **2006**, *2*, 9–21.
10. Dubinina, L. G. *Structural Mutations in the Experiments with Crepis capillaries;* Moscow, Nauka,– 1978 p 187.
11. Dubinin, N. P. Genetika. **1966**, *7*, 3–20.
12. Dubinin, N. P. *Proc. USSR*, **1971**, *2*, 165–178.
13. Dubinin, N. P. *Proc. USSR*, **1971**, *3*, 333–344.
14. Chernov, V. A. *Cytotoxic Substances in Chemotherapy of Malignant Tumors*; Moscow, Medicine, 1964, p 320.
15. Chernov, V. A.; Grushina A. A.; Lytkina L.G. *Pharmacol. Toxicol.*, **1963**, *26*(1), 102–108.
16. Weisfeld, L. I. *Genetika,–* **1965**, *4*, 85–92.
17. Weisfeld, L. I. *Genetika*, **1968**, *7*, 119–125.
18. Navashin, M. S. Nauka: Moscow, 1985, 349.
19. Protopopova, E. M.; Shevchenko, V. V.; Generalova, M. V. *Genetika*, **1967**, *6*, 19–23.
20. Sidorov, B. N.; Sokolov, N. N.; Andreev, V.A. *Genetika*, **1965**, *1*, 121–122.
21. Andreev, V. S.; Sidorov, B. N.; Sokolov, N. N. *Genetika*, **1966**, *4*, 28–36.
22. Sidorov, B. N.; Sokolov, N. N.; Andreev, V. A. *Genetika*, **1966**, *7*, 124–133
23. Rubtsov, N. B. *Priroda*, **2007**, *8*, 1–8.
24. Protopopova, E. M.; Shevchenko, V. V.; Grigorjeva, G. A. *Rep. Aca. Sci. USSR.* **1969**, *186*(2), 464–467.

BIOLOGY OF DEVELOPMENT OF PHYTOPATHOGENIC FUNGI OF FUSARIUM LINK AND RESISTANCE OF CEREALS TO IT IN CLIMATIC CONDITIONS OF TYUMEN REGION

N. A. BOME, A. JA. BOME, and N. N. KOLOKOLOVA

CONTENTS

12.1 Introduction.. 252
Keywords 257
References... 257

12.1   INTRODUCTION

The differences were discovered between the varieties of spring wheat for resistance to phytopathogenic fungi of the genus *Fusarium* in laboratory seed germination and seedling morphometric parameters. The effect of temperature (20°C, 10°C, and 5°C) was studied on the rate of development of *Fusarium nivale Ces.* (beginning of active growth, sporulation, and diameter of the colony). Infection load in a field experiment decreased the selection and valuable features. Crop growing conditions in different areas of the Tyumen region formed unevenly. The climate is influenced by cold arctic air masses of the Arctic Ocean, the Asian continent, as well as dry winds blowing from Kazakhstan and Central Asia. The climate is typically continental, and all the climatic factors vary greatly over the years, both in tension and in development time, creating a variety of combinations. There are elements of the climate reminding the western region (dry summer periods), circumpolar areas (very short and cold growing season) and the deserts of the south (dry, oppressive weather from spring to fall) [1].

The agricultural areas of the Tyumen region are characterized by harsh cold winters, relatively short summers, short springs and autumns, a small frost-free period, and sharp changes in temperature during the year and even during the day.

One of the causes of the yield decrease in agricultural crops, including cereals, is the growth of infection of the most dangerous diseases. Plants suffer both from pathogens belonging to soil pathological complex (root rot, *Fusarium* wilt, etc.) and from air-spread infections (rust, Septoria, smut disease, powdery mildew, etc.) [2].

Phytopathogenic fungi of the genus *Fusarium* belong to the most dangerous among more than 350 species of toxigenic fungi known in agriculture [3]. It is shown that the contamination of seeds of spring wheat can occur both in the hidden and explicit form, and to a large extent it is determined by the varietal characteristics [4].

According to the results of our research 7 genera of pathogenic fungi were singled out in the microflora of grains of spring wheat, barley, and rye varieties of different eco-geographical origin and different years of harvest. Of these genera, *Alternaria, Helminthosporium, Trichothecium, Tilletia, Fusarium* are representatives of field microflora, while *Mucor* and

Penicillium belong to mold species. The fungal spores of the genus *Alternaria* dominated on most grains of all varieties.

Pathogens from the genera *Helminthosporium* and *Fusarium* of the most harmful type, causing root rot and spot, were detected.

Taking into account the fact that pathogens of the genus *Fusarium* are common enough in the cereals (both spring and winter forms) in Tyumen region, and can cause significant yield losses, we have conducted laboratory and field studies on the biology of this genus. The experiments included method of phytopathological analysis of seeds with the calculation of the disease index [5–9].

In our experiment conducted in the laboratory on four varieties of soft spring wheat, dependence was observed of seeds' ability to normal germination from their contamination by pathogens. Cultivar "Tyumenskaya 80" had the lowest laboratory germination of seeds among varieties—88.2% at the maximal level of infection. In less contaminated varieties ("Saratovskaya 57," "Comet," "Mir 11") indicators of seeds germination ability were higher: "Mir 11"—98.5%, "Comet"—99.0%, "Saratovskaya 57"—99.3%.

Fungi of *Fusarium* genus are the most common pathogens among soil infections. They cause disease of roots and root collar, which leads to the death of productive stems, and the empty spike of infested plants.

According to the index of the disease of affected seedling varieties "Comet," "Tyumenskaya 80," "Mir 11" were classified as middle susceptible (RB = 28.15–30.45%, 21.59–28.43%, 20.40–29.50%, respectively), and variety "Saratovskaya 57"—as low susceptible (RB = 12.48–18.48%) (Table 1).

TABLE 1 Evaluation of spring wheat samples on infectious background for resistance to *Fusarium sp.*

Variety	Phytopathological analysis of seeds		Benzimidazole method		The rots of root		The score
	P_6, %	score	P_6,%	score	P_6, %	score	
Tyumenskaya 80	21.59	2	24.00	4	25.55	2	8
Comet	28.15	1	35.43	1	21.87	3	5
Saratovskaya 57	18.48	4	34.35	2	21.11	4	10
Mir 11	20.40	3	24.27	3	28.89	1	7

Note: >40%—susceptible, 20–40%—middle susceptible, 10–20%—low susceptible.

A stronger root growth and inhibition of vegetative parts were observed in the study of morphometric parameters of the background of the infected seedlings (Table 2).

TABLE 2   Quantitative traits indicators of spring wheat samples on infectious background of *Fusarium sp.*

Variety	Option	Length of sprout X±m$_x$, cm	Length of roots X±m$_x$, cm
Comet	Control	23.52±0.61	14.09±0.43
	Experiment	21.75±0.81	20.39±0.81*
Saratovskaya 57	Control	22.09±0.95	18.53±0.33
	Experiment	20.83±0.71	19.30±0.96
Mir 11	Control	24.25±0.75	19.28±0.71
	Experiment	19.63±0.75*	20.72±0.62
Tyumenskaya 80	Control	26.02±0.60	18.96±0.57
	Experiment	19.81±0.67*	24.96±0.42*

Note: * denotes the differences were statistically significant at $P < 0.05$.

Cases of stimulation of the growth processes of infected plants are described [10]. Often this phenomenon is temporary and connected to the physiological characteristics of the pathogen. Intensive growth of roots and lagging behind of overground parts of the plants can probably be explained by the fact that the introduction of the pathogen into the roots of the plants leads to blockage of vascular system, disrupts the transport of water and dissolved substances, reduces the rate of photosynthesis, and therefore, produces a delay of plant development.

Comprehensive assessment on the grounds that characterizes the intensity of seed germination, seedling variability of quantitative traits and primary root system, have allowed to identify varieties of spring wheat "Tyumenskaya 80," "Mir 11," "Saratovskaya 57," as the most resistant to infection.

Productivity of winter crops forms is dependent upon a number of biotic (pathogens) and abiotic (temperature, rainfall, etc.) factors. Pathogenic fungi that cause disease play negative role in plant growth and development.

In particular the snow mold, which is caused by *Microdochium nivale* (Fr.) Samuels and I.C. Hallett (*Fusarium nivale* Ces. ex Berl. and Voglino), is dangerous. It is widely specialized facultative parasite, always present in the soil.

One of the factors that determine the development of the fungus is the temperature. *Fusarium nivale* Ces. begins to develop at 5°C, the optimal growth is observed at 11–17°C [11, 12]. In our laboratory studies performed with U.B. Trofimova [13] the effect of temperature on the rate of development of the fungus was studied. By cultivating the fungus in the oven at 20°C, 10°C and 5°C on potato glucose agar in Petri dishes in the three-fold repetition we determined the diameter of the colony and especially sporulation.

The lowest rate of growth of the fungus was recorded at 5°C. Beginning of the growth in this variation was observed on the 8th day after sowing. Fungal colonies reached the diameter of Petri dishes on 42th day, with sporulation recorded only on 56th day (Fig. 1).

Beginning of the growth	2 days	4 days	8 days
	20°C	10°C	5°C
Start sporulation ➡	↑ 6 days ↓	↑ 16 days ↓	↑ 56 days ↓
The diameter of the colony	87.7±2.03 mm	33.0±2.00 mm	90.0±0.00 mm

FIGURE 1   Effect of temperature on the development of fungus Fusarium nivale Ces.

At a temperature of 10°C on 4th day of the experiment diameter of the colony was equal to 12.5 mm, and after 28 days Petri dish was completely occupied by the fungus. In this variant sporulation happened much earlier— on the 16th day. The fastest growth of *Fusarium nivale* Ces. colony was observed at 20°C. The active beginning of growth was already evident on 2nd day, sporulation was observed on the 6th, while on 8th day of the experiment the colony's diameter reached 90 mm.

Development of snow mold is determined by weather conditions of the spring period and isn't observed every year, so any conclusion on plants' resistance can only be made in the years of strong manifestation of the disease.

One of the conditions to obtain reliable results in the determination of resistance is the creation of an artificial background ensuring optimal infection load. This background on the experimental site was created by application into soil of an aqueous suspension of spores and mycelia of pure 14-day culture of *Fusarium nivale* Ces. Infectious load was 10^6 conidia/ml of inoculum (500 ml/m^2 of soil). Infection was carried out in autumn in the phase of bushing out before snow cover. Estimate of snow mold infection of plants was carried out 10 days after snow melting in the early resumption of the growing season according to methodical guidelines of V. D. Kobylyansky [14], on a scale worked out by V. I. Andreev and O. Molchanova [15].

A study of infection *in vivo* and hard infectious background revealed that harmfulness of snow mold manifested in the reduction of such morphometric characteristics of winter rye as plant height, leaf area, and productivity traits. Decline of more than 50% was noted in leaf area per 1 m^2, number of grains per plant, grain weight per spike and plant. There was a strong development of disease on the infectious background, which resulted in lower yields compared to the control samples on average by 38.1%.

In the growing season, characterized by a long warm autumn, conditions favorably evolved for active growth of the pathogen. Effect of pathogen was aggravated by soil and air drought in spring and summer. In the experimental variant with infectious load inhibition of growth processes was observed, which manifested in significant reduction in breeding-valuable features to 26.21–67.70%.

To the group of resistant varieties of winter rye belonged "Chulpan," "Ilmen" "Iset" and "Supermalysh 2," wave to middling susceptible – "Voshod 1," susceptible – "8s-191 Rossianka x Getera," "Desnyanka x Imerig," "Tetra" and "Siberia."

KEYWORDS

- **Fusarium nivale Ces**
- **Facultative fungi**
- **Pathogens**
- **Abiotic**

REFERENCES

1. Ivanov, P. K. Spring wheat. Moscow, 1968; 551 pp.
2. Kosogorova, E. A. Protection of field and vegetable crops from diseases. Tyumen: Publishing House of the Tyumen State University, 2002; 244 p.
3. Kudayarova, R. Mitotoksiny, R. Problems and prospects of the development of innovation in agricultural production. All-Russian Scientific and Practical Conference of the XVII specialized exhibition "AgroComplex-2007." Ufa: Bashkir State Agrarian University, 2007; Part 2. 79 p.
4. Khairulin, R. M.; Kutluberdina, D. R. The prevalence of fungi of the genus *Fusarium* in grain of spring wheat in the southern forest of the Republic Bashkortostan Last number, **2008**, *12*, 32–36.
5. Naumova, N. A. Analysis of seeds to fungal and bacterial infection. L.: Kolos, 1970; 32 p.
6. Evaluation of crops for resistance to diseases in Siberia. Guidelines. Novosibirsk, 1981; 48 p.
7. Mikhailin, N. I. Comparative evaluation of methods for determining the severity of root rot of spring wheat. *Agricultural Biology*, **1983**. *4*, 95 p.
8. Guidance on the study of the stability of the grass to the agents of diseases of the conditions for non-chernozem zone of the RSFSR. L.: WALS, 1977; 60 p.
9. Zrazhevskaya, T. G. Determination of the resistance of wheat to common root rot. Mycology and phytopathology. **1979**, *13(3)*, 58 p.
10. Rodigin, M. N. General phytopathology. Moscow High School, 1978; 365 p.
11. Rubin, A. Crop physiology. IV. Leguminous plants. Perennial grasses. Cereals (rye, barley, oats, millet and buckwheat). Moscow: Moscow State University, 1970; 654 p.
12. Yakovlev, N. Phytopathology. Programmed instruction. Moscow: Kolos, 1992; 384 p.
13. Trofimova, U. B.; Bome, N. A. Parameters of snow mold damage and resistance of winter rye to illness. Journal of Plant Protection, St. Petersburg: Pushkin, **2006**, *1*, 33–36.
14. Kobylyansky, V. D. (Eds.) Guidelines for the study of the world collection of rye. L., WRI, 1981; 20 p.
15. Andreev, V.; Molchanov, O. Snow mold of winter grains (Methods of study and control measures). M. Niitekhim, 1987; 46 p.

CHAPTER 13

REALIZATION OF POTENTIAL POSSIBILITIES OF A GENOTYPE AT LEVEL OF PHENOTYPE

N. A. BOME, S. A. BEKUZAROVA, L. I. WEISFELD, and I. A. CHERKASHINA

CONTENTS

13.1 Introduction 260
13.2 Material and Methods 263
13.3 Results 264
13.4 Discussion 268
13.5 Conclusion 269
Keywords 269
References 270

13.1   INTRODUCTION

Change of climatic conditions and growth of anthropogenic pressure demand the search of new paths and ways for optimization of plant metabolism.

Plant genetic resources serve as the base for improvement of environment's quality and ensure food and biological resources security.

In recent years, we get more alarming information on growing "genetic erosion." According to current forecasts, by the middle of XXI century up to 60% of plant species diversity can be lost [1]. Loss of plant resources' components is considered as one of factors of biosphere's ecological crisis [2].

Research of ways of action of biologically active substances, which are used in agriculture and don't cause mutation, plays an active role in development of ways of increasing productivity and adaptive properties in cultural plants.

Para-aminobenzoic acid (PABA) is widely used to increase adaptive properties of cultural plants.

Non-heritable character of PABA's action was discovered by Josef A. Rapoport. He revealed PABA's properties, which induce reparation and can decrease harmful action of mutagenic-active environmental pollutants.

PABA was first discovered by Rapoport as a modifier of living organisms' individual development already in 1940s. Using development of a model object – *Drosophila* – as an example, he showed that PABA evoked positive changes of non-heritable nature (i.e., it's no mutagen) [3]. Rapoport [4] proposed a scheme of relations between genotype, ferments and phenotype: "genes → their heterocatalysis (substrate: RNA molecules) → mRNA → catalysis of mRNA (substrate: amino-acids) → ferments → their catalysis → phenotype." According to Rapoport's theory [4], PABA unites with ferments into complexes. He called this phenomenon "phenotypic activation." He came to the conclusion that PABA induced higher rate of catalysis of certain ferments in plants and influenced positively on early stages of plant development. As a result plants become more adaptive to casual changes of environment and reveal to larger extent the hereditary potential of organism. In the series of works fulfilled under Rapoport's leadership and published in 1989 [5], it was shown on different agricultural plants in model and production experiments that PABA influenced positively on the productivity-defining complex of traits and improved adaptive properties of plants to unfriendly variation of external conditions. Studies on use of PABA in agriculture are going on and further in depth. It was shown on different cultivars of grain plants that PABA influenced positively some or other components, defining the yield of grain, in different years in different way [6].

Experiments in the Republic of North Ossetia – Alania [7–12] showed that addition of PABA to mixture of several nutrients (irlit, leskenit, maize extract, ambrosia sap and others) and microelements gave positive effect. Treatment by PABA of pea and bunias seeds before sowing [7], additional fertilizing of clover plants [8] and seed plantings of legumes [9] stimulated development of plants in greater extent than mixtures of nutrients without PABA. During the treatment of potato tuber sprouts such components as ambrosia sap, leskenit and PABA were added to water melted from snow [10]. This method led to growth of potato yield and decreasing of Fusarium rot infection. Adding of PABA to nutrient solution during the treatment of winter *Triticale* [11] had positive influence on protein content in the green mass. Long-time treatment of cornel cuttings by PABA [12] increased their acclimation rate. PABA was also used for receiving virus-free potatoes [13].

It was shown in joint studies of N.M. Emanuel Institute of Biochemical Physics of RAS and North Caucasian Institute for Mountain and Foothill Agriculture that PABA increased potato yield after tuber treatment with subsequent enveloping in ash [14]. Treatment of vegetable seeds and seedlings by PABA solution in mixture with boric acid and permanganate potassium [15] increases resistance to diseases of young seedlings of carrot, beet, cucumber, tomato, as well as further yields of those plants. After PABA treatment of binary seed mixture of early cultivars of winter wheat and winter vetch [16] their productivity and quality of green forage grew after skewing the mixture in the period from wheat's exit to the tube and beginning of vetch's budding to forming of wheat grain of milky ripeness.

In the northern forest-steppe of Tyumen Region, where saline soils are found along with fertile ones, joint treatment of seeds of three cultivars of barley with low salinity resistance by salt solution and PABA (0.01% solution) led to growth of germs' salinity resistance irrespective of NaCl concentration [17]. Positive results were obtained after spraying of inflorescences of barley mother plants by PABA solutions before crossing. In several hybrid combinations hybrids exceeded control in length and width of flag leaf, number of leaves per plant and plant height [18]. Spraying of PABA solutions on inflorescences of four amaranth samples increased seed productivity, and the strongest effect were achieved by using 0.02% concentration [18].

During the experiments data was received on realization of cultural plants' adaptation mechanisms under PABA influence. For practical needs PABA treatment is effective to limit the negative influence of agro-ecological factors, change growth and development processes in plants and ensure the maximal possible productivity.

In the present research effects of PABA treatment were studied on seeds, non-threshed inflorescences and vegetating plants in conditions of Northern Caucasus and Tyumen Region.

Method of sowing ears was preliminarily tested during two years on Tyumen home base of All-Russian N.I. Vavilov Institute for Plant Growing on more than 100 winter wheat samples from world collection. Productivity of winter forms of grain plants depends upon several biotic (pathogenic microorganisms) and abiotic factors. Positive results were obtained on such characters as winter hardiness, plant's resistance to snowy mold (the disease mainly caused by *Microdochium nivale* (Fr.) Samuels and I.C. Hallett, also known as *Fusarium nivale* Ces. ex Berl. and Voglino).

To increase manifestation of selection-valuable traits on phenotypic level it was proposed to use PABA solution in acid according to the method [19]. The trial was held in 2009 – 2010 on the experimental base of North Caucasian Institute for Mountain and Foothill Agriculture, where perennial legumes served as precursors. Unlike most above-mentioned studies, where PABA was dissolved in hot water according to the method of Rapoport et al. [5], in this research PABA was dissolved in acetic acid.

In the conditions of climatic change reaction of plant organism is determined by its ability to adaptation. In most cases sharp changes of growing conditions of such important crop as spring wheat suppress its economic valuable characters—germination of seeds, acclimation rate, growth and development of plants, photosynthetic activity. Cultivation of spring crops in Western Siberia depends upon vegetation period, conditions of harvest period, droughts and other stress influences. Activation of early stages of plant development leads to shortening of development periods from shoots to ripening and saves plants from death. PABA was used as phenotypic activator in 2009–2012 on Tobolsk complex scientific station of Ural Branch of RAS. Data on influence of PABA on development of spring wheat *Triticum aestivum* (L.) is represented lower.

13.2 MATERIAL AND METHODS

In experiments with ear treatment crops have been chosen as material, which are widely in Northern Caucasus for production and selection. The tested samples included winter wheat harvested in 2009 (cultivars Ivina, Vassa, Bat'ko, Don 107, Kollega, Kalym), winter wheat – cultivar Bastion, destined for Northern Caucasian Region, and selection samples of several millet species, introduced to North Ossetia: Japanese millet (*Echinochloa frumentacea*), panic (*Setaria italica* Panicumitalicum) and Italian millet (*Setaria italica*). All millet samples were harvested in 2010.

PABA is a powder consisting of small crystals, which easily and totally dissolve in 3% solution of acetic acid without warming. One teaspoon of dry powder of PABA (10 g) was dissolved in a small volume (20–25 ml) of 3% acetic acid. The received mixture was dissolved in 1 liter of water from water supply system under room temperature. PABA 0.1% concentration was received. To receive 0.2% concentration one should dissolve 20 g of dry powder.

We soaked 10 non-threshed inflorescences of each sample of cereals with mature seeds in PABA solution (0.1 or 0.2%) for 2–2.5 or 3–4 h. After this treatment inflorescences were planted in the soil in the field. Untreated inflorescences, soaked in the water under room temperature for 3–4 h, were applied as control (control I). There was also control II: PABA was dissolved in hot water, and inflorescences were soaked in this solution for 2–2.5 h.

Seeds from threshed inflorescences were preliminarily germinated in Petri dishes. If the seeds were soaked in PABA dissolved in acetic acid, their shoots emerged 3–4 d earlier than shoots of seeds soaked in water. Mold microflora developed on shoots from threshed seeds soaked in aqueous solution of PABA. But if PABA was dissolved in acetic acid, shoots grew for a long time without mold.

Experiments were conducted with participation of F.T. Tsomartova (PG student of Gorsky State Agrarian University, Vladikavkaz), and Luschenko G.V. (PG student of North Caucasian Institute for Mountain and Foothill Agriculture, m. Mikhaylovskoye).

Several ways of PABA use were studied in experiments, held in To-bolsk complex scientific station: presowing treatment of seeds for 8 h, spraying of vegetating plants in different ontogeny phases, complex treatment (soaking of seeds and spraying of plants). PABA was dissolved in hot water according to the standard method. The soil of experimental plot situated in sub-taiga zone in Tobolsk District of Tyumen Region belongs to sod-podzol type, it is cultivated, without signs of erosion, alkalescent and has pH 7.7. PABA was used in two concentrations: 0.001% for seed treatment for 8 h and 0.0001% for spraying of plants. Vegetating plants were sprayed in different ontogeny phases (germination, bushing out, exit to the tube, earing, flowering). Complex treatment was also used: it included seed treatment before sowing and plant spraying in every new ontogeny phase. Size of plots was 1 square meter, the experiment had 4 repeats. Rate of spring wheat's sowing was 650 seeds per square meter. Time of sowing was each year determined by weather conditions.

13.3   RESULTS

In contrast with the majority of above-mentioned studies, where PABA was dissolved in hot water according to the method worked out by Rapoport et al. [3], in this research PABA was dissolved in acetic acid [19]. The mixture of PABA and acetic acid creates acid medium and enables to prevent several fungal diseases with simultaneous preservation of propagating samples' genotypes.

Concentration of PABA solution 0.1 or 0.2% in acetic acid was sufficient for penetration of substance into seed embryos during 2–6 h in different variants of the experiment. High concentrations of PABA (9–11%) inhibit plant growth and development [10].

Lower we show average results of experiments on inflorescence treatment. After treatment inflorescences were planted to the field without threshing. Sowing qualities of seeds and wintering of winter crops were defined. Variants of the experiment are represented in Table 1.

TABLE 1 Variants of experiments description with PABA treatment of inflorescences ways of treatment of winter crops and millet inflorescences with para-aminobenzoic acid dissolved in acetic acid.

№ exp.	Experiments
0	Control I without treatment: soaking of inflorescences in water for 3–4 h
1	Control II: soaking of inflorescences in PABA (0.1–0.2%), dissolved in hot water, 2–2.5 h
2	Soaking of inflorescences in PABA (0.1%), dissolved in acetic acid, 2–2.5 h
3	Soaking of inflorescences in PABA (0.2%), dissolved in acetic acid, 2–2.5 h
4	Soaking of inflorescences in PABA (0.2%). dissolved in acetic acid, 3–4 h
5	Soaking of inflorescences in PABA (0.2%), dissolved in acetic acid, 5–6 h

Note: Average data is represented for control II with PABA concentrations 0.1 and 0.2%, because difference between them has been insignificant.

Efficiency of treatment by PABA solutions was connected with treatment duration and solution concentration. Dynamics of phenotype's activation under PABA influence is graphically shown on the histogram (Fig. 1), representing average indicators of addition levels for germinating capacity, energy of germination and frost hardiness of winter crops.

The histogram reflects the tendency of average additions to grow in each variant. After increase of PABA concentration from 0.01 to 0.02% average additions also increased from the 2nd experiment to the 3rd one. Dependence upon treatment duration is also shown under PABA concentration 0.2% (difference between the 3rd and the 4th variants). After longer treatment by PABA (for 5–6 h) additions were lower than in experiments 1–3 due to suppression of all indicators. The best result was achieved in experiment 4 (treatment for 3–4 h) on all indicators.

One can judge about possible phenotypic correction of morphological traits by PABA treatment, if leaf cover indicators of spring wheat cultivars are measured. Changes of flag leaf square of spring wheat were studied in earing phase after seed treatment or spraying of plants by PABA solutions. Table 2 includes average data for 3 years.

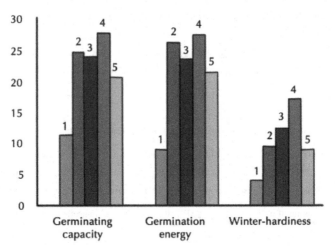

FIGURE 1 Comparison of averages (totally for all crops, %) of germinated capacity, energies of germination of seed for all crops and winter-hardness for winter wheat and winter barley at the different methods of treatment of ears of winter crops and panicles of kinds millet by para-aminobenzoic acid in the variants of experiments from 1 to 5 (see Table 1) by comparison to control of I (without treatment, in table 1 – a zero variant). Numbers on axis of ordinates are percent.

Square of flag leaf varied in control variants of tested cultivars of spring wheat from 28.5 sm² (Annet) to 39.03 sm² (Rix) (see Table 2).

TABLE 2 Variability of flag leaf square of spring wheat plants under PABA influence, sm² (2009–2011).

Cultivar	Variant	Min	Max	$X \pm S_x$
Annet	Control	–	–	28.50±2.60
	PABA	33.35±11.15	40.35±0.15	35.46±0.74*
Iren'	Control	–	–	30.30±0.69
	PABA	27.45±6.25	36.85±6.95	31.98±0.87
Ikar	Control	–	–	37.90±0.51
	PABA	37.65±6.25	42.45±13.45	39.33±0.51*
Rix	Control	–	–	39.03±0.86
	PABA	38.20±5.60	53.10±14.30	44.65±1.55*

Note: *differences between experimental variants and control are statistically significant.

Three cultivars had medium degree of trait's variation (CV = 11.48–15.81%), and one cultivar had high degree (CV=34.73%). Cultivars reacted in different ways on spraying of plants by PABA as it was proved by average data for three years of studies.

Significant increase of sheet plate under PABA influence was found in cultivars Annet, Ikar and Rix. Variability of the trait under PABA influence among cultivars was weak (CV=3.86–10.4%). In average during the years of research cultivar Rix formed the largest leaves after treatment of seeds and plants by PABA. In all variants PABA stimulated growth of blade irrespective of cultivar.

Concentration of pigments in leaf and efficiency of photosynthetic reactions have direct influence on productivity of plants. Quantity of pigments, their ratio in earing phase and their preservation in leaves of *Triticum aestivum* L. in the vegetation period serve as the most important characteristic of production process and photosynthetic activity of this crop.

The study of concentration of photosynthetic pigments in earing phase showed that cultivars Annet, Ikar, Rix, Iren' after treatment of plants by PABA had significant differences from control variant on quantity of chlorophyll a and b and carotenoids per unit of leaf square.

Treatment of spring wheat plants by PABA ensured significant increase of total pigment content in relation to control. This increase varied slightly among the cultivars: Annet – 53.4%, Ikar – 41.7%, Iren' – 50.0%, Rix – 44.7% (Table 3).

TABLE 3 Quantity and ratio of pigments in flag leaf of spring wheat plants under PABA influence, sm^2 (2009–2011).

Cultivar	Variant	Chlorophyll a, mg/sm^3	Chlorophyll b, mg/sm^3	Carotenoids, mg/sm^3	Total quantity of pigments, mg/sm^3
Annet	Control	90.1±0.31	51.4±0.23	76.7±0.24	218.2
	PABA	155.39±13.36*	71.23±3.10*	123.18±9.60*	334.78±21.83*
Ikar	Control	104.3±0.28	60.3±0.11	83.2±0.19	247.8
	PABA	155.66±13.25*	70.55±3.67*	123.05±9.66*	351.05±24.05*
Iren'	Control	92.1±0.11	53.5±0.08	78.7±0.09	224.3
	PABA	124.99±1.90*	78.51±4.44*	101.56±2.83*	338.3±32.31*
Rix	Control	90.2±0.11	50.1±0.09	73.1±0.10	213.4
	PABA	123.06±9.97*	83.74±10.68*	103.18±10.55*	308.85±3.17*

Note: *differences between experimental variants and control are statistically significant.

13.4 DISCUSSION

A significant role of PABA treatment is shown in this research for development of stable plant growing in Russia. PABA has positive influence on sowing qualities of seeds, morphological traits of plants and chemical composition of spring wheat's leaves. Increase of flaglist square and pigment content in it is especially important for spring crops due to short spring season in West Siberia, in particular in Tyumen Region. Ways of use and concentration of growth regulator are determined, which ensure fuller realization of ecological and biological potential of spring wheat cultivars in sub-taiga zone of Tyumen Region. Taking into account that the experiment has been held on soils with low content of micro- and macro-elements, humus and organic acids, one can assume that protective properties of the substance increase in extreme environmental conditions.

Study on PABA treatment of inflorescences was fulfilled due to the need of fastest selection of cultivars resistant to diverse environmental conditions. Annual cereals are needed on slope land of Northern Caucasus as high quality forage for cattle and as cover crops for intercropping of perennial legumes for creation of cultural pastures. Perennial grasses grow slowly in their first year. Taking into account their nitrogen-fixing capacity, legumes can already in the first year after sowing ensure additional fertilizing of perennial grasses with biological nitrogen. Annual millet crops are tolerant to drought and produce green mass and hay of good quality.

Advantages of dissolving PABA in acetic acid are shown. PABA improves significantly sowing quality of seeds in inflorescences planted in the field. The method of sowing non-threshed inflorescences is also important from the economic point of view. Sowing of inflorescences without threshing makes the breeder's work simpler both in scientific and organization relation: after selection of desired phenotypes in inflorescences propagation of best plants is fastened, and labor costs are cut. For the breeder it's also important that it becomes possible by sowing inflorescences to sort out the most promising families for the number of germinated seeds. It becomes possible to gather from each inflorescence more seeds for receiving posterity, periods of material's estimate and valuable samples' propagation are shortened. Usually germination of seeds and viability of plants decrease under unfavorable growing conditions, in case of plant's introduction to new environment, under high doses of mutagen or after remote hybridization of cultural plants with wild species. Treatment of inflorescences by PABA inhibits these negative trends [19]. The method is especially important for sowing recently harvested material of winter crops, when there is a lack of time for threshing ears and underworking grain. Such conditions occur often in the regions with short vegetation period, for example, in Tyumen Region.

13.5 CONCLUSION

1. A method of fastening of breeding process is proposed, which is based on making of nurseries not by seeds, but by ears (wheat) or panicles (millet) with germination of mature seeds in inflorescences.
2. This method includes soaking of inflorescences in para-aminobenzoic acid (concentrations 0.1 and 0.2%) dissolved in acetic acid.
3. Proposed method of treatment of seeds by PABA influences positively seed germination capacity and energy of germination as well as winter hardiness of plants.
4. High biological efficiency of complex and presowing treatment of spring wheat by para-aminobenzoic acid is shown, which reflects in growth of leaf cover square and increase of photosynthetic content of pigments in leaves. Cultivar's reaction on different ways of para-aminobenzoic acid treatment is defined.

The author is grateful to I. Smirnov and A. Bukonin for their assistance in the preparation of the text.

KEYWORDS

- **Flaglist square**
- **Genetic erosion**
- **Phenotypic activation**
- **Pigment content**

REFERENCES

1. Alexanyan, S. M. et al. Modern methods and international experience of preservation of wild plants' gene pool (on the example of wild fruits). Rauzin, E. G.; et al. *Almaty,* **2011,** 188 c. ISBN 978-601-7032-20-3.
2. United Nations Convention on Biological Diversity. Rio-de-Janeiro. 1992 (ratified by the Russian Federation in 1995).
3. Rapoport, I. A. *Proc. Inst. Cytol., Histol., Embryol. Acad. Sci.* **1948.** *2*(1), 3–128.
4. Rapoport, I. A. *Chemical Mutagens and Para-aminobenzoic Acid to Increase the Yield of Crops.* Nauka, Moscow, 1989, 3–37.
5. Rapoport, I. A.; *Editor of Chemical Mutagens and Para-aminobenzoic Acid in Enhancing the Productivity of Crops.* Nauka: Moscow. 1989, 253 ISBN 5-02-004014-2.
6. Eiges, N. S.; Weisfeld, L. I. *Regularities of the Action of Para-aminibenzoic Acid on the Cereals Cultures.* Chemical Mutagens and Problems of Agricultural Production. Nauka, Moscow, **1993,** 191–198. ISBN 5-02-004163-7.
7. Bekuzarova, S. A.; Abiyeva, T.S.; Tedeeva, A. A. *The Method of Pre-treatment of Seeds.* Patent 2270548. Published 27. 02. 2006. Bull. 67
8. Bekuzarova, S. A.; Farniyev, A.T.; Basiyeva, E. B.; Gaziyev, V. I.; Kalitseva, D.A. *The Method of Stimulation and Development of Clover Plants.* Patent 2416186. Published on 20.04.2011. Bull. 11.
9. Bekuzarova, S. A.; Shechedrina, D. I.; Farniyev, A. T.; Pliyev, M. A. *Method of Additional Fertilizing of Leguminous Grasses.* Patent 2282342. Published on 27.08.2006. Bull. 24.
10. Ikaev, B. V.; Marzoev, A. I.; Bekuzarova, S. A.; Basayev, I. B.; Bolieva, Z. A.; Kizinov, F. I. *The Method of Treatment of Pre-Plant Shoots of Potato Tubers.* Patent 2385558. Published on 10.04.2010. Bull. 10.
11. Bekuzarova, S. A.; Antonov, O. V.; Fedorov, A. K. *Method of Increase of Content of Protein In Green Mass of Winter Triticale.* Patent 2212777. Published on 27.09.2003.
12. Cabolov, P. H.; Bekuzarova, S. A.; Tigiyeva, I. F.; Tadtayeva, E. A.; Eiges, N. S. *The Method of Reproduction of Dogwood Drafts.* Patent 2294619. – Published on 10.03.2007. Bull. 7.
13. Shcherbinin, A. N.; Soldatova, T. B. *The Nutrient Medium for Micropropagation of Potato.* Patent 2228354. - Published on 10.05.2004.
14. Eiges, N. S.; Weisfel'd, L. I.; Volchenko, G. A.; Bekuzarova, S. A. *Method of Pre-Treatment of Tubers of Potatoes.* Patent 2202701. Published on 02.10.2007.
15. Eiges, N. S.; Weissfel'd, L. I.; Volchenko, G. A.; Bekuzarova, S. A. *The Method of Pre-treatment of Seeds and Seedlings of Vegetable Crops.* Patent 2200392. Published 20. 03. 2003. Bull. 8.
16. Eiges, N. S.; Weissfel'd, L. I.; Volchenko, G. A.; Bekuzarova, S. A.; Pliyev, M. A.; Hadarceva, M. V. *The Method of Receiving of Feeds in Green Conveyer.* Patent 2330410. Published 10. 08. 2008. Bull. 22.
17. Bome, N. A.; Govorukhina, A. A. *The Effectiveness of the Influence of Para-aminobenzoic Acid on the Ontogeny of Plants Under Stress.* **Bulletin of** Tyumen State University. Tyumen: Tyumen State University Publishing House, 1998, Vol. 2, pp 176–182.

18. Bome, N. A.; Bome, A. Ja.; Belozerova, A. A. *Stability of Crop Plants to Adverse Environmental Factors*. Monograph. Tyumen: Tyumen State University Publishing House, 2007, p. 192.
19. Bekuzarova, S. A.; Bome, N. A.; Weisfeld, L. I.; Tsomartova, F. T.; Lushchtnko, G. V. *Method of Presowing of Processings of Seeds of Selection Samples of Grain Crops*. Patent 2461185. Published on 20.09.2012. Bull. 26.

CHAPTER 14

COMMENTARY: COOPERATION OF HIGH SCHOOLS AND SCIENTIFIC INSTITUTIONS ON THE WAY OF EDUCATION OF SCIENTIFIC SHOTS TO MODERN CONDITIONS

N. I. KOZLOWA, A. E. RYBALKO, K. P. SKIPINA, and L. G. KHARUTA

CONTENTS

14.1 Introduction 274
Keywords 278
References 278

14.1 INTRODUCTION

The deep shocks connected with serious intervention in the unique nature of the Sochi Black Sea Coast, for long years will have negative consequences. In the created conditions the requirement for the ecologists capable competently to solve most different problem at creation of projects on restoration of ecologically safe environment, necessary for successful development of tourism and restoration of the resort industry raises. It is necessary to find ecologically comprehensible technical decisions connected with an active urbanization of territory, to keep the sites of the National park which has been not mentioned by creation of Olympic objects, to restore and create anew places of dwelling rare and vanishing species of representatives of flora and fauna. Experts IOC and our ecologists testify that in the conditions of the rough building limited to control terms of delivery in operation of objects, is very difficult to watch observance of positions ecologically a sustainable development of the Sochi region.

The requirement for the decision of the arisen environmental problems of region considerably will increase after end of all actions connected by carrying out Olimpiad-2014 when the period of search of the measures connected with transformation of region in vacation spot and tourism of world value will inevitably come, and also with development of the Sochi region as all-the-year-round resort. The success of these measures will be substantially provided by participation in the planned actions of highly skilled ecologists not only on a design stage, but also at realisation of the best in every respect projects, which should combine economic and ecological expediency. Only such approach will lead to occurrence of successful vacation spots, routes of ecological tourism, biodiversity preservation in region of Games.

In the developed conditions of natural increase of requirement of ecological shots the Sochi institute of RPFU, expanding preparation of experts of a biological direction, begins training on two new specialty having an ecological orientation – "Biology" and "Ecology and wildlife management".

The new status of institute (from 2011 year Sochi branch RPFU is transformed to the Sochi institute RPFU) gives possibility to raise vocational training level. Now the decision on creation of the scientifically-educational center (SEC), carrying out carrying out of researches and a professional training of the higher scientific qualification that will allows to carry out scientific researches and retraining of experts of the directions claimed in region is accepted. All scientific institutions with which the institute had creatively productive relations will take part in center work.

The major qualifying characteristics of the scientifically-educational center are high scientific level of carried out researches, high productivity of preparation of scientific shots, participation in preparation of students on a scientific profile of the scientifically-educational center, use of results of scientific researches in educational space of the developing biological direction connected with opening of new specialty—Biology, Ecology and wildlife management, Veterinary science and Veterinary and sanitary examination.

Within the limits of SEC implementation of possibility of improvement of professional skill of experts, in perspective area of a science—biotechnologies is planned. Researches on preservation of rare plants on which the Sochi Black Sea Coast is the most capacious enclave of Russia and to creation of possibility of use of these valuable plants in practice of the most various areas of plant growing on the basis of methods of biotechnology of new generation will be strengthened. The cellular engineering in a combination to modern methods of plant virology is an effective basis of introduction in culture of rare kinds, and also plants perspective for use in decorative gardening.

Ongoing research of students and teachers devoted to improving the educational process, the inclusion of biotechnology and to attract resources, academic institutions with which the Institute cooperates (1, 2, 4, 8, 15), improving the legal protection of the environment (3, 14). A number of papers devoted to the use of biotechnology in breeding of many rare plants for conservation and utilization in the economy (7, 9–13, 15, 16).

Scientific interests of teachers of chair the physiology directed on working out of methods of introduction in culture of disappearing representatives of local flora and their accelerated микроклонального of reproduction, along with search of optimum ways of increase of productivity of educational process, are the integral making preparation of the qualified experts. In laboratory of physiology of plants graduates of chair are prepared for the vigorous activity under the successful decision of similar problems under condition of development of a corresponding infrastructure. Therefore within the limits of created SEC there is a real possibility of participation of students in scientifically-practical activities on the accelerated reception of the improved landing material necessary for increasing requirement of the developing infrastructure of landscape gardening.

Themes of the degree works executed in laboratory of physiology of plants:

- Application of methods of biotechnology in preservation of a biodiversity of hand bells
  The Caucasian biospheric reserve;
- Preservation of a specific variety Campanulaceae the Western Caucasus Biotechnology methods
- Influence of plant growth regulator on growth and development of plants Gerbera the multiplied
- Features of micropropgation of a carnation acantolimonoides (Vanishing species)
- Reproduction conditions zephyrantes in culture in vitro.
- Micropropagation of lilies.
- High-quality features of reproduction of tulips methods of culture of fabrics.
- Studying of conditions of micropropagation of a potato of a cultivar Lugovsky.
- Working out of modes of regeneration virus free plants of Lisiantus (*Eustoma grandiflorum*) in meristem culture.
- Determination of the primary structure and characteristics of the new plant lipid transfer protein dill Anethum geaveolens.
- Assessment of the regeneration potential of different varieties of durum and soft wheat.
- Biotechnological process for the preparation of totally labeled with stable isotopes of lipid-transporting protein of lentils.
- Evergreen oaks Sukhumi subtropical arboretum.

Has received development the scientific direction connected with researches in area and problems of protection of an environment, in creation of model of the favorable environment in vacation spots and dwellings. Participation of students in these researches has come to the end with protection of degree works on problems of influence of anthropogenic changes of environment on health and social potential of the population of the Black Sea coast of Caucasus.

Results of scientific researches of students and teachers will be used for work on prospects of development of a sanatorium complex of Sochi which practical value is difficult for overestimating. On their basis it is necessary to define criteria of preservation and optimization of ecosystem, defining recreational possibilities of an environment. It can serve as a starting point on a way of maintenance steady and long-term development of territory of the Sochi region of the Black Sea coast of Krasnodar territory, especially with toughening of ecological requirements in connection with the forthcoming Winter Olympic Games in Sochi in 2014.

The themes of degree works executed under the guidance of science officers of National park:
- Ecological estimation of Navaginsky range of a firm field waste of Sochi;
- Dynamics of distribution of a HIV-infection in G. Sochi;
- Influence of anthropogenous loadings on a condition of water pool Red Glades;
- Receptions of preservation of the Nature sanctuary the Site from the sandy seaside
  "Vegetation" between bases of rest "Chernomorets" and "Energy" in Imeretinsky lowland;
- Birds of the Sochi Black Sea Coast;
- Birds of Imeretinsky lowland.

Thus, active participation of students of chair in expeditions under the account of a livestock of wild animals in places of dwelling and a condition of flora of the Sochi Black Sea Coast becomes a basis for the future teamwork of the Sochi institute with the Sochi national park and the Caucasian biospheric reserve within the limits of SEC.

Creative cooperation of the Sochi institute and integration of educational resources of chair of physiology with Institute medical приматологии the Russian Academy of Medical Science on problems of ecological physiology also began with the moment of opening of chair in 2000. The works of students connected with ecology of the person, and their direct participation in the scientific program of laboratories of Institute medical приматологии the Russian Academy of Medical Science in the field of virology, microbiology, immunology, studying of behavior of prima-cies have scientific and practical value. Participation of students under the guidance of visible scientists in priority directions of a medical science has allowed giving a scientific basis of a choice of the future trade to young scientists.

Now by request of Olympic committee the project on creation of an ecological track in places of growth of wild orchids of the Sochi Black Sea Coast is developed and there are begun works on its registration. The proj-ect of preservation rare and vanishing species of plants within the limits of the project on biodiversity preservation in region of Olympic games 2014 is submitted to consideration.

Thus, all above-stated opens prospects of development of activity SOC of the Sochi institute and reception of positive results of its activity on education of scientific shots in modern conditions of cooperation of high schools and scientific institutions.

KEYWORDS

- **Decorative gardening**
- **Landscape gardening**
- **Population**
- **Sochi Black Sea Coast**

REFERENCES

1. Skipina, K. P.; Rybalko, A. E. Training of personnel for introduction of biotechnologi-cal methods of preservation of a biodiversity in the suburb of Sochi. Proceeding The Moscow international scientific and practical conference (Moscow, on March, 15–17th, 2010) M: Joint-Stock Company "Ekspo-biochim-tehnologies," D.I. Mendeleyev University of Chemistry and Technology of Russia, 2010; p. 374.
2. Kozlowa, N. I.; Rybalko, A. E.; Skipina, K. P.; Kharuta, L. G. Biotechnological educa-tion problems in subdivision of Sochi institute Russian university of People's Friend-ship in view of preparation to Olympic games-2014. Proceeding of the Moscow in-ternational scientific and practical conference (Moscow, on March, 21–23st, 2011) M:

Joint-Stock Company "Ekspo-biochim-tehnologies," D.I. Mendeleyev University of Chemistry and Technology of Russia, 2011; p. 372.

3. Rybalko, A. A. Features of ecolaw training under implementation of biotechnological methods for preservation of biological diversity in Sochi Black Sea region. Ibid. p. 373.

4. Rybalko, A. E.; Tkachenko, V. P.; Rybalko, A. A. Biotechnology for landscape construction in subtropical zone – element for training of skilled personnel for landscape construction. Ibid. p. 371.

5. Bogdanov, I. V.; Finkina, E. I.; Balandin, S. V.; Rybalko, A. E.; Ovchinnikova, T. V. Biotechnological uniform stable isotope labeling of lentil lipid transfer protein. Ibid.

6. Pavlova, L. E.; Melnikova, D. N.; Finkina, E. I.; Balandin, S. V.; Rybalko, A. E.; Ovchinnikova, T. V. New lipidtransporting fiber from fruits *Citrus natsudaidai.* Ibid.

7. Rybalko, A. A.; Titova, S. M.; Rybalko, A. E.; Bogatyreva, S. N.; Kharuta, L. G. Micropropagation in Vitro of Plants of Family Gentiana and others Endemic of Plants of Northern Caucasus. Biotechnology and Ecology of Big Cities. Sergey D. Varfolomeev, G. E. Zaikov, L. P. Krylova, eds. Nova Science Publishers, Inc. 2011; pp. 61–69.

8. Kozlowa, N. I.; Rybalko, A. E.; Skipina, K. P.; Kharuta, L. G. Training of Personnel for Instillation of Biotechnological methods of Biovariety Conservation in the Suburbs of Sochi. Ibid. p. 145–150.

9. Maevskij, S. M.; Gubaz, S. L.; Rybalko, A. A. Introduction in culture in vitro three herbs of the North Caucasus (Mentha longifolia L., Origanum vulgare L., Thymus vulgaris L.). Proceeding of the Moscow international scientific and practical conference "Pharmaceutical and Medical Biotechnology" (Moscow, on March, 20–22th, 2012) M: Joint-Stock Company «Expo-biohim-technologies», D.I. Mendeleyev University of Chemistry and Technology of Russia, D.I. Mendeleyev University of Chemistry and Technology of Russia, 2012; pp. 456–457.

10. Gurchenkova, Y. A.; Rybalko, A. E.; Kharuta, L. G. Micropropagation of *Hypericum* for medicinal and ornamental purposes. Ibid. pp. 456–457.

11. Matskiv, A. O.; Arakeljan, M. A.; Rybalko, A. E. Introductions in culture in vitro rare bulbous plants of the Sochi Black Sea Coast (Scilla, Muscari, Galanthus) Ibid. p. 455

12. Urevich, I. A.; Petikjan, E. V.; Rybalko, A. E. Eustoma – new raw materials for pharmaceutics (eustomoside, eustoside and eustomorusside). Ibid. pp. 452–453.

13. Averyanova, E. A.; Rybalko, A. E.; Skipina, K. P.; Kharuta, L. G. Wild-growing kolchidsky orchids and their preservation as objects of education and producers of medicinal substances. Ibid. p. 453.

14. Rybalko A. A. New approaches to teaching of the ecological right in connection with an ecological condition of a biodiversity of the Sochi Black Sea Coast. Ibid. p. 436–437.

15. Kozlowa, N. I.; Rybalko, A. E.; Skipina, K. P.; Kharuta, L. G. Cooperation of high schools and scientific institutions on a way of education of scientific shots to modern conditions. Ibid. p. 435–436.

15. Shevlyakova, L. A.; Orlova, G. L.; Rybalko, A. A. Study of introduction in culture in vitro rare pharmaceutical plant Blackstonia perfoliata (L). Huds. (Working out of technology of reception of pharmaceutical raw materials). Ibid. p. 451–452.

16. Zagorodnuk, E. D.; Rybalko, A. E. Working out of a technique of micropropagation of pharmaceutical plant Lysimachia vulgaris L. Ibid. p. 450–451.

CHAPTER 15

FLUID FLOW AND CONTROL OF BENDING INSTABILITY DURING ELECTROSPINNING

A. POURHASHEMI and A. K. HAGHI

CONTENTS

15.1 Introduction..283
15.2 The Basics of Electrospinning Modeling.....................................289
    15.2.1 Viscoelastic Flow Analysis...291
        15.2.1.1 Governing Set of Equations..............................291
    15.2.2 Approaches to Viscoelastic Finite Element
        Computations..292
        15.2.2.1 Time Dependent Flows....................................292
    15.2.3 Basics of Hydrodynamics...292
    15.2.4 Electrohydrodynamic (EHD) Theory 293
    15.2.5 Electric Forces in Fluids..295
    15.2.6 Dimensionless Non-Newtonian Fluid Mechanics............295
    15.2.7 Detection of X-Ray Generated by Electrospinning..........296
15.3 Modeling Electrospinning of Nanofibers 297
    15.3.1 An Outlook to Significant Models....................................298
        15.3.1.1 Leaky Dielectric Model..................................298
        15.3.1.2 Whipping Model..299
        15.3.1.3 A Model for Shape Evaluation of Electrospinning
        Droplets 300

15.3.1.4 Nonlinear Model.. 300

15.3.1.5 A Mathematical Model For Electrospinning Process
Under Coupled Field Forces... 301

15.3.1.6 Slender-Body Model.. 301

15.3.1.7 A Model for Electrospinning Viscoelastic Fluids 304

15.3.1.8 Lattice Boltzmann Method (LBM).............................. 304

15.3.1.9 Mathematical Model for AC-Electrospinning 304

15.3.1.10 Allometry in Electrospinning..................................... 305

15.3.1.11 Multiple Jet Modeling.. 306

15.3.1.12 A Mathematical Model of the Magnetic Electrospinning
Process.. 307

15.3.1.13 Electrospinning Nanoporous Materials Model.......... 307

15.4 Summary and Outlook 308

Keywords 309

References... 310

15.1 INTRODUCTION

Electrospinning is a simple and relatively inexpensive mean of manufacturing high volume production of very thin fibers (more typically 100 nm to 1 micron) and lengths up to kilometers from a vast variety of materials including polymers, composites and ceramics [1–2]. Electrospinning technology was first developed and patented by Formhals [3] in the 1930s, and a few years later the actual developments were triggered by Reneker and co-workers [4]. These techniques are investigated and developed to satisfy the increasing needs for the refined nanosize hybrid fibers based on commercial polymers, various [5]. In this method, nanofibers produced by solidification of a polymer solution stretched by an electric field [6–8] , which can be applied in different areas including wound dressing, drug or gene delivery vehicles, biosensors, fuel cell membranes and electronics, tissue-engineering processes [7, 9–10].

The unique properties of nano-fibers is that they have extraordinarily high surface area per unit mass, very high porosity, tunable pore size, tunable surface properties, layer thinness, high permeability, low basic weight, ability to retain electrostatic charges and cost effectiveness [2]. Electrospinning has been proven to be the best nanofiber manufacturing process because of simplicity and material compatibility [11]. Table 1 illustrates the percentage of the study on different applications of electrospun nanofibers in various part of industry made by scopus site.

TABLE 1 Application of electrospun polymer nanofibers.

Polymer nanofibers application	Percent (%)
Tissue engineering	70
Nano-Sensors	10
Industrial application	8.4
Filter media	5
Protective clothing	4.3
Life science	1.3
Cosmetic	1

There are commonly two standards electrospinning setups, vertical and horizontal. With the development of this technology, several researchers are developed more intricate systems fabricating more complex nanofibrous structures in a more controlled and efficient way [10].

Generally in this process (Fig. 1), a polymer solution or melt is supplied through a syringe about 10–20 cm above a grounded substrate. The process is driven by an electrical potential of the order of kilovolts applied between the syringe and the substrate [9]. To a first approximation, Electrospinning of polymer solutions involves a rapid evaporation of the solvent. The evaporation of the solvent thus will happen on a time scale well below the second-range [12]. The elongation of the jet during electrospinning is initiated by the electrostatic force, gravity, inertia, viscosity and surface tension [1]. While, traditional spinning process principally uses gravity and externally applied tension, electrospinning applies externally applied electric field as driving force [5, 11].

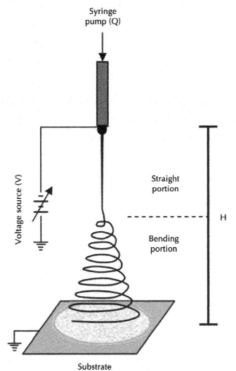

FIGURE 1   A scheme of electrospinning set up for nanofiber production.

The applied voltage induces a high electric charge on the surface of the solution drop at the needle tip. The drop now experiences two major electrostatic forces: The coulombic force, which is induced by the electrical field and the electrostatic repulsion between the surface charges. The intensity of these forces causes the hemispherical surface of the drop be elongated and form a conical shape also known as the Taylor cone. By further increasing the strength of the field, the electrostatic forces in the drop will overcome the surface tension of the fluid/air interface and therefore an electrically charged jet will be ejected [13]. As the jet diameter decreases, the surface charge density increases and the resulting high repulsive forces split the jet into smaller jets. This phenomenon may occur several times, leading to many small jets [14]. After the ejection, the jet elongates and the solvent evaporates, leaving an electrically charged fiber that, during the elongation process, becomes very thin [1–2, 12, 15].

The main characteristics of this process is the onset of a chaotic oscillation of the elongating jet which is due to the electrostatic interactions between the external electric field and the surface charges on the jet as well as the electrostatic repulsion of mutual fiber parts. The fiber can be spun directly onto the grounded (conducting) screen or on an intermediate deposit material. Because of the oscillation, the fiber is deposited randomly on the collector, creating a so called "non-woven" fiber fabric [1, 16].

Recent experiments demonstrate that an essential mechanism of electrospinning is a rapidly whipping fluid jet. The studies analyze the mechanics of this whipping jet by studying the instability of an electrically forced fluid jet with increasing field strength. Generally three different instabilities are identified: the classical (axisymmetric) Rayleigh instability and electric field induced axisymmetric and whipping instabilities. At increasing field strengths, the electrical instabilities are enhanced whereas the Rayleigh instability is suppressed. Which instability dominates depends strongly on the surface charge density and radius of the jet [16–17]. These instabilities depend on fluid parameters and equipment configuration such as location of electrodes and the form of spinneret [16].

In melt or dry/wet solution spinning, the shape and diameter of the die, as well as mechanical forces, inducing specific draw ratios and drawing speeds, highly determine dimensional and structural properties of the final fibers [18]; however, in electrospinning, the known effective properties are the polymer molecular weight, the molecular-weight distribution, the architecture (branched, linear, etc.) of the polymer, temperature and humidity and air velocity in the chamber and processing parameters (like applied voltage, flow rate, types of collectors, tip to collector distance) as well as the rheological and electrical properties of the solution (viscosity, conductivity, surface tension, etc.) and finally motion of target screen [2, 7, 10, 19]. Processing parameters which mentioned above can be manipulated such that a steady, electrostatically driven jet of fluid is drawn from the capillary tip and collected upon the grounded substrate [2, 9, 16]. The study of effects of various electrospinning parameters is important to improve the rate of nanofiber processing. In addition, several applications demand well-oriented nanofibers [11]. Several techniques such as dry rotary electrospinning [20] and scanned electrospinning nano-fiber deposition system [21] control deposition of oriented nanofibers.

Fiber diameter is the most important structural characteristic in electrospun nanofiber webs. Despite its importance, thus far, there is no successful method for determining fiber diameter. Indeed, a few studies have been conducted to develop a method for measuring fiber diameter [22]. Fiber diameter is usually determined from Scanning Electron Microscopy (SEM) images obtained from the electrospun webs. In consequence of the small fiber dimensions, high-quality images with appropriate magnifications are required. Another method for nanofiber diameter is applying an image analysis based on a binary image of the textile, which is used to create a distance map and skeleton. The fiber diameter may be determined from the values of the distance map at any pixel location on the skeleton [22–23]. The other common ways are using transmission electron microscopy (TEM) or atomic force microscopy (AFM) [22]. RSM (response surface methodology) also is used successfully for process optimization and obtaining a quantitative relationship between electrospinning parameters and the average fiber diameters and its distribution for some kind of nanofibers [24–25]. Artificial neural networks (ANN) also were successfully applied for modeling and controlling the electrospinning processes in recent years. ANN cannot create an equation similar to RSM, but it works like human brain, estimating the response based on the trained data in the inquired range [22, 24, 26]. Rabbi et al. employed RSM and ANN methods to analyze polyurethane (PU) and polyacrylonitrile (PAN) nanofibers morphology which is synthesized by electrospinning so as to determine the parameters affecting the nanofibers morphology and optimizing the process [25, 27]. Porosity is another important property of electrospun webs which can also be measured by using SEM, TEM and FEM image analysis [28].

Frenot et al. studies indicated some important findings including: (i) fibers of different sizes, i.e., consisting of different numbers of parent chains, exhibit almost identical hyperbolic density profiles at the surfaces, (ii) the end beads are predominant and the middle beads are depleted at the free surfaces, (iii) there is an anisotropy in the orientation of bonds and chains at the surface, (iv) the center of mass distribution of the chains exhibits oscillatory behavior across the fibers, and (v) the mobility of the chain in nanofiber increases as the diameter of the nanofiber decreases [19].

The major weak point of electrospinning method is a convective instability in the elongating jet. The jet will rapidly start whipping while it travels towards the collector. Therefore, during the electrospinning process, the whole substrate is covered with a layer of randomly placed fiber. The created fabric has a chaotic structure which is difficult to characterize its properties [1, 11]. Another weak point is that electrospun fibers often have beads as "by products" [29]. Some polymer solutions are not readily electrospun when the polymer solution is too dilute, due to limited solubility of the polymer. In these cases, the lack of elasticity of the solution prevents the formation of uniform fibers; instead, droplets or necklace-like structures known as 'beads-on-string' are formed [30]. The electrospun beaded fibers are related to the instability of the jet of polymer solution. The bead diameter and spacing were related to the fiber diameter, solution viscosity, net charge density carried by the electrospinning jet and surface tension of the solution [29–30].

A striking feature of the electrospinning process is that jets can be launched from any liquid surface in principle. Thus, a variety of configurations were reported that produce jets from free liquid surfaces, without using a spinneret. These contain the use of a magnetic liquid in which "spikes" can be formed to concentrate field lines at points on a liquid surface, liquid-filled trenches, wetted spheres, cylinders and disks, conical wires, rotating beaded wires and gas bubbles rising through the liquid surface [31]. One of the perceived drawbacks of the electrospinning for industrial purposes is its low production rate. The nozzle-less (free liquid surface) technology makes possibilities to produce nanofibers layers in a mass industrial scale. Nanofibers nonwoven structured layers are ideal for creating composite materials (Table 3) [32].

TABLE 2   Comparison of Nozzle vs. Nozzle-Less electrospinning.

Production variable	Nozzle	Nozzle-Less
Mechanism	Needle forces polymer downwards. Drips and issues deposited in web.	Polymer is held in bath; even distribution is maintained on electrode via rotation.
Hydrostatic Pressure	Production variable – required to be kept level across all needles in process.	None.
Voltage	5–20 kv	30–120 kv
Taylor cone Separation	Defined mechanically by needle distance.	Nature self-optimizes distance between Taylor cones.
Polymer Concentration	Often 10% of solution.	Often 20% or more of solution.
Fiber Diameters	80, 100, 150, 200, 250 and higher. Standard deviation likely to vary over fiber length.	80, 100, 150, 200, 250 and higher. Standard deviation of ±30%

By referring to Wikipedia, a model is a schematic description of a system, theory, or phenomenon that accounts for its known or inferred properties, which may be used for further study of its characteristics, and a mathematical modeling is a representation in mathematical terms of the behavior of real devices and objects. Since the modeling of devices and phenomena is essential to both engineering and science. Engineers and scientists have very practical reasons for doing mathematical modeling. In this way, models are classified into three groups, models describing the behavior; models explaining why that behavior and results occurred as they did and models that allow us to predict future behaviors or results that are as yet unseen or unmeasured.

Despite the simplicity of the electrospinning technology, industrial applications of electrospinning are still relatively rare, mainly due to the unresolved problem of very low fiber for existing devices and difficulties in controlling the process. Collection of experimental data and their confrontation with simple physical models appear as an effective approach towards the development of practical tools for controlling and optimizing the electrospinning process [33]. In the other hand, it's necessity to develop theoretical and numerical models of electrospinning demands different optimization procedure for each material [7]. Utilizing a model to express the effect of electrospinning parameters will assist researchers to make an easy and systematic way for presenting the influence of variables. Additionally, it causes to predict the results under a new combination of parameters. Therefore, without conducting any experiments, one can easily estimate features of the product under unknown conditions [34]. In general, modeling and simulation of electrospinning process is useful to understand the following:

(a) The cause for whipping instability.
(b) The dependence of jet formation and jet instability on the process parameters and fluid properties, for better jet control and higher production rate.
(c) The effect of secondary external field on jet instability and fiber orientation [2, 11].

In the following section, some basic and necessary theories for electrospinning modeling are reviewed.

15.2   THE BASICS OF ELECTROSPINNING MODELING

The modeling of the electrospinning process will be useful for the factors perception that cannot be measured experimentally [35]. Although electrospinning gives continuous fibers, mass production and the ability to control nanofibers properties are not obtained yet. In electrospinning, the nanofibers for a random state the collector plate; while, in many users of these fibers such as tissue engineering, well-oriented nanofibers are needed. Modeling and simulations will give a better understanding of electrospinning jet mechanics [11]. The development of a relatively simple model of the process was prevented by the lack of systematic, fully characterized experimental observations, suitable to lead and to test the theoretical development [16]. The governing parameters on electrospinning process investigated by the modeling are solution volumetric flow rate, polymer weight concentration, molecular weight, the applied voltage and the nozzle to ground distance [1, 6, 36]. The macroscopic nanofiber properties can be determined by multi scale modeling approach. For this purpose, at first, the effective properties were determined by using modified shear lag model, then by using volumetric homogenization approach, the macro scale properties concluded [37].

So far, two important modeling zones have been introduced which are: (a) The zone close to the capillary (jet initiation zone) outlet of the jet and (b) The whipping instability zone in which the jet spirals and accelerates towards the collector plate [11, 35, 38].

Depending on solution concentration during experiments, various impacts for the electric field were observed. Since the effects of solution concentration and electric field strength on mean fiber diameter changed at different spinning distances, some interactions and coupling effects are present between the parameters [34].

The parameters influence the nature and diameter of the final fiber to obtain the ability to control them. It is a major challenge. For selected applications, it is desirable to control not only the fiber diameter, but also the internal morphology [39]. An ideal operation would be: the nanofibers diameter to be controllable, the surface of the fibers to be intact and a single fiber would be collectable. The control of the fiber diameter can be affected by the solution concentration, the electric field strength, the feeding rate at the needle tip and the gap between the needle and the collecting screen (Fig. 2.) [1, 12, 40].

Control over the fiber diameter remains a technological bottleneck. However a cubic model for mean fiber diameter was developed for samples by A. Doustgani and et al. A suitable theoretical model of the electrospinning process is one that can show a strong-moderate-minor rating effects of these parameters on the fiber diameter. Some disadvantages of this method are low production rate, non oriented nanofiber production, difficulty in diameter prediction and controlling nanofiber morphology, absence of enough information on rheological behavior of polymer solution and difficulty in precise process control that emphasis necessity of modeling [36, 40–41].

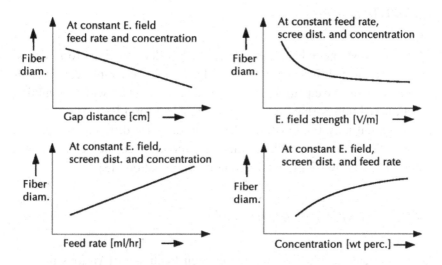

FIGURE 2   Effect of process parameters on fiber diameter.

15.2.1   VISCOELASTIC FLOW ANALYSIS

Recently, significant progress is made in the development of numerical algorithms for the stable and accurate solution of viscoelastic flow problems, which exits in processes like electrospinning process. A limitation is made to mixed finite element methods to solve viscoelastic flows using constitutive equations of the differential type [42].

15.2.1.1   GOVERNING SET OF EQUATIONS

The analysis of viscoelastic flows includes the solution of a coupled set of partial differential equations: The equations depicting the conservation of mass, momentum and energy, and constitutive equations for a number of physical quantities present in the conservation equations such as density, internal energy, heat flux, stress, etc., depend on process [42].

15.2.2 APPROACHES TO VISCOELASTIC FINITE ELEMENT COMPUTATIONS

There are fundamentally different approaches like: A mixed formulation may be adopted different parameters like velocity, pressure, etc., including the constitutive equation, is multiplied independently with a weighting function and transformed in a weighted residual form [42]. The constitutive equation may be transformed into an ordinary differential equation (ODE). For transient problems this can, for instance, be achieved in a natural manner by adopting a Lagrangian formulation [43].

15.2.2.1 TIME DEPENDENT FLOWS

By introducing a selective implicit/explicit treatment of various parts of the equations, a certain separating at each step of the set of equations may be obtained to improve computational efficiency. This suggests the possibility to apply devoted solvers to sub-problems of each fractional time step [42].

15.2.3 BASICS OF HYDRODYNAMICS

As nanofibers are made of polymeric solutions forming jets, it is necessary to have a basic knowledge of hydrodynamics [44]. According to the effort of finding a fundamental description of fluid dynamics, the theory of continuity was implemented. The theory describes fluids as small elementary volumes, which still are consisted of many elementary particles.

The equation of continuity:

$\frac{\partial \rho_m}{\partial t} + div(\rho_m v) = 0$ (For incompressible fluids $div(v) = 0$)

The Euler's equation simplified for electrospinning:

$$\frac{\partial v}{\partial t} + \frac{1}{\rho_m} \nabla p = 0$$

The equation of capillary pressure:

$$P_c = \frac{\gamma \partial^2 \zeta}{\partial x^2}$$

The equation of surface tension:

$$\Delta P = \gamma \left(\frac{1}{R_X} + \frac{1}{R_Y} \right)$$ R_x and R_y are radii of curvatures

The equation of viscosity:

$$\tau_{ij} = \eta \left(\frac{\partial v_i}{\partial x_j} + \frac{\partial v_j}{\partial x_i} \right)$$ (For incompressible fluids, $\tau_{i,j}$ = Stress tensor)

$$V = \frac{\eta}{\rho_m}$$ (kinematic viscosity)

15.2.4 ELECTROHYDRODYNAMIC (EHD) THEORY

In 1966, Taylor discovered that finite conductivity enables electrical charge to accumulate at the drop interface, permitting a tangential electric stress to be generated. The tangential electric stress drags fluid into motion, and thereby generates hydrodynamic stress at the drop interface. The complex interaction between the electric and hydrodynamic stresses causes either oblate or prolate drop deformation, and in some special cases keeps the drop from deforming [45–46].

Feng used a general treatment of Taylor-Melcher for stable part of electrospinning jets by one-dimensional equations for mass, charge and momentum. In this model, a cylindrical fluid element is used to show electrospinning jet kinematic measurements [9].

In (Fig. 3) the essential parameters are: radius, R, velocity, v_z, electric field, E_z, total path length, L, interfacial tension, γ, interfacial charge, σ, tensile stress, τ, volumetric flow rate, Q, conductivity, K, density, ρ, dielectric constant, ε, and zero-shear rate viscosity, η_0. The most important equation that Feng used are [9, 47]:

$$\tilde{R}^2 \tilde{v}_z = 1$$

$$\tilde{R}^2 \tilde{E}_z + Pe_e \tilde{R} \tilde{v}_z \tilde{\sigma} = 1$$

$$\tilde{v}_z\tilde{v}'_z = \frac{1}{Fr} + \frac{\tilde{T}'}{Re_j\,\tilde{R}^2} + \frac{1}{We}\frac{\tilde{R}'}{\tilde{R}^2} + \varepsilon(\tilde{\sigma}\tilde{\sigma}' + \beta\tilde{E}_z\tilde{E}'_z + \frac{2\tilde{\sigma}\tilde{E}_z}{\tilde{R}})$$

$$\tilde{E}_z = \tilde{E}_0 - \ln\chi\left[(\tilde{\sigma}\tilde{R})' - \frac{\beta}{2}(\tilde{E}\tilde{R}^2)''\right]$$

$$E_0 = \eta_0 v_0 \Big/ R_0 \quad \beta = (\varepsilon\!\big/\!\tilde{\varepsilon}) - 1 \quad \tau = R^2(\tilde{\tau}_{zz} - \tilde{\tau}_{rr})$$

Feng solved equation under different fluid properties, particularly for non-Newtonian fluids with extensional thinning, thickening, and strain hardening, but Helgeson et al. developed a simplified understanding of electrospinning jets based on the evolution of the tensile stress due to elongation [9].

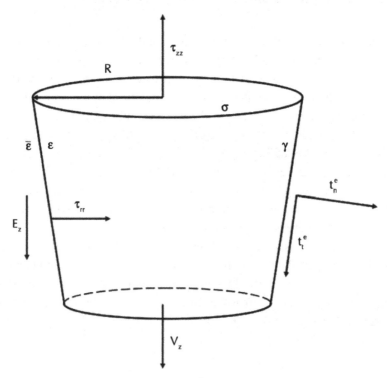

FIGURE 3 Scheme of the cylindrical fluid element used in electrohydrodynamic modeling [9].

15.2.5 ELECTRIC FORCES IN FLUIDS

The initialization of instability on the surfaces of liquids should be done by applying the external electric field, inducing electric forces on surfaces of liquids. A localized approximation was developed to calculate the bending electric force, which acts on an electrified polymer jet, as an important element of the electrospinning process for manufacturing of nanofibers. Using this force, a far-reaching analogy between the electrically driven bending instability and the aerodynamically driven instability was established. The description of the wave's instabilities is expressed by equations called dispersion laws. The dependence of wavelength on the surface tension γ is almost linear and the wavelengths between jets are a little bit smaller for lower depths. Dependency of wavelength on electric field strength is exponential. The dispersion law is identified for four groups of dielectrics liquids with using of Clausius-Mossotti and Onsager's relation (nonpolar liquids with finite and infinite depth and weakly polar liquids with finite and infinite depth). According to these relations relative permittivity is a function of parameters like temperature, square of angular frequency, wave length number and reflective index [2, 44, 48–49].

15.2.6 DIMENSIONLESS NON-NEWTONIAN FLUID MECHANICS

The best way for analyzing fluid mechanics problems is to convert parameters to dimensionless form. By using this method the numbers of governing parameters for given geometry and initial condition reduce. The non-dimensionalization of a fluid mechanics problem generally starts with the selection of a characteristic velocity, then because of the flow of non-Newtonian fluids, the stress depends non-linearly on the flow kinematics, the deformation rate is a main quantity in the analysis of these flows. Next step after determining different parameters is to evaluate Characteristic values of the parameters. The non-dimensionalization procedure is to employ the selected characteristic quantities to obtain a dimensionless version of the conservation equations and to get to certain numbers like Reynolds number and the Galilei number. The excessive number of governing dimensionless groups poses certain difficulties in the fluid mechanics analysis. Finally by using these results in equation and applying boundary conditions it can be achieved to study different properties [50].

15.2.7 DETECTION OF X-RAY GENERATED BY ELECTROSPINNING

Electrospinning jets that produce nanofibers from a polymer solution by electrical forces are fine cylindrical electrode creating extremely high electric-field strength in their vicinity at atmospheric conditions. However, this quality of electrospinning is only scarcely investigated, and the interactions of the electric fields generated by them with ambient gases are nearly unknown. Pokorny et al. reported on the discovery that electrospinning jets generate X-ray beaming up to energies of 20 keV at atmospheric conditions. [51]. Mikes et al. investigated on the discovery that electrically charged polymeric jets and highly charged gold-coated nanofibrous layers in contact with ambient atmospheric air generate X-ray beams up to energies of hard X-rays [52–53].

The first detection of X-ray produced by nanofiber deposition was observed using radiographic films. The main goal of using these films understands Taylor cone creation. The X-ray generation is probably dependent on diameters of the nanofibers affected by the flow rate and viscosity. Thus, it is important to find the ideal concentration (viscosity) of polymeric solutions and flow rate to spin nanofibers as thin as possible. The X-ray radiation can produce black traces on the radiographic film. These black traces were made by outer radiation generated by nanofibers and the radiation has to be in the X-ray region of electromagnetic spectra, because the radiation of lower energy is absorbed by the shield film. Radiographic method of X-ray detection is efficient and sensitive. It is obvious that this method didn't tell us anything about its spectrum, but it can clearly show its space distribution. The humidity, temperature and rheological parameters of polymer can affect on the X-ray intensity generated by nanofibers [44]. The necessity of modeling in electrospinning process and a quick outlook of some important models will be discussed as follows.

15.3   MODELING ELECTROSPINNING OF NANOFIBERS

Using theoretical prediction of the parameter effects on jet radius and morphology can significantly reduce experimental time by identifying the most likely values that will yield specific qualities prior to production [36]. All models start with some assumptions and have shortcomings that need to be addressed [15]. The basic principles for dealing with electrified fluids that Taylor discovered are impossible to account for most electrical phenomena involving moving fluids under the seemingly reasonable assumptions, which the fluid is either a perfect dielectric or a perfect conductor. The reason is that any perfect dielectric still contains a nonzero free charge density. During steady jetting, a straight part of the jet occurs next to the Taylor cone in which only axisymmetric motion of the jet is observed. This region of the jet remains stable in time. However, further along its path, the jet can be unstable by non-axisymmetric instabilities such as bending and branching, where lateral motion of the jet is observed in the part near the collector [9].

Branching as the instability of the electrospinning jet can happen quite regularly along the jet if the electrospinning conditions are selected appropriately. Branching is a direct consequence of the presence of surface charges on the jet surface as well as of the externally applied electric field. The bending instability leads to drastic stretching and thinning of polymer jets towards nano-scale in cross-section. Electrospun jets also creat perturbations similar to undulations, which can be the source of various secondary instabilities leading to nonlinear morphologies developing on the jets [18]. The bending instabilities occurring during electrospinning was studied and mathematically modelled by Reneker et al by viscoelastic dumbbells connected together [54]. Both electrostatic and fluid dynamic instabilities can contribute to the basic operation of the process [19].

Different stages of electrospun jets were investigated by different mathematical models during last decade by one or three dimensional techniques [7, 55]. Physical models, studying the jet profile, the stability of the jet and the cone-like surface of the jet, was developed due to significant effects of jet shape on fiber qualities [6]. Droplet deformation, jet initiation and, in particular, the bending instability which control, to a major extent, fiber sizes and properties are controlled apparently predominantly by charges located within the flight jet [18]. An accurate, predictive tool using a verifiable model which accounts for multiple factors would provide a means to run many different scenarios quickly without the cost and time of experimental trial-and error [36].

15.3.1 AN OUTLOOK TO SIGNIFICANT MODELS

The models typically treat the jet mechanics by using the localized-induction approximation by analogy to aerodynamically driven jets. They include the viscoelasticity of the spinning fluid and were also augmented to account for solvent evaporation in the jet. These models can describe the bending instability and fiber morphology, because of difficulty in measure model variables they may not accurately design and control the electrospinning process. [9]. Here are some current and more discussed models:

15.3.1.1 LEAKY DIELECTRIC MODEL

The principles for dealing with electrified fluids were summarized by Melcher and Taylor [48]. Their research showed that it is impossible to explain the most of the electrical phenomena involving moving fluids given the hypothesis that the fluid is either a perfect dielectric or a perfect conductor, since both the permittivity and the conductivity affect the flow. An electrified liquid always includes free charge. Although the charge density may be small enough to ignore bulk conduction effects, the charge will accumulate at the interfaces between fluids. The presence of this interfacial charge will result in an additional interfacial stress, especially a tangential stress, which in turn will modify the fluid dynamics [48, 56].

The electrohydrodynamic theory proposed by Taylor as the leaky dielectric model is capable of predicting the drop deformation in qualitative agreement with the experimental observations [45–46].

Although Taylor's leaky dielectric theory provides a good qualitative description for the deformation of a Newtonian drop in an electric field, the validity of its analytical results is strictly limited to the drop experiencing small deformation in an infinitely extended domain. Extensive experiments showed a serious difference in this theoretical prediction [45].

Some investigations were done to solve this defect. For example, to examine electrokinetic effects, the leaky dielectric model was modified by consideration the charge transport [56–57]. When the conductivity is finite, the leaky dielectric model can be used [57]. Saville indicated that the solution is weakly conductive so the jet carries electric charges only on its surface [56–57].

15.3.1.2 WHIPPING MODEL

Whipping model is a model for the electrospinning process that mathematically depicts the interaction between the electric field and fluid properties. This model may be applied to predict "terminal" jet diameter. The mathematically derived limiting diameter is equated with experimentally measured fiber diameter [17, 58].

The diameter depends only on the flow rate, electric current, and surface tension:

$$h_t = c\left(\frac{I}{Q}\right)^{-2/3} \gamma^{1/3}$$

where c is constant [59].

The important assumptions in the derivation of this "terminal" diameter contain uniform electric field. There is no phase change and no inelastic stretching of the jet. The experimental research showed that this model can be qualitatively valid for selected polymeric solutions including polyethylene oxide (PEO) and polycaprolactone (PCL) in low concentrations [17, 58–59].

The simplified version of the model predicts that
- at the early stages of whipping behavior, the amplitude of the whipping jet h is controlled by charge repulsion and inertia, and h grows exponentially with time in agreement with the earlier linear instability analysis.
- at the late stages of whipping, the inertia effects are damped by viscosity, and the envelope of the whipping jet is controlled by the interplay of charge repulsion and stretching viscosity.

- at the end of the whipping/stretching the surface tension balances the charge repulsion and sets the terminal diameter [17, 59].

This model basically works by comparing theory with experimental data. The critical point in this model is determining an accurate treatment for more understanding physical mechanism of charge transport near the nozzle to have a good prediction, which it is not accurately conceivable especially for highly conductive fluids. By giving the shape and charge density of a jet, this formula can predict both when and how the jet will become unstable [17].

15.3.1.3   A MODEL FOR SHAPE EVALUATION OF ELECTROSPINNING DROPLETS

Comprehension the drops behavior in an electric field is playing a critical role in practical applications. The electric field-driven flow is of practical importance in the processes in which improving the rate of mass or heat transfer between the drops and their surrounding fluid [45]. Numerically, investigations about the shape evolution of small droplets attached to a conducting surface depends on strong electric fields (weak, strong and superelectrical) are done and indicate that three different scenarios of droplet shape evolution are distinguished, based on numerical solution of the Stokes equations for perfectly conducting droplets by investigation of Maxwell stresses and surface tension [13, 60]. The advantages of this model are that the non-Newtonian effect on the drop dynamics is successfully identified on the basis of electrohydrostatics at least qualitatively. In addition, the model shows that the deformation and breakup modes of the non-Newtonian drops are distinctively different from the Newtonian cases. There is a restriction in this model, because regarding the electrohydrostatic, the drop phase should be highly conductive and much less viscous [45].

15.3.1.4   NONLINEAR MODEL

During large-strain stretching of the polymer nonlinear viscoelasticity models must be used by understanding the role of rheology [61]. For example, a simple two-dimensional model can be used for describing formation of barb electrospun polymer nanowires with a perturbed swollen cross-section and the electric charges "frozen" into the jet surface. This model is integrated numerically using the Kutta-Merson method with the adoptable time step. The result of this modeling is explained theoretically as a result of relatively slow charge relaxation compared to the development of the secondary electrically driven instabilities, which deform jet surface locally. When the disparity of the slow charge relaxation compared to the rate of growth of the secondary electrically driven instabilities becomes even more pronounced, the barbs transform in full-scale long branches. The competition between charge relaxation and rate of growth of capillary and electrically driven secondary localized perturbations of the jet surface is affected not only by the electric conductivity of polymer solutions but also by their viscoelasticity. Moreover, a nonlinear theoretical model was able to resemble the main morphological trends recorded in the experiments [18].

15.3.1.5   A MATHEMATICAL MODEL FOR ELECTROSPINNING PROCESS UNDER COUPLED FIELD FORCES

There is not a theoretical model to describe the electrospinning process under the multi-field forces; therefore, a simple model might be very useful to indicate the contributing factors. Modeling this process can be done in two ways: (a) the deterministic approach using classical mechanics like Euler approach and Lagrange approach. (b) The probabilistic approach uses E-infinite theory and quantum like properties. Many basic properties are harmonious by adjusting electrospinning parameters such as voltage, flow rate and others, and it can offer in-depth inside into physical understanding of many complex phenomena which cannot be fully explain [8].

15.3.1.6   SLENDER-BODY MODEL

One-dimensional models for inviscid, incompressible, axisymmetric, annular liquid jets falling under gravity are obtained by means of methods of regular perturbations for slender or long jets, integral formulations, Taylor's series expansions, weighted residuals, and variational principles [38, 62].

Using Feng's theory some familiar assumptions in modeling jets and drops are applied: The jet radius R decreases slowly along z direction while the velocity v is uniform in the cross section of the jet so it is lead to the nonuniform elongation of jet. According to the parameters can be arranged into three categories: process parameters (Q, I and E_∞), geometric parameters (R_0 and L) and material parameters (ρ, η_0 (the zero-shear-rate viscosity), $\varepsilon, \bar{\varepsilon}$, K, and γ). The jet can be represented by four steady-state equations: the conservation of mass and electric charges, the linear momentum balance and Coulomb's law for the E field.

Mass conservation can be stated by:

$$\pi R^2 v = Q$$

R: Jet radius

The second equation in this modeling is charge conservation that can be stated by:

$$\pi R^2 KE + 2\pi R v \sigma = I$$

The linear momentum balance is:

$$\rho v v' = \rho g + \frac{3}{R^2}\frac{d}{dz}(\eta R^2 v) + \frac{\gamma R'}{R^2} + \frac{\sigma \sigma'}{\bar{\varepsilon}} + (\varepsilon - \bar{\varepsilon})EE' + \frac{2\sigma E}{R}$$

The Coulomb's law for electric field:

$$E(z) = E_\infty(z) - \ln \chi(\frac{1}{\bar{\varepsilon}}\frac{d(\sigma R)}{dz} - \frac{\beta}{2}\frac{d^2(ER^2)}{dz^2})$$

L: The length of the gap between the nozzle and deposition point
R_0: The initial jet radius

$$\beta = \frac{\varepsilon}{\bar{\varepsilon}} - 1$$

$$\chi = \frac{L}{R_0}$$

By these four equations, the four unknown functions R, v, E and σ are identified.

At first step, the characteristic scales such as $\mathbf{R_0}$ and $\mathbf{v_0}$ are denoted to format dimensionless groups. Inserting these dimensionless groups in four equations discussed above the final dimensionless equations is obtained.

The boundary conditions of four equations that became dimensionless can be expressed (see Fig. 4) as:

$$\ln z = 0 \quad R(0) = 1 \quad \sigma(0) = 0$$

$$\ln z = \chi \quad E(\chi) = E_\infty$$

***The first step is to write the ODE's as a system of first order ODE's by a numerical relaxation method as e.g., *solvde* from numerical recipes. The basic idea is to introduce new variables, one for each variable in the original problem and one for each of its derivatives up to one less than the highest derivative appearing. For solving this ODE the Fortran program is used, in the first step an initial guess uses for χ and the other parameters would change according to χ [1, 63]. The limitation of slender-body theory is: avoiding treating physics near the nozzle directly [38].

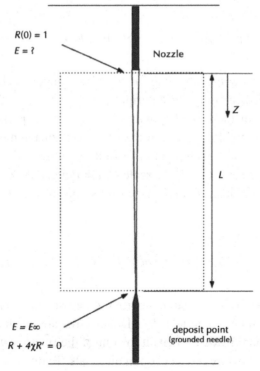

FIGURE 4 Boundary conditions [1].

15.3.1.7   A MODEL FOR ELECTROSPINNING VISCOELASTIC FLUIDS

When the jet thins, the surface charge density varies, which in turn affects the electric field and the pulling force. Roozemond combined the "leaky dielectric model" and the "slender-body approximation" for modeling electrospinning viscoelastic jet [53]. It must be considered that all variables are uniform on the cross-section of the jet, and vary only along z [61]. The jet could be represented by four steady-state equations: the conservation of mass and electric charges, the linear momentum balance and Coulomb's law for the electric field, with all quantities depending only on the axial position z. The equations can be converted to dimensionless form using some characteristic scales and dimensionless groups like most of the models. These equations could be solved by converting to ODE's forms and using suitable boundary conditions [38, 63].

15.3.1.8   LATTICE BOLTZMANN METHOD (LBM)

An efficient numerical modeling for complex structure web with a reliable prediction needs a complete description of the structure that it is highly difficult. One of these models with agreement by experiments is lattice Boltzmann method [64]. Developing lattice Boltzmann method instead of traditional numerical techniques like finite volume, finite difference and finite element for solving large-scale computations and problems involving complex fluids, colloidal suspensions and moving boundaries is so useful [11].

15.3.1.9   MATHEMATICAL MODEL FOR AC-ELECTROSPINNING

Much of the nanofiber research reported so far was about nanofibers which was made from DC potential [6]. In DC-electrospinning, the fiber instability or 'whipping' made it difficult to control the fiber location and the resulting microstructure of electrospun materials. To overcome these limitations, some new technologies were applied in the electrospinning process.

The investigations proved that the AC potential resulted in a significant reduction in the amount of fiber 'whipping' and the resulting mats exhibited a higher degree of fiber alignment but are observed to contain more residual solvent. In AC-electrospinning, the jet is inherently unsteady due to the AC potential unlike DC ones, so all thermal, electrical, and hydrodynamics parameters were considered to be effective in the process [65–66].

The governing equations for an unsteady flow of an infinite viscous jet pulled from a capillary orifice and accelerated by an AC potential can be expressed as follows:
1) The conservation of mass equation
2) Conservation of charge
3) The Navier-Stokes equation

Using these governing equations, final model of AC electrospinning is able to find the relationship between the radius of the jet and the axial distance from nozzle, and a scaling relation between fiber radius and the AC frequency [66].

15.3.1.10   ALLOMETRY IN ELECTROSPINNING

Electrospinning applies electrically generated motion to spin fibers; therefore, it is difficult to predict the size of the produced fibers, which depends on the applied voltage in principal. Therefore the relationship between radius r of jet and the axial distance z from nozzle was the subject of investigation [67]. It can be described as an allometric equation by using the values of the scaling exponent for the initial steady stage, instability stage, and terminal stage [68].

The relationship between r and z can be expressed as an allometric equation of the form

$$r \approx z^b$$

where b is the power exponent. When b = 1 the relationship is isometric and when b ≠ 1 the relationship is allometric [67, 69].

Assume that the volume flow rate (Q) and the current (I) remains unchanged during the electrospinning procedure, we have the following scaling relations: $Q < r0$, and $I < r0$ [67].

Many experiment data shows a power law relationship between electric current and solution flow rate under the condition of fixed voltage: $I \approx Q^b$ where b is the same as the previous equation [70]. Due to high electrical force acting on the jet, it can be illustrated [67]

$$\frac{d}{dz}\left(\frac{v^2}{2}\right) = \frac{2\sigma E}{\rho r}$$

Conservation of mass [67, 69-70]: $\pi r^2 \rho v = Q$

Conservation of charge [67, 69]: $2\pi r \sigma v + k\pi r^2 E = I$

From the above Equations it can be seen that [67]

$$r \approx z^b, \sigma \approx r, E \approx r^{-2}, \frac{du^2}{dz} \approx r^{-2}, r \approx z^{-\frac{1}{2}}$$

Allometric scaling law:

The charged jet can be considered as a one-dimensional flow like mentioned. Then the mass and charge conservation equation consider like above. If the equations modified, they can be changed [67]

$$2\pi r \sigma^\alpha v + k\pi r^2 E = I$$

$$r \approx z^{-\alpha/(\alpha+1)}$$

where α is a surface charge parameter, the value of α depends on the surface charge in the jet.

When $\alpha = 0$ no charge in jet surface, and in $\alpha=1$ use for full surface charge [67].

15.3.1.11   MULTIPLE JET MODELING

It was experimentally and numerically exhibited that the jets from multiple nozzles expose higher repulsion by another jets from the neighborhood by Columbic forces than jets spun by a single nozzle process [5]. Yarin and Zussman achieved upward electrospinning of fibers from multiple jets without the use of nozzles; instead using the spiking effect of a magnetic liquid [71]. For large-scale nanofibre production and the increase in production rate, multi-jet electrospinning systems can be designed to increase both productivity and covering area [72–73]. The linear Maxwell model and nonlinear Upper-Convected Maxwell (UCM) model were used to calculate the viscoelasticity. By using these models, the primary and secondary bending instabilities can be calculated. Maxwell model and the non-linear UCM model lead to rather close results in the flow dominated by the electric forces. In a multiple-nozzle set up, not only the external applied electric field and self-induced Coulombic interactions influence the jet path, but also mutual-Coulombic interactions between different jets contribute [73].

15.3.1.12   A MATHEMATICAL MODEL OF THE MAGNETIC ELECTROSPINNING PROCESS

Various forces which make the jet flow cause the whipping circle increasing larger and larger [74]. For controlling the instability, magnetic electrospinning is proposed. For describing the magnetic electrospun jet, it can be used Reneker's model [54]. This model may not consider the coupling effects of the thermal field, electric field and magnetic field. Therefore, the momentum equation for the motion of the beads is [75]

$$m\frac{d^2 r_i}{dt^2} = F_C + F_E + F_{ve} + F_B + F_q$$

Wan et al. also established an electro-magnetic model was to explain the bending instability of the charged jet in the electrospinning process. In order to verify the validity of the model, the motion path of the jet was then simulated numerically. The results showed that the magnetic field generated by the electric current in the charged polymer jet might be one of the main reasons which caused the helix motion of the jet during electrospinning [76].

15.3.1.13   ELECTROSPINNING NANOPOROUS MATERIALS MODEL

The electrospun nanoporous materials received much attention recently because of the potential for a broad range of applications in widely different areas such as photonic structures, microfluid channels (nanofluidics), catalysis, sensors, medicine, and pharmacy and drug delivery [28]. Due to this importance, some theoretical models were focused on preparing electrospun nanoporous materials with controllable pore sizes and numbers.

Xu et al. applied the following theorical model for electrospinning dilation.

Conservation of mass gives: $\pi r^2 \rho u = Q$

Where Q is the volume flow rate, ρ is the liquid density, u is the velocity and r is radius of the jet, the radius of the jet decreases with the increase of the velocity of the incompressible charged jet. The critical jet velocity (maximum velocity before jet changing into unstable form) is calculated as: $u_{cr} = Q / \pi \rho r_{cr}^2$, where r_{cr} is radius of jet before changing into unstable form.

The velocity can exceed this critical value if a higher voltage is applied. In cases the radius of the jet reaches the value of the critical value, and the jet speed increases up to its critical value, in order to keep the conservation of mass equation, the jet dilates by decreasing its density, leading to porosity of the electrospun fibers; we call this phenomenon electrospinning-dilation.

The properities of electrospun nanoporous microspheres can be predicted using the following equation: $\pi r_{cr}^2 \tilde{\rho} u_0 = \pi R^2 \tilde{\rho} u_{min} = Q$

Where $\tilde{\rho}$ is the density of dilated microsphere, u_0 is the velocity of the charged jet at $r = r_{cr}$, R is the maximal radius of the microsphere, u_{min} is the minimal velocity. Higher voltage means higher value of the jet speed (u_0) at $r = r_{cr}$, and a more drastic electrospinning-dilation process happens, resulting in a lower density ($\tilde{\rho}$) of dilated microsphere, smaller size (R) of the microsphere and smaller pores as well [77–78].

15.4 SUMMARY AND OUTLOOK

Electrospinning is a very simple and versatile method of creating polymer-based high-functional and high-performance nanofibers that can revolutionize the world of structural materials. The process is versatile in that there is a wide range of polymer and biopolymer solutions or melts that can spin. The electrospinning process is a fluid dynamics related problem. In order to control the property, geometry, and mass production of the nanofibers, it is necessary to understand quantitatively how the electrospinning process transforms the fluid solution through a millimeter diameter capillary tube into solid fibers, which are four to five orders smaller in diameter. When the applied electrostatic forces overcome the fluid surface tension, the electrified fluid forms a jet out of the capillary tip towards a grounded collecting screen. Although electrospinning gives continuous nanofibers, mass production and the ability to control nanofibers properties are not obtained yet. Combination of both theoretical and experimental approaches seems to be promising step for better description of electrospinning process. Applying simple models of the process would be useful in atoning the lack of systematic, fully characterized experimental observations and the theoretical aspects in predicting and controlling effective parameters. The analysis and comparison of model with experiments identify the critical role of the spinning fluid's parameters. The theoretical and quantitative tools developed in different models provide semi- empirical methods for predicting ideal electrospinning process or electrospun fiber properties. In each model, researcher tried to improve the existing models or changed the tools in electrospinning by using another view. Therefore, it was attempted to have a whole view on important models after investigation about basic objects. A real mathematical model, or, more accurately, a real physical model, might initiate a revolution in understanding of dynamic and quantum-like phenomena in the electrospinning process. A new theory is much needed which bridges the gap between Newton's world and the quantum world.

KEYWORDS

- **Beads-on-string**
- **By products**
- **Non-woven**
- **Scopus site**
- **Spikes**
- **Taylor cone**
- **Whipping**

REFERENCES

1. Solberg, R. H. M. *Position-controlled Deposition for Electrospinning*. In *Department Mechanical Engineering*. 2007, Eindhoven University of Technology: Eindhoven. 67.
2. Chronakis, S. I. *Processing, Properties and Applications*, in *Micro-Nano-Fibers by Electrospinning Technology*. 264–286.
3. Formhals, A. *Process and Apparatus for Preparing Artificial Threads*, U.S. Patent, Editor. 1934: Germany.
4. Reneker, D. H.; Chun, I. *Nanotechnology*, **1996**, *7*, 216–223.
5. Kim, G.; Cho, Y. S.; Kim, W. D. *Eur. Polym. J.*, **2006**, *42*, 2031–2038.
6. Theron, S. A.; Zussman, E.; Yarin, A. L. *Polymer*, **2004**, *45*: 2017–2030.
7. Kowalewski, T. A.; Barral, S.; Kowalczyk, T. *Modeling Electrospinning of Nanofibers*. IUTAM Symposium on Modelling Nanomaterials and Nanosystems, **2009**, *13*, 279–293.
8. Xu, L. *Chaos, Solitons and Fractals*, **2009**, *42*, 1463–1465.
9. Helgeson, E. M., et al. *Polymer*, **2008**, *49*, 2924–2936.
10. Bhardwaj, N.; Kundu, S. C. *Biotechnol. Adv.*, **2010**, *28*, 325–347.
11. Karra, S. *Modeling Electrospinning Process and a Numerical Scheme Using Lattice Boltzmann Method to Simulate Viscoelastic Fluild Flows*. In *Mechanical Engineering* 2007, Indian Institute of Technology Madras: Chennai, 60.
12. Bognitzki, M., et al. *Adv. Mater.*, **2001**, *13*, 70–73.
13. Basaran, O. A.; Suryo, R. *Nat. Phys.*, **2007**, *3*, 679–680.
14. Mottaghitalab, V.; Haghi, A. K. *Korean J. Chem. Eng.*, **2011**, *28*, 114–118.
15. Titchenal, N.; Schrepple, W. *Mater. Sci. Eng.*
16. Shin, Y. M., et al. Polymer, **2001**, *42*, 9955–9967.
17. Hohman, M. et al., *Phys. Fluid*, **2001**, *13*, 2201–2220.
18. Holzmeister, A.; Yarin, A. L.; Wendorff, J. H. *Polymer*, **2010**, *51*, 2769–2778.
19. Frenot, A.; Chronakis, I. S. *Curr. Opin. Coll. Interf. Sci.*, **2003**, *8*, 64–75.
20. El-Auf, A.K. *Nanofibers and Nanocomposites Poly (3,4-ethylene dioxythiophene)/ Poly(styrene sulfonate) by Electrospinning*. In *Department of Materials Science and Engineering*. Drexel University: Philadelphia, 2004, p 261.
21. Czaplewski, D.; Kameoka, J.; Craighead, H. G. *J. Vacuum Sci. Technol. B (Microelectronics and Nanometer Structures)*, **2003**, *21*, 2994–2997.
22. Ziabari, M.; Mottaghitalab, V.; Haghi, A. K. *Korean J. Chem. Eng.*, **2008**, *25*, 905–918.
23. Ziabari, M.; et al., *Nanoscale Res. Lett.*, **2007**, *2*, 597–600.
24. Abdul Karim, S.; et al., *J. Appl. Sci. Res.*, **2012**, *8*, 2510–2517.
25. Nasouri, K.; et al., *J. of Appl. Poly. Sci.*, **2012**, *126*, 127–135.
26. Faridi-Majidi, R.; et al., *J. of Appl. Pol. Sci.*, **2012**, *124*, 1589–1597.
27. Rabbi, A.; et al., *Fibers Polym.*, **2012**, *13*, 1007–1014.
28. Ziabari, M.; Mottaghitalab, V.; Haghi, A. K. *Korean J. Chem. Eng.*, **2008**, *25*, 923–932.
29. Fong, H.; Chun, I.; Reneker, D. H. *Polymer*, **1999**, *40*, 4585–4592.
30. Yu, H.; Fridrikh, J., S. V.; Rutledge, G. C. *Polymer*, **2006**, *47*, 4789–4797.
31. Forward, M. K.; Rutledge, G. C. *Chem. Eng. J.*, **2012**, *183*, 492–503.

32. Petrik, S.; Maly. M. *Production Nozzle-Less Electrospinning Nanofiber Technology*, 2009.
33. Kowalewski, T. A.; Blonski, S.; Barral, S. *Bull. Polish Acad. Sci. Tech. Sci.*, **2005**, *53*, 385–394.
34. Ziabari, M.; Mottaghitalab, V.; Haghi, A. K. *Korean J. Chem. Eng.*, **2010**, *27*, 340–354.
35. Patanaik, A.; Jacobs, V.; Anandjiwala, R. D. *Experimental Study and Modeling of the Electrospinning Process*. In *86th Textile Institute World Conference*. Hong Kong, 2008, pp 1160–1168.
36. Thompson, C. J.; et al., *Polymer*, **2007**, *48*, 6913–6922.
37. Agic, A. *Mater. Science Forum* **2012**, *714*, 33–40.
38. Feng, J. *J. Phys. Fluid*, **2002**, *14*, 3912–3926.
39. Helgeson, M. E.; Wagner, N. J. *Ame. Ins. Chem. Eng.*, **2007**, *53*, 51–55.
40. Doustgani, A.; et al., *Composites: Part B*, **2012**, *43*, 1830–1836.
41. Fridrikh, V. S.; et al., *Phys. Rev. Lett.*, **2003**, *90*, 144502–4.
42. Baaijens, P.T. F. *Mixed Finite Element Methods for Viscoelastic Flow Analysis: A Review*. In *Faculty of Mechanical Engineering*. Eindhoven University of Technology Center for Polymers and Composites: Eindhoven, 2001, p. 37.
43. Rasmussen, H. K.; Hassager, O. *J. Non-New. Fluid Mech.*, **1993**, *46*, 298–305.
44. Mikeš, I. P. *Physical Principles of Electrostatic Spinning*. In *Physical Engineering*, Technical University in Liberec Liberec, 2011, p. 122.
45. Ha, J. W.; Yang, S. M. *J. Fluid Mech.*, **2000**, *405*, 131–156.
46. Taylor, G. I. *Proc. Roy. Soc. London. Ser. A, Math. Phys. Sci.*, **1966**, *291*, 159–166.
47. Wan, Y. Q. Guo, Q.; Pan, N. *Int. J. Non. Sci. Num. Sim.*, **2004**, *5*, 5–8.
48. Melcher, J. R.; Taylor, G. I. *Annu. Rev. Fluid Mech.*, **1969**, *1*, 111–146.
49. Yarin, A. L.; Koombhongse, S.; Reneker, D. H. *J. Appl. Phy.*, **2001**, *89*, 3018–3026.
50. de Souza Mendes, R. P. *J. Non-New. Fluid Mech.*, **2007**, *147*, 109–116.
51. Pokorńy, P.; Mikes, P.; Lukáš, D. *A Lett. J. Exploring Frontiers Phys.*, **2010**, *92*, 47002–47007.
52. Mlikes, P.; et al., *High Energy Radiation Emitted from Nanofibers*. In *7th International Conference - TEXSCI 2010*. Liberec, Czech Republic, 2010, p 3.
53. Kornev, K. G. *J. Appl. Phy.*, **2011**, *110*, 124910–124915.
54. Reneker, D. H.; et al., *J. Appl. Phy.*, **2000**, *87*, 4531 –4547.
55. He, J. H.; et al., *Polym. Int.*, **2007**, *56*, 1323–1329.
56. Saville, D. A. *Annu. Rev. Fluid Mech.*, **1997**, *29*, 27–64.
57. Parageorgiou, T. D; V.Broeck, J. M. *IMA J. Appl. Math.*, **2007**, *72*, 832–853.
58. Sarkar, K.; et al., *J. Materials Pro. Tech.*, **2009**, *209*, 3156–3165.
59. Rutledge, C. G.; Warner, S. B. *Electrostatic Spinning and Properties of Ultrafine Fibers*, in *Materials Competency*. National Textile Center Research Briefs, 2003.
60. Reznik, S. N.; et al., *J. Fluid Mech.*, **2004**, *516*, 349–377.
61. Feng, J. J. *J. Non-New. Fluid Mech.*, **2003**, *116*, 55–70.
62. Ramos, J. I. *Appl. Math. Model.*, **1996**, *20*, 593–607.
63. Roozemond, P. C. *A Model for Electrospinning Viscoelastic Fluids*. In *Department of Mechanical Engineering*. Eindhoven University of Technology: Eindhoven, 2007, p 26.
64. Wang, M., et al., *Int. J. Therm. Sci.*, **2007** *46*, 848–855.

65. Shin, Y. M., et al., *Appl. Phys. Lett.*, **2001**, *78*, 3–7.
66. Ji-Huan, H.; Yue, W.; Ning, P *Int. J. Nonlinear Sci. Numerical Simul.*, **2005**, *6*, 243–248.
67. He, J. H.; Liu, H. M. *Nonlinear Anal.*, **2005**, *63*, e919–e929.
68. He, J. H.; Wan, Y. Q.; Yu, J. Y. *Int. J. Nonlinear Sci. Numerical Simul.*, **2004**, *5*, 243–252.
69. He, J. H.; Wan, Y. Q. *Polymer*, **2004**, *45*, 6731–6734.
70. He, J. H.; Wan, Y. Q.; Yu, J. Y. *Polymer*, **2005**, *46*, 2799–2801.
71. Yarin, A. L.; Zussman, E. *Polymer*, **2004**, *45*, 2977–2980.
72. Varesano, A.; Carletto, R. A.; Mazzuchetti, G. *J. Mat. Pro. Tech.*, **2009**, *209*, 5178–5185.
73. Theron, S. A., et al., *Polymer*, **2005**, *46*, 2889–2899.
74. Wu, Y., et al., *Chaos, Solitons Fractals*, **2007**, *32*, 5–7.
75. Xu, L.; Wu, Y.; Nawaz, Y. *Com. Math. App.*, **2011**, *61*, 2116–2119.
76. Wan, S., et al., *Adv. Sci. Lett.*, **2012**, *10*, 566–569.
77. Xu, L.; Liu, F.; Faraz, N. *Com. Math. App.*, **2012**, *64*, 1017–1021.
78. He, J. H., et al., *Polym. Int.*, **2007**, *56*, 1323–1329.

CHAPTER 16

RELAXATION PARAMETERS OF POLYMERS

NINEL N. KOMOVA, GENNADY E. ZAIKOV,
and ALFONSO JIMENEZ

CONTENTS

16.1 Introduction 314
16.2 Theoretical Part 315
16.3 Experimental 322
16.4 Concluding Remarks 322
Keywords 327
References 327

16.1 INTRODUCTION

Any measurement of physical system is made by means of some device (in the more general case – the measuring environment). Thus there is an interaction of the device to measured system therefore the system condition to some extent changes depending on intensity of influence from the device. At measurements of classical system quite pertinently to suppose, that the measurement doesn't change a condition of measured system at all. If describe a condition of measured system and measurement procedure make so in details that features of influence of measuring system that a situation cardinally are shown changes. It appears that owing to the quantum nature of physical systems during the measurement the conditions of measured system are changing. These changes are so more than more information received by measurement. It is necessary to pay for the information. So in the theory of measurements the information increase corresponds to reduction of entropy [1]:

$$S = -\sum_{i}^{n} p_i \ln p_i ,$$

where p_i – aprioristic probabilities of various conditions of system, n – quantity of conditions.

Thus, increasing accuracy of measurement, we necessarily increase also return influence of measuring procedure by a condition of measured system.

John Neumann for quantum system has proved a background and mathematically has strictly formulated a postulate of a reduction. According to this postulate at measurement of some observable size the system condition changes in such a manner that in a new condition the measured observable has already another certain value, and it has turned out as a result of measurement. Occurrence of this condition is called as a condition reduction of a system.

In the theory of measurements it is considered two types of measuring systems: passive and active [2]. In the passive measuring system there is a comparison of the defined size with the standard without any active influence on the system, which parameters define. The feature of active measuring system is influence on characterized system, and the response of system to this influence gives the information for calculation of demanded parameters.

As the active measuring system assumes a certain influence on characterized object in the course of this influence the object can undergo changes. Therefore for reception of the most exact value of the defined parameter in the theory of measurements perform the operation of coordination between measuring system and the measured object, consisting in reduction, and at the best data dissipation, influences of entrance influence on measured object.

At the measurements concerning difficult systems or objects, the measured size often depends on set of various circumstances. Usually the nature and quantitative characteristics of these dependences are unknown. The circumstances influencing on result of measurement don't remain constants during carrying out of measurement, there fore it's impossible to correct this or that error of measurement. It means that measurement isn't selective, and the result of measurement comprises also other factors.

In the greatest measure these principles are important for sizes which assume measurements in which basis the difficult physical and mathematical models demanding a certain sort of updating according to conditions of measurements lie. In the mechanic of polymers such sizes are the parameters characterizing relaxation properties of materials. These sizes and dependences corresponding to them give the chance to judge structure of polymers, to find temperatures of structural transitions and service conditions of corresponding materials [3, 4].

16.2 THEORETICAL PART

One of widely used methods in research elastic and relaxation properties of polymers in the block at periodic sinusoidal loadings is the method of Aleksandrova-Lazurkina [6]. Unlike resonant this method is applied for high elasticity deformations of polymers in the field of the frequencies lying considerably below own frequency of the sample – far from resonant area. In this case, the phase relations, phase lag of deformation from stress-relaxation time is determined only by or through the appropriate range of relaxation time and elastic material. In this method phase parities don't depend on the form, the size and density of the sample that allows to find relaxation time of a material from measurements.

The method is based idea of rubbery (*high elasticity*) deformation as a reflection of the deformation of flexible macromolecules, and the appearance of the elastic forces of deformation and shape recovery after unloading – the result of thermal motion of parts of macromolecules. However, all the patterns that underlie the method refer to the equilibrium states of the body under load. The study of temporal patterns of rubbery deformation in a regime of constant stress or strain, as well as in periodic loads confirmed the significant role of the kinetics of rubbery deformation relaxation phenomena in the behavior of polymeric materials under mechanical stress, and in the process of vitrification of polymers [5, 7].

Depending on the interim regime changes impact the behavior of the material. At a constant temperature with increasing speed or increasing the frequency of impacts observed so-called effect of "hardening" of the material [8].

Total deformation of the polymer is composed of an elastic, rubbery (high elastic) flow and deformation. When considering the polymer in the rubberlike (high elasticity) state accepts that the macroscopic viscosity of the material is great and flows absent.

To obtain the dependence of rubbery component of the strain on the applied stress using the simplest model for these conditions [9]. In this case, such a model is a three-element model representing a model of Kelvin (parallel connected spring and damper), connected in series with a spring. The equation describing the relationship between stress and strain of this model is as follows:

$$\frac{d\sigma}{dt} + \frac{E_0 + E_1}{\eta}\sigma = E_0\frac{d\varepsilon}{dt} + \frac{E_0 E_1}{\eta}\varepsilon, \tag{1}$$

where σ – stress acting on the system being studied; ε-deformation occurring in the system under the applied stress, E_0 – the module of elasticity; E_1 – high elasticity module; η – micro viscosity.

Then the deformation of the polymer is made up of elastic strain $\varepsilon_0 = \sigma/E_0$ and high elasticity ε_1 parts. Rewriting Eq. (1) relative rates of change of strain and isolating highly elastic component of deformation, we can received:

$$\frac{d\varepsilon_1}{dt} + \frac{E_1}{\eta}\varepsilon_1 = \frac{\sigma}{\eta} \qquad (2)$$

If the stress varies with time harmonically with frequency ω:

$$\sigma = \sigma_0 \cos \omega t, \qquad (3)$$

full deformation is described by the equation:

$$\varepsilon = Ce^{-\frac{t}{\tau}} + \sigma_0 \left\{ \left(\frac{1}{E_0} + \frac{1}{E_1}\frac{1}{1+\omega^2\tau^2} \right) \cos \omega t + \frac{1}{E_0}\frac{\omega\tau}{1+\omega^2\tau^2}\sin \omega t \right\}, \qquad (4)$$

where the parameter $\tau = \eta / E_1$ is called as *relaxation time*. In some works [10, 11] this parameter is named delay time, and relaxation time is named the parameter, which proportional to it:

$$\tau_1 = \tau\frac{E_1}{E_1 + E_0}. \qquad (5)$$

The first exponential member of the Eq. (4) contains constant C, witch depending on initial conditions, and defines an unsteady part of deformation fading in due course. Therefore, if from the beginning of carrying out of measurement has passed enough time $t \gg \tau$ (the transients which have arisen at the moment of a start of motion, have already faded and takes place the established conditions) it is possible to neglect this member and consider only that part of Eq. (4), which is concluded in braces. This part describes the stationary oscillations, which are studied on experience. They consist of oscillations in a phase with the pressure, described by expression in parentheses and consisting of elastic and *high elastic* components, and the oscillations, which are lagging behind pressure on a phase on $\pi/2$. As these two harmonious oscillations are directed along one axis (a vector of their speeds are collinear) the amplitude of deformation is expressed by the equation:

$$\varepsilon_0 = \sigma_0 \sqrt{\left(\frac{1}{E_0} + \frac{1}{E_1}\frac{1}{1+\omega^2\tau^2} \right)^2 + \frac{1}{E_1^2}\frac{\omega^2\tau^2}{\left(1+\omega^2\tau^2\right)^2}} \qquad (6)$$

Using condition $E_0 \gg E_1$: as the high elasticity module for polymeric materials on some orders less than the module of elasticity, it is possible to receive dependence of deformation on pressure and frequency of loading (ω):

$$\varepsilon_0 = \frac{\sigma_0}{E_1} \frac{1}{\sqrt{1+\omega^2\tau^2}}. \tag{7}$$

The received expression can be transformed as:

$$\frac{\varepsilon_0}{\sigma_0} = \frac{1}{E_1\sqrt{1+\omega^2\tau^2}}. \tag{8}$$

Parameter ε_0/σ_0 is dynamic compliance (I) and equal to the inverse dynamic module. The compliance makes sense strain in a single strain.

Using complex representation of harmoniously changing deformation: $\varepsilon(t) = \varepsilon_0 e^{i\omega t}$, strain rate will have an expression $d\varepsilon(t)/dt = \omega \, \varepsilon 0 e i (\omega t + \pi/2)$. Substituting this expression in the differential Eq. (1) and reducing on $\varepsilon_0 e^{i\omega t}$, we received:

$$(i\eta\omega + E)E^*(i\omega) = iE\eta\omega, \tag{9}$$

where E * is a complex dynamic modulus, which can be represented as:

$$E^*(i\omega) = \frac{\eta^2\omega^2 E}{E^2 + \eta^2\omega^2} + i\frac{\eta\omega E^2}{E^2 + \eta^2\omega^2}. \tag{10}$$

The first term is a real, and the second – the imaginary part of the complex dynamic modulus (E *= E `+ iE ``), which is proportional to E and depends on the frequency. E `` (ω) determines the losses at harmonic deformation and is the module of losses.

Similarly complex dynamic module E * (i ω) can be represented by a complex dynamic compliance I * as the sum of the imaginary and real I `` I `parts. Considering that I * (i $\omega$) E * (i $\omega$) = 1, we can provide the relevant expressions in the form: I * (i$\omega$) = I `($\omega$) + i I `` (ω), where

$$I' = \frac{1}{E_0} + \frac{1}{E_1}\frac{1}{1+\omega^2\tau^2} = I_0 + \frac{I_1}{1+\omega^2\tau^2} \tag{11}$$

$$I'' = \frac{1}{E_1} \frac{\omega\tau}{1 + \omega^2\tau^2} = I_1 \frac{\omega\tau}{1 + \omega^2\tau^2}, \tag{12}$$

where $I_0 = 1/E_0$, and $I_1 = 1/E_1$. Absolute measured deformation looks like:
$I = \sqrt{I'^2 + I''^2}$

From condition $I_1 \gg I_0$ (as $E_0 \gg E_1$):

$$I \approx \frac{I_1}{\sqrt{1 + \omega^2\tau^2}}. \tag{13}$$

The phase angle δ between the I and I `` , i.e., between strain and stress is defined as:

$$tg\delta = \frac{I''}{I'} = \frac{I_1\omega\tau}{I_1 + I_0(1 + \omega^2\tau^2)} \tag{14}$$

In essence the angle δ describes the mechanical loss, i.e., share of mechanical energy, which came into heat, or the proportion of dissipated energy per cycle of deformation per unit volume. A measure of this transformation may be an area corresponding to the hysteresis loop formed by the dependence of deformation on the voltage in the cycle of periodic actions (between the curve of loading and unloading).

At low frequencies, when you can measure the hysteresis loop and hysteresis loss coefficient is used mechanical loss [3]: $\chi = \Delta W/W$, where W- total work force for a series of mechanical deformation, and ΔW – dissipated energy per cycle of deformation, which is proportional to the square hysteresis loop.

Between χ and tgδ there is a dependence at all frequencies in terms of linear viscoelasticity Thus, for asymmetric vibrations from 0 to $2\varepsilon_0$ according to work [12] such dependence is found:

$$\chi = \frac{2\pi tg\delta}{4\sqrt{1 + tg^2\delta} + \pi tg\delta}. \tag{15}$$

The decision of this equation concerning parameter tg δ gives dependence:

$$tg\delta = \frac{4\chi}{\sqrt{4\pi^2(1-\chi) - 6\chi^2}}. \tag{16}$$

This expression can be represented as: $\text{tg}\delta = \psi$. In expression (13) from condition $E_0 \gg E$ and $I_1 \gg I_0$ at low frequencies in a first approximation, we obtain:

$$\text{tg}\delta = \omega\tau \qquad (17)$$

Equating last two expressions, we receive: $\psi = \omega\tau$, whence $\tau = \psi / \omega$.

According to the second postulate of the Boltzmann adopted in his theory of the elastic aftereffect, and the underlying Boltzmann-Volterra model that describes the relaxation phenomena, using a function of heredity [13]: action occurred in the past few strains on the stresses caused by deformation of the body at any given time, do not depend on each other and therefore algebraically added. This position has received also the name of a principle of the Boltzmann's superposition. It should be noted that the polymer body superposition principle holds in the upper-bounded the range of deformation, stress and rate of change.

Given this principle, considering the dissipative processes occurring during application of periodic voltage to the material in the rubberlike state for a long time, we can conclude that there is an accumulation of mechanical energy dissipation in each cycle. Then, if the energy is transformed into heat during one cycle is determined by parameter χ_1, then in a low heat with the environment during N cycles of the energy dissipated during the time t will be: $\chi_{com} = \chi_1 t \nu$, where $t \nu = N$.

Stored energy in the sample is converted into heat, which should lead to an increase in temperature. The principle of temperature-time superposition [14], which establishes the equivalence of the effect of temperature and duration of exposure on the relaxation properties of polymers, we can assume that the increase in the impact load on the material is proportional to the action of temperature. Empirical dependence of the temperature ΔT of the exposure time t and the intensity (frequency) exposure to ν in the first approximation be written as: $\Delta T = bt\nu$, where the b-parameter, taking into account the characteristics of energy conversion, depending on the structure of the material.

Relaxation time of the supplied periodic voltage decreases with increasing temperature and obeys the Arrhenius equation:

$$\tau = \tau_0 e^{U/RT} \qquad (18)$$

For elastic-plastic bodies similar dependence follows from Aleksandrova-Gurevich's equation [15] and has the form

$$\tau = \tau_0 \exp[(U_0 - a\sigma)/RT], \tag{19}$$

where U_0 – activation energy of relaxation process, the constant of the material.

This equation takes into account the dependence of the relaxation of the load. If we assume that $U_0 - a\sigma \approx U$ and to determine the relative relaxation time as τ_t/τ_1 (the ratio of the current value of the relaxation time of the initial value of you during the load application), then, on the basis of Eq. (18), we can represent this value as an expression:

$$\frac{\tau_t}{\tau_1} = \exp\left(\frac{U}{R(T_1 + \Delta T)} - \frac{U}{RT_1}\right) \tag{20}$$

where the temperature T_1 corresponds to the beginning of load application in the relaxation time τ_1, and the increment of the ΔT is the temperature change in the impact load.

After application of rather simple algebraic transformations the Eq. (19) will become:

$$\frac{1}{\ln\frac{\tau_1}{\tau_t}} = \frac{RT_1}{U} + \frac{RT_1^2}{U}\Delta T \, . \tag{21}$$

If instead the increment of temperature ΔT would use the proposed higher proportion of the value of the exposure time t and frequency of the applied load v, then the Eq. (20) becomes:

$$\left(\ln\frac{\tau_1}{\tau_t}\right)^{-1} = \frac{RT_1}{U} + \frac{RT_1^2}{U}(bvt)^{-1} \tag{22}$$

Using this expression and taking into account the approach adopted, it is possible to experimental data on changes in the mechanical loss factor (tangent of mechanical losses) over time, the impact loads to find the estimates of the activation energy of relaxation process, determine the extent to which the process is stationary (steady), the degree of linearity of the relaxation processes and the range of conditions and the regime correct determination of relaxation parameters for a periodic load.

16.3 EXPERIMENTAL

In the present work as object of research polyethylene of low density (*LDPE*)). Samples in the form of the cylinder: diameter (d) from 8 mm at a parity h/d = 1.5 made pressing at temperature 180°C, pressure of 150 kgs/sm². To obtain a homogeneous sample produced an extract of the pressure and temperature 180°C at least 10 minutes with pre-pressing for the release of air located between the grains of the original polymer.

Samples subjected to periodic monoaxial deformation of compression on the installation described in [16] at room temperature (293 K). As a result of the periodic action of the voltage on the sample received the stress-strain during loading and unloading of the form a hysteresis loop. The study used three discrete frequencies of loading: 0.017, 0.17 and 1.7 Hz. Under each of these frequencies is tested at least three samples for 30 min., Taking readings every 5 min. Results cheated, define the parameters of the mechanical losses as the ratio of the hysteresis loop to the area between the curve of loading and the axis of strain ($\chi = \Delta W/W = S_{loop}/S_{hole}$). The results of measurements of at least three samples were averaged and subjected to further processing in accordance with those presented in the theoretical part of the calculations.

16.4 CONCLUDING REMARKS

Figure 1 shows kinetic curves of variation of the mechanical loss by prolonged exposure of three frequencies: 0.017, 0.17 and 1.7 Hz. It is seen that with increasing time of deformation coefficient of mechanical losses vary, but relationships have different characteristics

So for low frequencies 0.017Гц (curve 1) and 0.17 Hz (curve 2) the initial value of this parameter is higher than the next. In all probability this is due to the fact that during the reduction of χ is the system output at steady state, i.e., where the constant C in Eq. (4) becomes equal to 0. For higher frequency – 1.7 Hz (curve 3) the establishment of this regime is much faster. For higher frequency – 1.7 Hz (curve 3) the establishment of this regime is much faster.

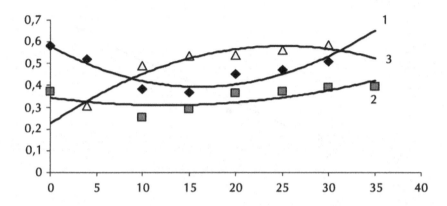

FIGURE 1 Change of the mechanical losses coefficient eventually at influence of periodic loadings with frequency: 1 – 0.017 Hz, 2 – 0.17 Hz, 3 – 1.7Hz.

Figure 2 shows the inverse of the logarithm of the relative relaxation time in degree –1 (which corresponds to the left side of the Eq. (21)) the reciprocal of the time of impact load on the sample with a frequency of 0.017 Hz. Dependence is well approximated by a straight line, i.e., the found coordinates dependence of relative time of a relaxation and time of influence of loading is directly proportional. Meaningfully the value found at the intersection of this dependence with the vertical axis and bearing T_1 equal to ambient temperature (293 K) can determine the activation energy. In these conditions (Fig. 2) it is equal to 4.9 kJ/mol. The slope in Fig. 2 provides an estimate of the value of the parameter "b" in formula (21). The calculation shows that the frequency of 0.017 Hz, b = 11.88. Since the dependence is linear in a rather wide time interval, this gives reason to conclude that the activation energy of relaxation process with periodic loading of the solid LDPE under these conditions virtually unchanged.

Using the representation of the relative relaxation time of the duration of the periodic effects in the corresponding coordinates for the frequency of 0.17 Hz (Fig. 3) makes it possible to calculate the activation energy and the parameter b for the relaxation process of solid LDPE under these conditions. With accuracy to the experimental errors (for a test frequency of 0.17 Hz), the activation energy is 4.9 kJ/mol and b = 0.414.

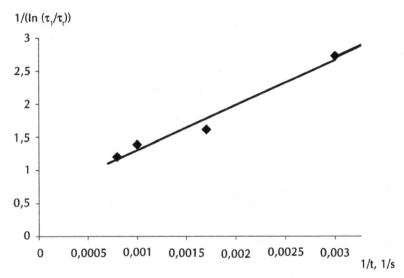

FIGURE 2 Dependence of relative time of a relaxation (τ_t/τ_0) on duration of influence with frequency of 0.017 Hz.

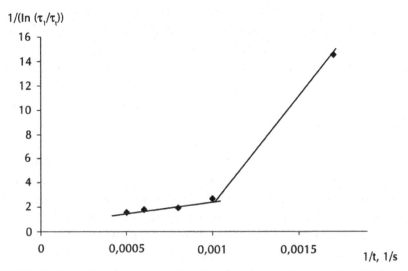

FIGURE 3 Dependence of relative time of a relaxation (τ_t/τ_0) on duration of influence with frequency of 0.17 Hz.

In Fig. 4 a similar dependence is shown for the frequency of 1.7 Hz. The calculated value of activation energy is the 2.4 kJ/mol. The value b =0.04.

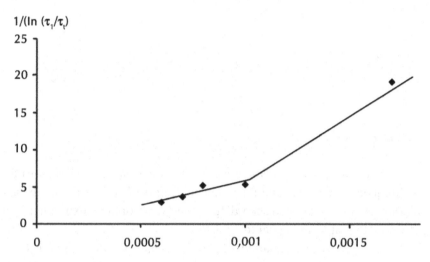

FIGURE 4 Dependence of relative time of a relaxation (τ_t/τ_0) on duration of influence with frequency of 1.7 Hz.

Analyzing the obtained values, we should note a decrease of b with increasing frequency (Fig. 5), indicating that the difference in the relaxation processes occurring at different frequencies. Another interesting fact is that for frequencies 1.7 and 0.17 Hz sampling rate on the parameter b is the same and equal to 0.07, while the frequency of 0.017 Hz (three orders of magnitude smaller than the largest) is the product of three times and equals 0.202. It should be noted the difference in the nature of the drawings: Fig. 2 – to $v = 0.017$ Hz and Figs. 3 and 4, respectively for 0.17 and 1.7 Hz.

In the Fig. 2 no jumps in the dependence, in Figs. 3 and 4, the values for the initial periods of exposure time are several times higher than those in the subsequent course of dependencies. Perhaps this difference is caused by various structural transformations under mechanical loading with different frequencies.

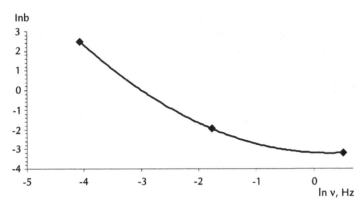

FIGURE 5   Dependence of parameter "b" from frequency of loading.

With regard to the activation energy, it is the smallest (2.4 kJ/mol) for the frequency of 1.7 Hz and for frequencies of 0.17 and 0.017 Hz, the activation energy, calculated according to the results of these experiments, it turns out the same and equal to 4.9 kJ.

To paraphrase the equation of the Aleksandrov-Gurevich, Eq. (18), where instead of the stress (σ) using the frequency ν, instead of the coefficient "a" use the "b," then we can define a certain characteristic value, similar to U_0, the initial activation energy of relaxation process, a constant:

$$U_0 = U + b\nu$$

Analysis of the dependence of the initial activation energy U_0 of the frequency (Fig. 6) shows that with increasing frequency ν decreases linearly energy U_0 frequency of loading, Hz.

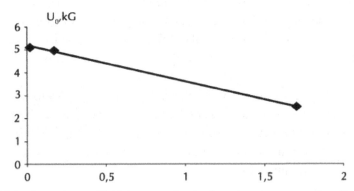

FIGURE 6   Dependence of initial energy of activation relaxation process from frequency of loading.

Thus, using the principle of temperature-time superposition and kinetic coefficient of mechanical losses can be at different intensities of loads to determine the time interval in which the measurement of relaxation parameters will be most correct. Besides using the above approximation, we can give a preliminary assessment of the relaxation parameters and analyze the nature of relaxation processes taking measurements without changing the initial temperature.

KEYWORDS

- **Boltzmann-Volterra model**
- **Monoaxial deformation**
- **Relaxation time**
- **Rubbery**

REFERENCES

1. Nikolis, G.; Prigogine. I. *Exploring Complexity. An introduction.* W. H. Freeman and Company: New York, 1997.
2. Luce, R. D.; Krantz, D. H.; Suppes, P.; Tversky, A. Foundations of measurement. (Representation, xiomatization and invariance). New York: Academic Press, 1998.
3. Turner, R. B. *Pol. Com.,* **1984,** *5*(2), 151–154.
4. Bartenev, R. M.; Baglyuk, S. V.; Tulinova, V. V. *Polym. Sci.,* **1990,** *32*(7), 1364–1371.
5. Shershnev, V. A.; Emeljanov, S. V. *Review MITH. T.,* **2006,** *1*(5), 3–19.
6. Malkin, A .J. *Vysokomolek. Soedin.,* *51*(1), 106–136.
7. Malkin, A. J. *Methods of Measurement of Mechanical Properties of Polymers.* Chemistry Moscow, **1978,** p 336
8. Askadsky, A. A.; Markov, V. M.; Kovriga, O. V. *Vysokomolek. Soed.,* **2009,** *51*(5), 576–582.
9. Nechaev, S. K.; Khokhlov, A. R. *Physics Lett.,* **1988,** *126*(7), 431–433.
10. Lukovkin, G. M.; Arzhakov, M. S.; Salko, A. E.; Arzhakov, S.A. *Deformation Destruct. Mater.,* **2006,** *6,* 18–24.
11. Gotlib, Y. Y.; Torchinskii, I. A.; Toshchevikov, V. P.; Shevelev, V. A. *Polymer Sci.,* **2010,** *52*(1), 82–93.
12. Irzhak, V. I. *Russ Chem. Rev.,* **2000,** *69*(3), 261–265.
13. Ferry, J. *Viscoelastic Properties of Polymer,* 3 edn. Wiley, 1980, p 672.

14. Doi, M.; Edwards, S. F. *The Theory of Polymer Dynamics,* Oxford University Press: Oxford, 1986.

15. Suen, J. K. C.; Joo, Y. L.; Armstrong, R. C. *Annu. Rev. Fluid Mech.*, **2002**, *34*, 417–444.

16. Lomovsky, V. A.; Fomkina, Z. I.; Bulba, V. L. et al. *Research Technique of the Relaxation Phenomena of Polymers in a High Elastic Condition.* (Dynamic methods). MITHT. 5, 2010.

CHAPTER 17

THE MORPHOLOGICAL FEATURES OF POLY(3-HYDROXYBUTYRATE) WITH AN ETHYLENE-PROPYLENE COPOLYMER BLENDS

A. A. OL'KHOV, L. S. SHIBRYAEVA, YU. V. TERTYSHNAYA, A. L. IORDANSKII, and G. E. ZAIKOV

CONTENTS

17.1 Introduction..330
17.2 Experimental..330
17.3 Results and Discussion 331
Keywords 337
References..338

17.1   INTRODUCTION

The methods of DSC and IR spectroscopy were used to study various blends of poly(3-hydroxybu-tyrate) with ethylene-propylene copolymer rubber (EP). When the weight fractions of the initial polymers are equal, a phase inversion takes place; as the blends are enriched with EP, the degree of crystallinity of poly(3-hydroxybutyrate) decreases. In blends, the degradation of poly(3-hydroxybutyrate) begins at a lower temperature compared to the pure polymer and the thermooxidative activity of the ethylene-propylene copolymer in the blend decreases in comparison with the pure copolymer.

Composite materials based on biodegradable polymers are currently evoking great scientific and practical interest. Among these polymers is poly(3-hydroxybu-tyrate) (PHB), which belongs to the class of poly(3-hydroxyalkanoates). Because of its good mechanical properties (close to those of PP) and biodegradability, PHB has been intensely studied in the literature [1]. However, because of its significant brittleness and high cost, PHB is virtually always employed in the form of blends with starch, cellulose, PE [2], etc., rather than in pure form.

This work is concerned with the study of the structural features of PHB-EPC blends and their thermal degradation.

17.2   EXPERIMENTAL

The materials used in this study were EPC of CO-059 grade (Dutral, Italy) containing 67.4 mol % ethylene units and 32.6 mol % propylene units. PHB with $М_ц = 2.5 \times 105$ (Biomer, Germany) was used in the form of a fine powder. The PHB : EPC ratios were as follows: 100 : 0, 80 : 20, 70 : 30, 50 : 50, 30 : 70, 20 : 80, and 0: 100 wt%.

The preliminary mixing of the components was performed using laboratory bending microrolls (brand VK-6) under heating: the microroll diameter was 80 mm, the friction coefficient was 1.4, the low-speed roller revolved at 8 rpm, and the gap between the rolls was 0.05 mm. The blending took place at 150°C for 5 min.

Films were prepared by pressing using a manual heated press at 190°C and at a pressure of 5 MPa; the cooling rate was ~50°C/min.

The thermophysical characteristics of the tested films and the data on their thermal degradation were obtained using a DSM-2M differential scanning calorimeter (the scanning rate was 16 K/min); the sample weight varied from 8 to 15 mg; and the device was calibrated using indium with Tm = 156.6°C. To determine the degree of crystallinity, the melting heat of the crystalline PHB (90 J/g) was used [2]. The Tm and Ta values were determined with an accuracy up to 1°C. The degree of crystallinity was calculated with an error up to ±10%. The structure of polymer chains was determined using IR spectroscopy (Specord M-80). The bands used for the analysis were structure-sensitive bands at 720 and 620 cm^{-1}, which belong to EPC and PHB, respectively [3]. The error in the determination of reduced band intensities did not exceed 15%.

17.3   RESULTS AND DISCUSSION

The melting endotherms of PHB, EPC, and their blends are shown in Fig. 1. Apparently, all the first melting thermograms (except for that of EPC) show a single peak characteristic of PHB.

The thermophysical characteristics obtained using DSC for blends of various compositions are listed in Table 1. As is apparent from this table, the melting heat $\triangle H_{ml}$ of PHB during first melting changes just slightly in comparison with the starting polymer. During cooling, only a single peak corresponding to the crystallizing PHB additionally appears.

However, the repeated melting endotherms of some blends (70% PHB + 30% EPC, 50% PHB + 50% EPC) display a low-temperature shoulder. Note that the melting enthalpy significantly changes as one passes from an EPC-enriched blend to a composition where PHB is predominant. When the content of EPC is high, the melting heat $\triangle H_{m2}$ of the recrystallized PHB significantly decreases. This effect should not be regarded as a consequence of the temperature factor, because the material was heated up to 195°C during the DSM-2M experiment and the films were prepared at 190°C; the scanning rate was significantly lower than the cooling rate during the formation of the films (50 K/min). Thus, the state of the system after remelting during DSC measurements is close to equilibrium.

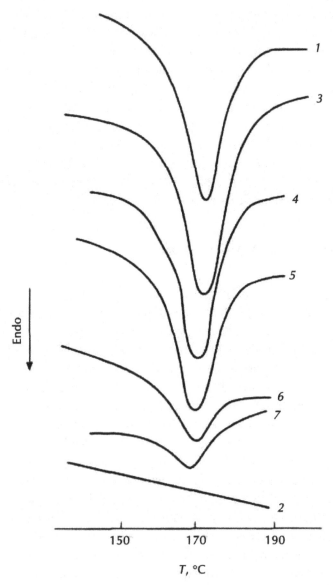

FIGURE 1 The melting endotherms of (1) PHB, (2) EPC, and their blends with compositions (3) 80 : 20, (4) 70 : 30, (5) 50: 50, (6) 30: 70, and (7) 20: 80 wt %.

TABLE 1   The thermophysical properties of PHB-EPC films.

PHB : EPC, wt %	T_m, °C	ΔH^*_{m1}	ΔH^*_{m2}	T_{cr} °C	Degree of crystallinity**, %
100 : 0	174	88.3	88.9	64	98
80 : 20	173	75.8	76.2	64	84
70 : 30	172	59.5	60.1	60	66
50 : 50	172	56.4	52.1	62	63
30 : 70	172	29.3	20.5	60	33
20 : 80	171	22.3	15.7	–	25
0 : 100	–	–	–	–	–

* Calculated as areas under the melting curves: the first melting and the melting after the recrystalization, respectively.
** Calculated according to the DH*ml values.

These results make it possible to assume that the melting heat and the degree of crystallinity of PHB decrease in EPC-enriched blends due to the mutual segmental solubility of the polymers [4] and due to the appearance of an extended interfacial layer. Also note that the degree of crystallinity may decrease because of the slow structural relaxation of the rigid-chain PHB. This, in turn, should affect the nature of interaction between the blend components. However, the absence of significant changes in the T_m and T_a values of PHB in blends indicates that EPC does not participate in nucleation during PHB crystallization and the decrease in the melting enthalpy of PHB is not associated with a decrease in the structural relaxation rate in its phase. Thus, the crystallinity of PHB decreases because of its significant amorphization related to the segmental solubility of blend components and to the presence of the extended interfacial layer.

Figure 2 shows the IR spectra for two blends of different compositions. As is known, the informative structure-sensitive band for PHB is that at 1228 cm^{-1} [5]. Unfortunately, the intensity of this band cannot be clearly determined in the present case, because it cannot be separated from the EPC structural band at 1242 cm^{-1} [3]. The bands used for this work were the band at 620 cm^{-1} (PHB) and the band at 720 cm^{-1} (EPC) [6], which correspond to vibrations of C–C bonds in methylene sequences (CH$_2$), where n > 5, occurring in the trans-zigzag conformation. The ratios between the optical densities of the bands at 720 and 620 cm^{-1} (D_{720}/D_{620}) are transformed in the coordinates of the equation where (5 is the fraction of EPC and W is the quantity characterizing a change in the ratio between structural elements corresponding to regular methylene sequences in EPC and PHB.

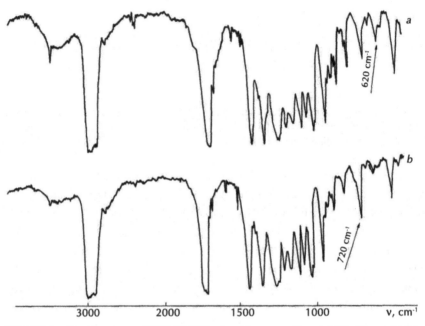

FIGURE 2 The IR spectra of PHB-EPC blends with compositions (a) 80 : 20 and (b) 20 : 80 wt %.

Figure 3 demonstrates the value of W plotted as a function of the blend composition. Apparently, this dependence is represented by a straight line in these coordinates but shows an inflection point. The latter provides evidence that phase inversion takes place and that the nature of intermolecular interactions between the polymer and the rubber changes.

$$W = \log[D_{720}\beta / D_{620}(1-\beta)] + 2$$

where (5 is the fraction of EPC and W is the quantity characterizing a change in the ratio between structural elements corresponding to regular methylene sequences in EPC and PHB.

Figure 3 demonstrates the value of W plotted as a function of the blend composition. Apparently, this dependence is represented by a straight line in these coordinates but shows an inflection point. The latter provides evidence that phase inversion takes place and that the nature of intermolecular interactions between the polymer and the rubber changes.

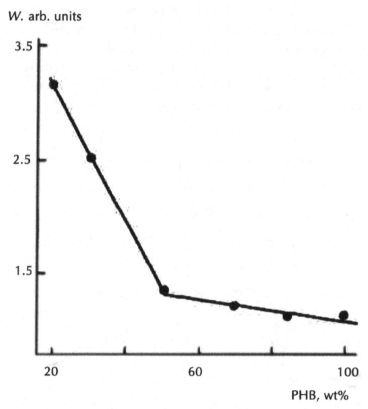

FIGURE 3   Plot of W vs. the content of PHB in the blend.

The phase inversion causes the blends in question to behave in different ways during their thermal degradation. The DSM-2M traces (Fig. 4) were measured in the range 100–500°C. The thermograms of the blends display exothermic peaks of the thermal oxidation of EPC in the range 370–400°C and endothermic peaks of the thermal degradation of PHB at T > 250°C. For the pure PHB and EPC, the aforementioned peaks are observed in the ranges 200–300°C and 360–430°C, respectively. The blend samples studied in this work display two peaks each, thus confirming the existence of two phases. Note that the peak width increases (curves 3, 4 in Fig. 4), and the heat Q of the thermal degradation of PHB changes in all the blends studied here (Table 2). This effect is apparently determined by the blend structure rather than by its composition. In blends, PHB becomes more active compared to the pure polymer and the rate of its thermal degradation increases. The temperature corresponding to the onset of thermal degradation 7^ decreases from 255°C; the value characteristic of the pure PHB, to 180°C (Table 2). The structure of the polymer becomes less perfect in this case; two likely reasons for this are a change in the morphology and the appearance of an extended interfacial layer.

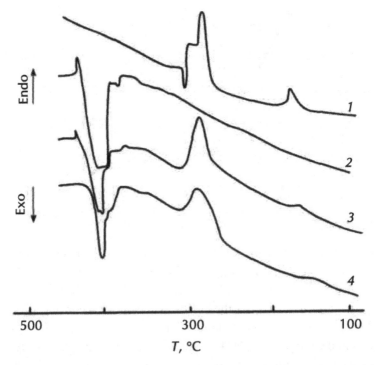

FIGURE 4   The DSC traces of (1) PHB, (2) EPC, and their blends with compositions (3) 70 : 30 and (4) 30 : 70 wt %.

As to EPC, it acquires a higher thermal stability in the blends under examination, as indicated by the increase in the temperature corresponding to the onset of its thermal oxidation T° (Table 2). The position of the exothermic peaks on the temperature scale characteristic of EPC indicates that its activity in blends is lower than that in the pure sample. The low-temperature shoulder of the exothermic EPC peak in the range 360–380°C (Fig. 4) decreases with increasing content of PHB. Apparently, this effect is due to a change in the copolymer structure related to the interpenetration of PHB and EPC segments.

TABLE 2   The parameters of the thermal degradation process.

PHB : EPC, wt %	T_{ad}(EPC), °C	T_{ad}(PHB), °C	Q^*(PHB), kJ/g
100 : 0	–	2553	0.53
70 : 30	370	180	1.38
30 : 70	380	250	0.51
0 : 100	360	–	–

* The specific heat of thermal degradation per g of PHB

Thus, the existence of two peaks in DSC thermograms of the blends indicates the presence of two phases in the PHB-EPC blends. The phase inversion takes place in the vicinity of the composition with equal component weights. The components influence each other during film formation, and, hence, the appearance of the extended interfacial layer is presumed for samples containing more than 50% EPC. A change in the structure of the blends affects their thermal degradation. The degradation of PHB in blends is more pronounced than that in the pure PHB, but the thermal oxidation of EPC is retarded.

KEYWORDS

- **Biodegradability**
- **EPC**
- **PHB**
- **Thermal oxidation**

REFERENCES

1. Seebach, D.; Brunner, A.; Bachmann, B. M.; Hoffman, T.; Kuhnle, F. N. M.; Lengweier, U. D. *Biopolymers and Biooligomers of (R)-3-Hydroxyalkanoic Acids: Contribution of Synthetic Organic Chemists*, Zurich: Edgenos-sische Technicshe Hochschule,
   1996.
2. Ol'khov, A. A.; Vlasov, S. V.; Shibryaeva, L. S.; Litvi-nov, I. A.; Tarasova, N. A.;
   Kosenko, R. Yu.; Lordan-skii, A. L.; *Polym. Sci., Ser. A,* **2000**, *42(4)*, 447.
3. Elliot, A. In *Infra-Red Spectra and Structure of Organic Long-Chain Polymers*, London: Edward Arnold, 1969.
4. Lipatov, Yu. S. In *Mezhfaznye yavleniya v polimerakh (Interphase Phenomena in
   Polymers)*, Kiev: Naukova Dumka, 1980.
5. Labeek, G.; Vorenkamp, E. J.; Schouten, A. J. *Macromolecules*, **1995**, *28*(6), 2023.
5. Painter, P. C.; Coleman, M. M.; Koenig, J. L. *The Theory of Vibrational Spectroscopy
   and Its Application to Polymeric Materials*, New York: Wiley, 1982.

CHAPTER 18

PERSPECTIVES OF APPLICATION MULTI-ANGLE LASER LIGHT SCATTERING METHOD FOR QUALITY CONTROL OF MEDICINES

A. V. KARYAKIN, E. D. SKOTSELYAS, and P. A. FLEGONTOV

CONTENTS

18.1 Introduction 340

18.2 Materials, Instruments, and Methods 341

18.3 Results 342

    18.3.1 Molecular Weight Distribution in Polysaccharides 342

    18.3.2 Molecular Weight Distribution in Albumin and
            Immunoglobulin Products 343

18.4 Conclusion 350

Keywords 351

References 351

18.1   INTRODUCTION

Multi-angle laser light scattering detector (MALLS) method was validate for the analyses molecular mass distribution of hydroxylethyl starch and was used for the determination percentage and molecular mass of polymers and aggregates in albumin's and immunoglobulin's products obtained from different manufacturers. Calculated percentages of aggregates in albumin products by both European Pharmacopoeia and MALLS methods were compared.

Application of light scattering theory has become quite widely used in the chemistry of polymers and biochemistry for the last 15–20 years. Development of the multi-angle laser light scattering detector (MALLS) was one of the breakthrough achievements in the former field. MALLS is constituted of a flow-through cell surrounded by 18 photodiodes registries at different angles vertically polarized monochromatic laser light scattering. Collected data are used to calculate molecular mass (MM) of proteins [1–3] and polysaccharides [4–6], sometimes size of the molecules, and, if combined with size-exclusion chromatography (SEC), molecular mass distribution (MMD).

MALLS signal is proportional and directly reflects the size of molecules and their concentrations, which makes it more sensitive for detection of large molecules and their complexes. The combination of MALLS and ultraviolet (UV) or refractive index (RI) detectors with SEC not only provides information about sample concentration but also the sample molecular weight, regardless of the conformation of the molecules detected in the fractions at the time of the analysis [7].

MALLS is used to quickly and accurately determine MMD of polysaccharide based plasma surrogates and European Pharmacopoeia (EP) recommend this method for the analyses of hydroxylethyl starch (HES) [8] and dextran products. There is no such recommendation provided in Russian Pharmacopoeia.

The MALLS-RI-SEC approach was implemented to validate the method MMD for the analyses of hydroxylethyl starch and MALLS-UV-SEC – to investigate the commercial manufacturing of the albumin and immunoglobulin products.

18.2 MATERIALS, INSTRUMENTS, AND METHODS

- Chromatographic system Knauer: isocratic pump Smartline 1000, injector with a loop volume 20 mkl;
- Smartline UV detector 2500 (Knauer), the wavelength of 280 nm;
- Refractive index detector Optilab rEX (Wyatt Technology);
- MALLS DAWN HELEOS-II (Wyatt Technology);
- Columns combination: one column TSKgel G3000PW 300 × 7.5 mm, 12 µm and two columns TSKgel G5000PW 300 × 7.5 mm, 17 µm (Tosoh Bioscience) for HES analysis, molecular weights ranging from 10 to 2000 kDa;
- Mobile phase for HES analysis – 0.05 M $NaNO_3$, 1.54 mM NaN_3 in water;
- Molecular weight standards set (American Polymer Standards corporation);
- Column Protein Pak 300SW (Waters) for separation of native globular proteins with the molecular weights ranging from 10 to 400 kDa;
- Mobile phase for protein analysis - 0.2 M phosphate buffer pH 7.0;
- Flow rate – 1 ml/min.

Molecular weight was calculated using Astra 5.3.4.14 software (Wyatt Technology), according to the basic light scattering theory equation, in a simplified form:

$$M = (K{\times}c / R(\Theta)_{{\to}0})^{-1}$$

where: M – weight-average molecular weight;

K – an optical parameter, depending from refractive index and refractive index increment;

c – concentration in mg/ml;

R(Θ) – intensity of light scattering at angle Θ.

R(Θ) is being determined for different angles and then extrapolate these values to zero angle. This approach allows eliminating the influence of molecular structure [7].

Distribution of the protein fractions percentage was calculated by the method of EP, as well as on the basis of MALLS detector signal. In this case, peak area was reduced to the unit molecular weight and further calculation was similar to the first option.

For the analysis were used:

- Commercially manufactured ReoHES 130, Vilana, Russia and sub-stance of HES, Serumwerk Bernburg AG, Germany;

- 40 series commercially manufactured products of albumin for infu-sion (Tambov Region Blood Transfusion Station (RBTS), Russia, 10 series; RBTS Lipetsk, Russia, 8 series; Blood Transfusion Sta-tion (BTS) Department of Health in Moscow № 1, Russia, 8 series; Baxter, Germany, 4 series; Oktapharma, Austria, 5 series; Biotest, Austria, 5 series, and albumin from human serum, lyophilized pow-der, fatty acid free, globulin free (Sigma, A3782);

- 9 series commercially manufactured products of immunoglobulin for intramuscular and intravenous injection, compliant EP (RBTS Nizhny Novgorod, Russia, 2 series; Oktapharma, Austria, 3 series; Talecris Biotherapeutics, USA, 1 series and Complex immunoglob-ulin preparation (CIP) Immuno-Gem, Russia, 3 series).

All protein products were diluted by eluent to a final concentration 1% (w/v).

For thermal denaturation 1% (w/v) albumin's solution was incubated at 75°C.

18.3 RESULTS

18.3.1 MOLECULAR WEIGHT DISTRIBUTION IN POLYSACCHARIDES

Typical chromatogram of HES (Vilana, Russia) is presented on Fig. 1. The validation results of polysaccharides MMD determination:

Linearity. Linear correlation coefficient was $r = -0.9989$ for the curve logarithm of the molecular weight versus retention time in the range MM from 10 to 2000 kDa;

The accuracy of the method was characterized by the "recovery." It's range was 98.2–100.3. Such a narrow range suggests that the systematic error is not significant;

Robustness. The relative standard deviation was less than 1% for the series of analyzes at different column's temperatures and mobile phase velocity. This indicates the stability of the chromatographic system to the influence of external factors;

Precision. The relative standard deviation was less than 1% for the repeatability and intermediate precision.

This method is universal for different polymers, and it can be used for MMD analysis of blood substitutes based on dextran instead of requiring a lengthy and complicated procedure recommended by EP.

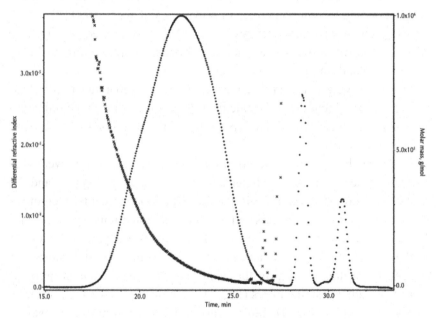

FIGURE 1   Typical chromatogram of ReoHES 130 (Vilana, Russia). "+" – Refractive index data, "x" – molar mass plot.

18.3.2   MOLECULAR WEIGHT DISTRIBUTION IN ALBUMIN AND IMMUNOGLOBULIN PRODUCTS

The MMD distribution in albumin and immunoglobulin products manufactured from the blood plasma according to the EP requirements is characterized by the method of SEC with detection of the protein absorption at

280 nm UV. Distribution of the fraction percentage in a sample is evaluated based on the relative percentage of each fraction peak calculated from the total area of all peaks except for those reflecting peptide fragments and other low molecular weight substances.

When the samples of the human albumin products are analyzed according to the EP regulations, the peaks detected in the void volume of the column are considered to be reflecting polymers and aggregates and contribution of those peaks reflecting absorption of a stabilizing additive is not considered. Usually, the area of the peak(s) due to polymers and aggregates is not greater than 10 percent of the total area of the chromatogram (corresponding to about 5 percent of polymers and aggregates) [8].

The comparative analyses protein's MMD was carried out by both MALLS and EP methods in commercial products of albumin from different Russian and foreign manufacturers.

The following detected fractions were clearly separated in all chromatograms and included in the calculations: fractions in the void reflecting polymers and aggregates as well as the fractions corresponding to dimers, monomers and fragments.

In the analysis of the albumin product samples all detectors were able to register peaks of the monomer and dimer fractions (Figs. 2, and 3). There were no peaks revealed in the 200–400 kDa molecular weight regions corresponding to trimers and tetramers of albumin in any of the analyzed commercial products. Fraction of aggregates was practically absent in the production of Lipetsk and Tambov RBTS. Peak of high molecular weight aggregates, eluting in the dead volume, were found in preparations from other manufacturers (Fig. 2). Peak of this fraction was divided in several samples (Fig. 3). Table 1 shown the range of molecular weight and radii mean square of the aggregates fractions obtained in the analysis of various series albumin's products. Percentage of fraction of aggregates was determined by the calculated values of molecular weight and peak's areas recorded by MALLS (MALLS's method), as well as the peak areas recorded by UV detector (EP's method). In all cases, the EP's method gives higher results than MALLS (Figs. 4 and 5). The percentage ratio of the aggregates calculated by these two methods, about the same for different lots of the drug from one manufacturer and varies among different firms.

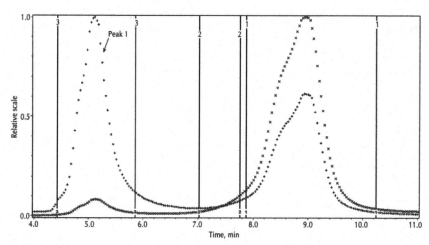

FIGURE 2 Typical chromatogram of albumin with one peak of aggregates (Biotest, Austria). Areas selected on the chromatogram: 1 – monomers, 2 – dimers, 3 – aggregates. "+" – MALLS, "x" – UV-detector.

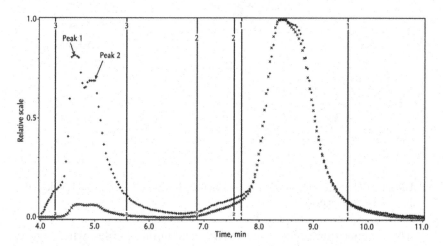

FIGURE 3 Typical chromatogram of albumin with two peaks of aggregates (Baxter, Germany). Areas selected on the chromatogram: 1 – monomers, 2 – dimers, 3 – aggregates. "+" – MALLS, "x" – UV-detector.

TABLE 1   Molecular parameters of aggregates in the analyzed products of albumin.

Manufactures	MM, MDa		Radii of aggregates molecules, nm	
	Peak 1	Peak 2	Peak 1	Peak 2
Moscow BTS	4.4–5.0	2.3–3.2	19–82	18–60
Baxter	5.8–10.0	2.1–4.3	44–137	35–63
Biotest	3.5–4.6	–	30–55	–
Oktapharma	2.8–4.2	–	30–90	–
Sigma	0.6–1.3	–	30–40	–

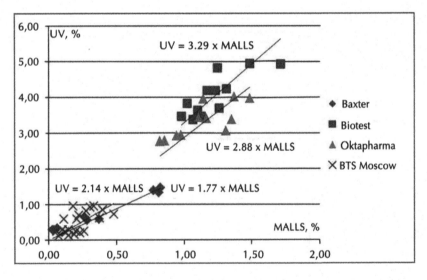

FIGURE 4   Relative content of aggregates calculated by both MALLS and UV detection for the one aggregates peak (Fig.2).

Thermal denaturation of Lipetsk RBTS albumin samples, where aggregates were absented, revealed large aggregates with a molecular mass of 15 MDa and the mean square radius of 120 nm (Fig. 6a, b, c). Intermediate oligomer peaks of albumin were not founded.

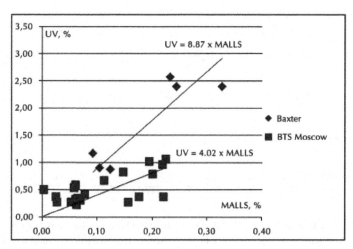

FIGURE 5 Relative content of the aggregates calculated by both MALLS and UV detection for the divided aggregates peak.

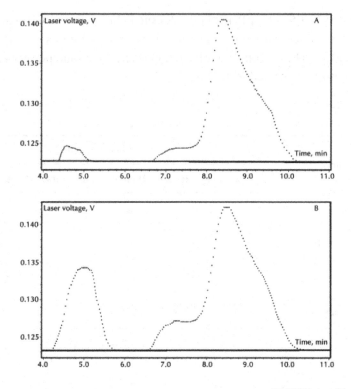

FIGURE 6 *(Continued)*

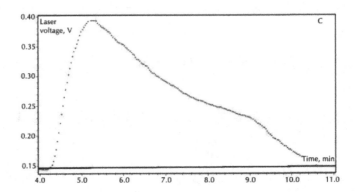

FIGURE 6 Chromatograms of thermal treatment albumin's solution (Lipesk RBTS, Russia): A – before treatment; B – 1 hour of thermal treatment; C – 2 hours of thermal treatment.

Both MALLS and EP methods detected the fraction of fragments, monomers, dimers and aggregates in all immunoglobulin's preparations (Fig. 7a, b, c). The calculated molecular mass fractions and percentages of aggregates are presented in Tables 2 and 3.

TABLE 2 Molecular weight distribution in immunoglobulin's products.

Manufactures	Molecular mass, kDa			
	Aggregates	Dimers	Monomers	Fragments
Octapharma	1,216	307	157	–
N. Novgorod RBTS	1,636	298	157	119
Talecris Biotherapeutics	1,396	308	154	89
Immuno-Gem	5,471 692	332	173	64

TABLE 3   Distribution of the fraction percentage in immunoglobulin's products.

Manufactures	Distribution of the fraction percentage, %					
	Method	Aggregates		Dimers	Monomers	Fragments
Octapharma	EP	1.1		18.0	79.9	1.0
	MALLS	0.1		16.8	82.0	1.1
N. Novgorod RBTS	EP	1.1		16.3	81.5	1.1
	MALLS	0.7		14.8	83.4	1.1
Talecris Biotherapeutics	EP	0.4		4.6	93.4	1.6
	MALLS	1.8		5.9	89.1	3.2
Immuno-Gem	EP	21.4	6.6	7.7	58.8	5.5
	MALLS	16.2	8.5	9.2	57.1	9.0

In the preparations for intravenous and intramuscular injection (Fig. 7a, b) MALLS detected an additional peak of high molecular weight fraction. On the chromatograms of the complex immunoglobulin product (CIP), which consists of different classes of immunoglobulins (Fig. 7C), UV detector revealed two resolved peaks corresponding to the aggregates and monomers of IgG, and two unresolved fractions with molecular weight values of approximately 700 and 330 kDa, corresponding to IgA and IgM. MALLS reveals two peaks: with the elution times of 5 min for the first fraction (the molecular weight more then 5,000 kDa) and 7 minutes for the second fraction (MM 176 kDa).

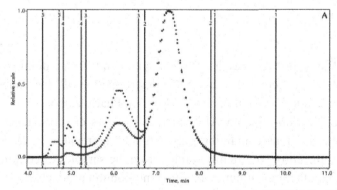

FIGURE 7   *(Continued)*

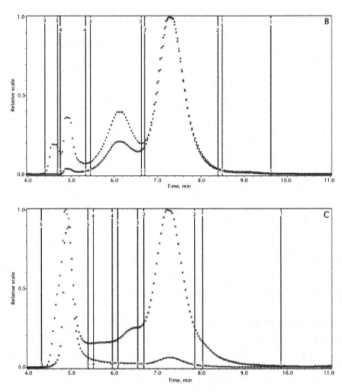

FIGURE 7 Typical chromatograms of immunoglobulins. Areas selected on the chromatogram: 1 – fragments, 2 – monomers, 3 – dimers, 4 and 5 – aggregates. "+" – MALLS, "x" – UV-detector. A - Immunoglobulin Human Antirhesus Rh0(D) 'Rezonativ' (Octapharma, Austria); B - Immunoglobulin Human Normal (N.Novgorod RBTS, Russia); C – Complex immunoglobulin preparation (Immuno-Gem, Russia).

18.4 CONCLUSION

1. The validation results allow recommending including MALLS detector method for the MMD determination of blood substitutes based on HES and dextran in Russian Pharmacopoeia.

2. Utilization of the MALLS detector allows effective detection of the fractions containing polymers and aggregates in the analysis of the chromatograms obtained from the samples of commercially manufactured albumin and immunoglobulin products.

3. Utilization of the MALLS detector showed lower concentration of aggregates in the samples of the albumin products as compare with the calculations derived from conventional EP analysis.
4. The percentage of macromolecular complexes of albumin increased as the result of thermal inactivation.
5. Application of the MALLS detector containing method to the samples of the immunoglobulin samples allowed registering macromolecular impurities in the fraction of polymers and aggregates, which were not detected by just UV detector.

KEYWORDS

- **Laser light scattering**
- **Polysaccharides**
- **Recovery**
- **Trimmers**

REFERENCES

1. Rongxin, S. U.; Wei, Q. J.; Zhimin, H. E.; Yubin, Z. G.; Fengmin, J. N. *Food Hydrocolloids*, **2008**, *22*, 995–1005.
2. Barackman, J.; Prado, I.; Karunatilake, C. *J. Chro.*, **2004**, *1043*, 57–64.
3. Sjekloca, L.; Konarev, P. V.; Eccleston, J.; Taylor, I. A.; Svergun, D. I.; Pastore, A. *Biochem. J.*, **2009**, *419*, 347–357.
4. Baggenstoss, B. A.; Weigel, P. H. *Ana. Biochem.*, **2006**, *352*(2), 243–251.
5. Ming-Hsuan, C.; Bergman, C. J. *Cardohydrate Polym.*, **2007**, *69*, 562–578.
6. Gidh, A. V.; Decker, S. R.; Vinzant, T. B.; Himmela, M. E.; Williford, C. *J. Chromatogr.*, **2006**, *1114*, 102–110.
7. Wen, J. I. E; Arakawa, T.; Philo, J. S. *Ana. Biochem.*, **1996**, *240*, 155–166.
8. European Pharmacopoeia 7.0, 2010.

CHAPTER 19

INTERACTION AND STRUCTURE FORMATION OF GELATIN TYPE A WITH THERMO AGGREGATES OF BOVINE SERUM ALBUMIN

Y. A. ANTONOV and I. L. ZHURAVLEVA

CONTENTS

19.1 Introduction.. 354
19.2 Objectives 355
19.3 Materials and Methods.. 356
     19.3.1 DLS... 357
     19.3.2 CD Measurements 357
     19.3.3 Fluorescence Measurements... 358
     19.3.4 Rosenberg's Method 358
19.4 Results and Discussion 359
     19.4.1 Characterization of the Native BSA and Thermally
            Aggregated BSA Sample by DLS 359
     19.4.2 Limited Thermo Aggregation as a Factor Determining
            Molecular Sizes of BSA-Gelatin Complexes and Stability
            BSA Structure in Complex 361
19.5 Conclusion 370
Keywords 370
References.. 370

19.1   INTRODUCTION

Intermacromolecular interactions are very important in natural biological systems [1–3] as well as in biotechnological applications [4–6]. For example, many molecules including biopolymers participate in biological functions as a molecular assembly or tissue: the self-assembly of the bacterial flagella, antigen-antibody reactions, the high activity and selectivity of enzymes, etc., is accurately achieved by intermacromolecular interactions [1]. The binding of proteins to nucleic acids, which are natural polyelectrolyte, is an integral step in gene regulation [1, 7, and 8]. A weak intermacromolecular interactions are responsible for a dramatically changes in thermodynamic compatibility of biopolymers [9–11]. Intermacromolecular interactions can be utilized for isolation of proteins [4, 5, and 12] and enzymes [13], enzyme immobilization [3, 14], encapsulation [15] and drug delivery [14, 16]. The most intriguing type of complexes is that containing two proteins [1]. When two or more protein binds together, often to carry out their biological function [2]. Proteins might interact for a long time to form part of a protein complex, [17] or a protein may interact briefly with another protein just to modify it [18]. Therefore understanding of the effect of protein structure on protein-protein interactions, for example, of smooth and skeletal muscle proteins permit the manipulation of protein side chains in order to enhance gelation properties. Of particular interest would be studies on the mechanism of complexation as well as on the molecular characteristic of resulting complexes. These studies have also aided in the understanding of biological systems. This application note highlights the use of the Malvern Zetasizer Nano ZS for characterization of interbiopolymer complexes [19]. However, data are lacking for understanding of how structural and aggregation properties of the interacting proteins affect structure formation, of complexes.

This work studies the effect of the limited thermo aggregation of BSA on the structure development in the complex forming water–acid gelatin-BSA system above the temperature of the conformation transition of gelatin, using dynamic light scattering (DLS), fluorescence measurements, and circular dichroism.

It is known [20,21], that thermoaggregation of BSA in water at moderate temperatures changes thermodynamic properties of the protein decreasing significantly thermodynamic compatibility of BSA with other proteins, for example, with ovalbumin. Since the limited thermoaggregation of the globular protein can lead to changes its conformation state, and electrical properties, as well as activity of the protein in its saturated solutions, we assumed that such thermomodification of BSA molecules can affect on the molecular and structural properties of the inter protein complexes.

BSA or plasma albumin is a well-known globular protein (M_w = 67 kDa) that has the tendency to aggregate in macromolecular assemblies. Its three-dimensional structure is composed of three domains, each one formed by six helices. 17 disulphide bonds are located in BSA molecule. The most common molecular form is ellipsoid (4.1 nm × 14.1 nm) [22]. Gelatin is a protein derived by partial hydrolysis of collagen. The protein is unique in that it is made up of triplets of amino acids, gly-X-Y. The X and Y can be any amino acid but the most common are proline and 4-hydroxyproline, which have a five member ring structure. Helix–coil transition of gelatin and gelation of gelatin solutions has been studied extensively in past [23]. Gelatin is well known, widely used in industry for its textural and structuring properties [24], and capacity to form interpolymer complexes with polyelectrolyte (e.g., [5, 25, 26]). Gelatin-BSA mixtures are well known and used in biomedical and controlled release areas [27].

19.2 OBJECTIVES

This work studies the effect of the limited thermo aggregation of BSA on the structure development in the complex forming water-gelatin-BSA system above the temperature of the conformation transition of gelatin, using dynamic light scattering, circular dichroism, and fluorescence measurements. It was established that the structure of the complexes formed is different in the case of the native and thermally aggregated BSA. Intermacromolecular interaction gelatin with native BSA leads to the collapse gelatin macromolecules and formation compact (30 nm in radius) BSA-gelatin complexes. Preheating of the BSA solutions at 57°C (below the denaturation temperature of BSA) for 10 min. leads to formation of the protein thermo aggregates with the average radius approximately 100 nm. Interaction of gelatin with the thermally aggregated BSA results in formation of the large complex particles with the middle radius ~1500 nm. This process accompanies the partial destabilization of the secondary and tertiary structures of the protein and an additional exposure of the hydrophobic triptophan residues to the surface of the globule. It is shown that the electrostatic interaction of the oppositely charged groups of BSA and gelatin is responsible for formation of such complex particles, whereas the secondary forces (hydrophobic interaction and hydrogen bonds) play an important role in stabilisation of the complex particles. The zeta potentials of the native and the thermally aggregated BSA samples were determined, and the solvent quality has been quantified by determining the activity of the protein samples in their saturated solutions. It was shown that the values of the zeta potential of the complex BSA-gelatin particles and a large size of the thermally aggregated BSA are a key factors in determining the structure formation, while the level of activity of the BSA samples have a smaller effect on the structure of complex particles.

19.3   MATERIALS AND METHODS

The BSA Fraction V, pH 5 (Lot A018080301), was obtained from Across Organics Chemical Co. (protein content = 98–99%; trace analysis, Na < 5000 ppm, CI < 3000 ppm, no fat acids were detected). The isoelectric point of the protein is about 4.8–5.0, and the radius of gyration at pH 5.3 is equal to 30.6 Å [22]. The water used for solution preparation was distilled three times. Most measurements were performed at pH 5.3. The extinction of 1% BSA solution at 279 nm was $A^{1cm}_{279}= 6.70$, and that value is very close to the tabulated value of 6.67 [28]. The gelatin sample used is gelatin type A 200 Bloom PS 8/30 (Lot 09030) produced by SBW Biosystems, France. The Bloom number, weight average molecular mass and the isoelectric point of the sample, as reported by the manufacturer, are, respectively, 207, 99.3 kDa and 8–9. Since the commercial sample contained traces of peptides and various substances regarded as impurities, an additional purification by washing with deionised water for 3 h at 5°C, was used. The major characteristics of the purified gelatin were described in a previous paper [29]. To prepare molecularly dispersed gelatin solutions, deionised water was gradually added to the gelatin, and stirred first at 60°C for 20 min and then at 40°C for 1 h. The required pH value of the solutions (5.4) was adjusted by addition of 0.1–0.5 M NaOH or HCl. The resulting solutions were centrifuged at 50,000 g for 1 h at 40°C to remove insoluble particles.

The ternary water-gelatin-BSA systems containing gelatin in coil conformation and native BSA sample were prepared by mixing solutions of each biopolymer at 40°C. All measurements were performed after previous holding of biopolymer solutions and their mixtures during 12–15 h. Previous experiments shown that longer store of solutions before measurement do not change the results of measurements.

Solutions of BSA aggregates with the middle size 100 nm were obtained by heating 5 g of 5 wt % solution of the native BSA sample at 57°C for 10 min and subsequent dilution to the desired concentrations.

19.3.1 DLS

Determination of Intensity-size distribution, and volume-size distribution functions, as well as zeta potentials of BSA, gelatin, and BSA+gelatin particles were performed, by the Malvern Zetasizer Nano instrument (England), using a rectangular quartz capillary cell. The concentration of gelatin in the water-gelatin-BSA mixtures was kept 0.25% (w/w). The intensity-size distribution functions were determined for mixtures with different BSA/gelatin weight ratio (q). For each sample the measurement was repeated 3 times. The samples of native BSA and gelatin were filtered before measurement through DISMIC-25cs (cellulose acetate) filters (sizes hole of 0.22 μm). Subsequently the samples were centrifuged for 30 sec at 4000 g to remove air bubbles, and placed in the cell housing. Solutions of BSA aggregates were used without filtration. The detected scattering light intensity was processed by Malvern Zetasizer Nano software.

19.3.2 CD MEASUREMENTS

CD measurements of BSA solution alone and in the presence of gelatin were performed using a Chiroscan Applied Photophysics instrument.

Far-UV CD spectra were measured in 1-mm cells at 20°C and 40°C. The solutions were scanned at 50 nm min^{-1} using 2 s as the time constant with a sensitivity of 20 mdeg and a step resolution of 0.1.

19.3.3   FLUORESCENCE MEASUREMENTS

Fluorescence emission spectra between 280 and 420 nm were recorded on a RF 5301 PC Spectrofluorimeter (Shimadzu, Japan) at 25°C and 40°C with the excitation wavelengths set to 250, 270, and 290 nm, slit widths of 3 nm for both excitation and emission, and an integration time of 0.5 s. The experimental error was approximately 2%.

19.3.4   ROSENBERG'S METHOD

The method of Middaugh et al., [30] (hereafter called Rosenberg's method) has been used to quantify the solvent quality. The method consists in determining the dependence of protein solubility in the given aqueous solvent on the concentration of PEG in the Solvent (1)–Protein (2)–PEG (3) system. Extrapolation of this dependence to $C_{PEG} = 0$ gives the value for the effective activity of the protein in its saturated solution (log $C_{biopolymer}$). Evidence for the validity of this extrapolation includes (a) the experimentally observed linearity of log solubility versus PEG concentration plots, (b) the extrapolation of such plots to correct activities in the situation where protein activities can be experimentally determined, and (c) the independence of the extrapolated activities on protein concentration over a wide range. A more detailed analysis [31] makes it possible to relate the activity to the value of the second virial protein coefficient characterizing the protein–solvent interaction. The data obtained represents the activity of BSA in its saturated solution and gives – in the case of water as a solvent – an indication of the hydrophily of the biopolymer. In thermodynamic terms, Rosenberg's methods relate the activity of the protein to the second virial coefficient A_{12} that characterizes the water (1)-biopolymer (2) interaction [31].

Experimentally, the method consists of preparing binary solutions of PEG and BSA at the required temperature, and pH. After mixing for 1 h, a separation of phases is established by means of centrifugation at 50,000 g for 30 min. The weight concentration of BSA in the supernatant is determined by determination of absorption at 279 nm.

19.4 RESULTS AND DISCUSSION

19.4.1 CHARACTERIZATION OF THE NATIVE BSA AND THERMALLY AGGREGATED BSA SAMPLE BY DLS

The intensity size distributions and the volume size distributions functions for 0.25 wt % solutions native BSA and the BSA^{TA} samples preheated at different temperatures and pH 5.4 are shown in Fig. 1. Our preliminary experiments shown that the intensity size distributions for BSA does not depends on the concentration, at least in the range from 0.1 wt% till 0.5 wt% (data not shown). One can see from Fig. 1 that approximately 90% of all BSA particles have the average radius 3.7 nm that is in accordance with known data [19]. Preheating the BSA sample at 54°C do not leads to appreciable changes in the intensity size distributions and the volume size distributions functions. At 55°C a small part of the protein undergo to aggregation (Fig. 1b). At the temperature 57°C all BSA molecules undergo to the limited aggregation with the average radius of particles 110 nm. At 58°C the average radius of particles increases up to 300 nm. Further increase the temperature leads to the beginning of thermal denaturation and irreversible precipitation of BSA [32].

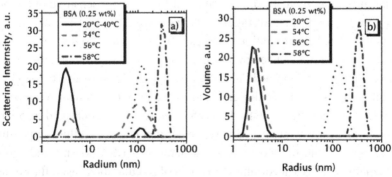

FIGURE 1   The intensity size distributions function (a) and the volume size distributions function (b) for 0.25 wt% solutions native BSA and the BSA^{TA} sample preheated at different temperatures. pH 5.4; 40°C.

The intensity size distributions for BSA and BSA^{TA} samples at pH 5.4 and different concentrations of the protein and at 40°C (before beginning of thermo aggregation of BSA) are presented in Fig. 2.

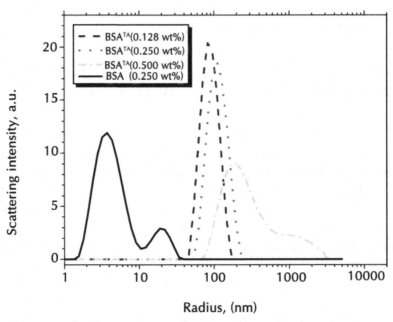

FIGURE 2   The intensity size distributions functions for BSA and BSA^{TA} samples at different concentrations of the protein. pH 5.4;40°C.

On the contrary to the native BSA, the average size of BSA^{TA} and the width of it's the intensity size distribution function are sensitive to the protein concentration. The average radius increases (from 84 nm to 110 nm) when the BSA^{TA} concentration changes from 0.128 wt% to 0.25 wt%, and it reach 180 nm at the concentration of 0.5 wt%. Since thermo aggregation of BSA at the concentrations higher then 0.25 wt % leads to formation of aggregates with a larger polydispersity, all subsequent experiments with BSA^{TA} sample were performed with the 0.25 wt% solutions preheated at 56.5°C for 10 min. as it was described in the experimental part. At this temperature the secondary and the local tertiary structures of BSA doesn't undergo an appreciable changes according to the data of circular dichroism, fluorescent spectroscopy, and differential scanning microcalorymetry [32].

19.4.2   LIMITED THERMO AGGREGATION AS A FACTOR DETERMINING MOLECULAR SIZES OF BSA-GELATIN COMPLEXES AND STABILITY BSA STRUCTURE IN COMPLEX

The intensity size distribution functions in water-BSA-gelatin and water-BSA^{TA} –gelatin systems at different q values are shown correspondingly in Figs. 3 and 4.

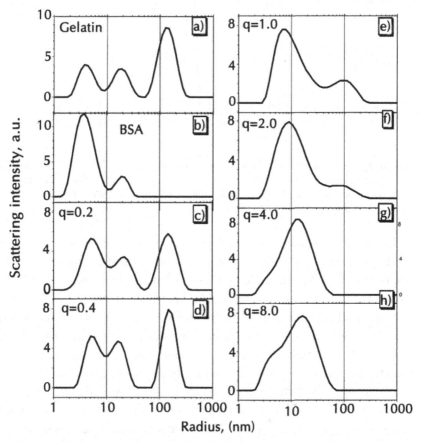

FIGURE 3   The intensity size distributions function for gelatin (0.25 wt%) (a), BSA (0.25 wt%) (b), and their mixtures at the concentration of gelatin 0.25 wt% and different q values (c–h). pH 5.4;40oC. The overlap concentrations (C*) for BSA and gelatin are 1.25 wt% and 3.0 wt%, correspondingly [38].

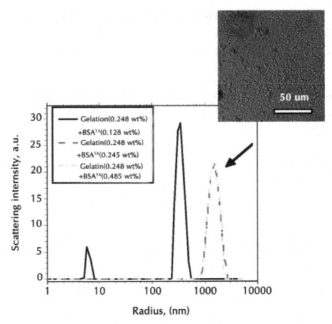

FIGURE 4 The intensity size distributions function for gelatin (0.25 wt%) (a), BSA (0.25 wt%) (b), and their mixtures at the concentration of gelatin 0.25 wt% and different q values (c–h). pH 5.4; 40°C. The overlap concentrations (C*) for BSA and gelatin are 1.25 wt% and 3.0 wt%, correspondingly [38].

The gelatin sample was strongly polydisperse. The main peak (about 50%) has the average radius 140 nm. A small parts of the low molecular weight fractions with the average sizes 3.8 nm (27%) and 9 nm (23%) correspondingly were also detected. As can be seen from Fig. 3, the presence of small amount of BSA in the gelatin solution (at q = 0.2 and q = 0.4) do not effect on the scattering intensity and position of the main gelatin peak. At a higher q values (1.0) the intensity of this peak decrease considerably. At q = 4 this peak completely disappears and the new dominant peak appears with the average radius 14 nm. At a higher content of BSA in the mixture (at q = 8) the appearance of shoulder was detected with the average radius ≅ 4 nm. The shoulder reflect "free" BSA molecules, and its becomes larger at q = 8. The main conclusion from the data presented in Fig. 2 is that above the temperature of the conformational transition intermacromolecular interaction of gelatin with native BSA leads to collapse gelatin macromolecules and formation compact (30 nm in radius) BSA-gelatin complexes. In the presence of high ionic strength (0.25, NaCl), when electrostatic interactions were suppressed the size distribution function of gelatin becomes insensitive to the presence of BSA (data not not presented). Taking into account that the molecular weight of gelatin is 243,000 Da and that of BSA is equal to 67,000 Da, we can roughly evaluate the "molar" ratio BSA/gelatin in the water insoluble complex. A simple calculation shows that this ratio is ~6/1.

The behaviour of gelatin in the presence of BSA^{TA} is completely different (Fig. 4). The interaction of gelatin with BSA^{TA} leads to formation of the large complex particles. At q = 0.5 the average size of the particlesis 356 nm that is 3 times higher then those of the binary water-gelatin and water-BSA^{TA}. At a higher q values (q = 1.0) the average size of the particles sharply increases up to 1500 nm.

An addition of NaCl in the ternary water-gelatin-BSA^{TA} system results in the partial dissociation of the complex aggregated, and the absence of free BSA molecules (Fig. 5). On the other hand, an addition of NaCl in the in the initial binary solutions of BSA and gelatin (0.25 M NaCl) leads to the partial dissociation of the complex aggregated, and the absence of free BSA molecules (Fig. 6).

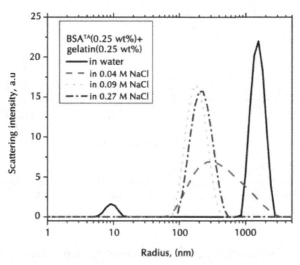

FIGURE 5 The effect of an addition NaCl in the ternary gelatin (0.25 wt%)-BSATA (0.25 wt%) system on the intensity size distributions functions. 5.4; 40oC. The overlap concentrations (C*) for BSA and gelatin are 1.25 wt% and 3.0 wt % correspondingly [38].

These facts shows that with one hand the complexes are formed via electrostatic interaction, rather than through hydrogen bonds formation or hydrophobic interaction. The role of salt is to "soften" the interactions, which is equivalent to making the electrostatic binding constant smaller. On the other hand, non electrostatic forces play an important role in stabilization of the formed complexes (Fig. 6).

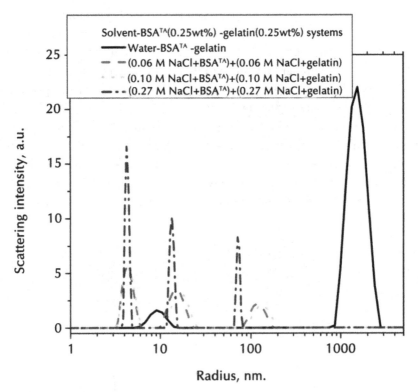

FIGURE 6   The effect of an addition of NaCl in the binary water-gelatin and water-BSA^{TA} system on the intensity-size distribution functions of the ternary water-gelatin (0.25 wt%)-BSA^{TA} (0.25 wt%) system, obtained by mixing of the binary salt solutions of gelatin and BSA^{TA}.

The question is arises, what is the reason of the great difference in structure formation of the native and BSA^{TA} samples with gelatin? We consider three possible reasons for such behaviour of BSA: 1. Change in the solvent quality of BSA molecules after their limited thermoaggregation.

2. Possible changes in structure and in the charge of the thermally aggregated BSA alone and in the presence of gelatin compared with native BSA sample. 3. Steric reasons affecting on complex formation.

Figure 7 shows the effect of PEG on the solubility of BSA in water before and after thermally induced aggregation at pH 5.4. The dependences obtained proved to be rectilinear. Extrapolation of this dependence to $C_{PEG} = 0$ gives the value for the effective activity of the protein in its saturated solution (log $C_{biopolymer}$) [30].

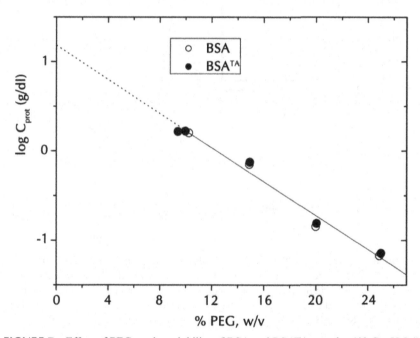

FIGURE 7   Effect of PEG on the solubility of BSA and BSATA samples (40°C, pH 5.4). The solid line is a least-square fit of the data to a straight line. The dotted line is a linear extrapolation of the solubility to zero PEG concentration.

The results obtained show that thermally induced aggregation of BSA does not affect appreciably on the activity of saturated BSA solutions, and therefore solvent quality of the BSA before and after the limited aggregation is almost the same.

It is known that at a relatively high temperature, polyelectrolyte can effect on the secondary, and the local tertiary structures of globular proteins, elaborating part of the charged functional groups of the protein from inside of globule to the globule surface [32]. Such structural changes can affect on electrostatic interaction of the oppositely charged proteins. In order to examine such possibility, the CD and fluorescence spectra of BSA samples were obtained in the presence of gelatin. The data obtained are shown in Figs. 8 and 9.

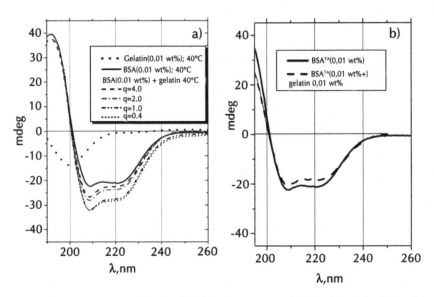

FIGURE 8 Comparison of the circular dichroism spectra of 0.01 wt% gelatin, 0.01 wt% BSA, and the water-gelatin (var) –BSA (0.01 wt%) mixtures at different q values after subtraction of the gelatin spectrum (a); comparison of the circular dichroism spectra of 0.01 wt% BSA^{TA} and 0.01 wt% BSA^{TA}+ 0.01 wt% gelatin after substraction of the gelatin spectrum (b). 40°C. pH = 5.4.

Figure 8 a shows far-UV CD spectra at 40°C for BSA and gelatin in water. Gelatin at +40°C shows the coil conformation. The triple helix can partly reform when the temperature of the sample falls below the gelation temperature [33]. At 40 °C BSA contains 51.4 % of the alpha helical structure. The average helix content, f_H (in %) for BSA molecules was calculated according to [34, 35]. The presence of gelatin in the BSA solution at q = 2 leads to significant increase in the negative band at 222 nm and significant increase in the negative maximum at 209 nm as shown in Fig. 8a. At a higher gelatin content in the mixture (at q = 1.0) this effect was more pronounced. Such increase in the negative band at 222 nm corresponds to an increase in the content of alpha-helical structures from 51.4% to 62%. Introduction of NaCl in the water results in full insensitivity of CD spectra of gelatin to the presence of BSA in all the q range studied (data not shown). In contrary of the native BSA, the presence of gelatin in the BSA^{TA} sample results in an appreciable decrease in the negative band at 222 nm and decrease in the negative maximum at 209 nm as shown in Fig. 8b. Increase gelatin content in the mixture from q = 1 to q = 2.0 and 4.0 do not effect on CD spectra of the BSA^{TA} sample (data not shown). Figure 9 shows the effect of gelatin on the spectra of fluorescence intensity of the native BSA sample (Fig. 9a) and the BSA^{TA} sample (Fig. 9b) at the excitation wavelength (λex) = 270 nm.

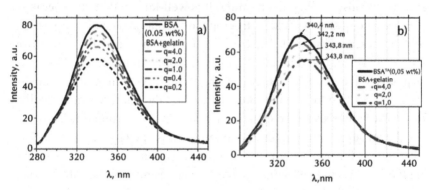

FIGURE 9 Fluorescence spectra of 0.05 wt% BSA and BSA-gelatin systems at different q values (a); Fluorescence spectra of 0.05 wt% BSA^{TA} and BSA^{TA}-gelatin systems at different q values (b). Exc λ=270 nm.40°C. pH 5.4.

The wavelength of maximum emission (λ_{max}) for the both BSA samples was about 339–340 nm, regardless of λ_{exc} in the range of 250–290 nm. The presence of gelatin in the native BSA sample leads to a visible decrease in fluorescence intensity. The decrease in the intensity was also observed at the λ_{exc} = 250 nm and λ_{exc} = 290 nm (data not shown) although to a lesser degree than at λ_{exc} = 270 nm. The fluorescence quenching was especially remarkable at a highest content of gelatin in the BSA solution (at q = 0.4). At the same time we did not observe any appreciable shift the maximum of the fluorescence intensity at q values from 0.4 to 4.0 that indicates the absence of an appreciable unfolding of BSA molecules in the presence of gelatin molecules. Moreover, CD data shows even on the significant ordering of the secondary structure of the native BSA in the presence of gelatin.

Unsimilar to the system BSA-gelatin containing native BSA sample, the results of fluorescence analysis for the system containing BSA^{TA} sample shows a shift in the wavelength of the maximum of the intrinsic fluorescent emission with quenching of fluorescence, indicating changes in protein conformation [34]. In proteins that contain all three aromatic amino acids, fluorescence is usually dominated by the contribution of the tryptophan residues, because both their absorbency at the wavelength of excitation and their quantum yield of emission are considerably greater than the respective values for tyrosine and phenylalanine. With proteins containing tryptophan, the changes in intensity and shifts in wavelength are usually observed upon unfolding. It is well-known that, in general, the fluorescence quantum yield of tryptophan decreases when its exposure to solvent increases [36, 37]. The emission maximum is usually shifted from shorter wavelengths to about 350 nm upon protein unfolding, which corresponds to the fluorescence maximum of pure tryptophan in aqueous solution. According to this, we take into account the fluorescence quenching in terms of local rearrangements of the tryptophan surroundings.

The apparent wavelength of maximum emission (i_{max}) for pure BSA is 339 nm (i_{max} for pure tryptophan in aqueous solution is 354 nm). Comparing these values indicates that the tryptophan residues of the native protein are only partially located on the surface and exposed to water. Significant parts are located inside the globule. An excess of gelatin results in a decrease in the fluorescence intensity and in a small shift of λ max from 339 to 344 nm. These facts indicate a partial unfolding of BSA with an additional exposure of the hydrophobic tryptophan residues to the surface.

If the partial destabilization of the secondary structure of BSA^{TA} and an additional exposition of the hydrophobic tryptophan residues to the surface in the presence gelatin takes place it may increase the charge of the interacting molecules and effect on the structure of the complexes formed. Really, the z- average potential on the BSA (z^{za}_{BSA}) is equal to 9.27±0.8 mV), whereas $z^{za}_{BSA}{}^{TA} = -12.31\pm0.8$. At the compositions corresponding to the maximal binding of BSA by gelatin z average negative potential of the complex particles gelatin-BSA ($z^{za}_{BSA-Gel}$) $\cong$ 0, whereas the values of the same parameter for the BSA^{TA} sample is –6.16. Therefore, the charges of the interacting BSA and BSA^{TA} molecules and the complex particles formed are different. These factors can affect on the sizes of complexes gelatin with BSA.

On the other hand, it is reasonably to suggest that formation of the large aggregates of BSA^{TA} sample leads to the steric difficulty in their interaction with gelatin with subsequent collapse of the later. In this case it is more probable that formation of the complex particles results from the simple joining of the BSA^{TA} to gelatin without significant changes of the sizes of interacting biopolymers. The relatively higher negative charge of the complex BSA^{TA}/gelatin particles then that of BSA-gelatin particles is also do not promote to collapse of the gelatin molecules.

Figure 10 shows schematically the main structural changes observed during interaction BSA or BSA^{TA} with gelatin in coil conformation. The interaction of BSA with gelatin leads to the collapse of the coil. The collapsed gelatin molecule wraps six BSA molecules. BSA^{TA} interacts with gelatin molecules results in formation the large macromolecular assembly. The interaction leads to the partial unfolding globular protein and formation of the charged complex particles.

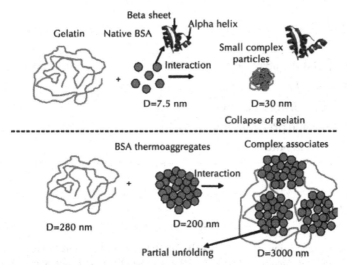

FIGURE 10 Schematic representation of the main structural changes observing during interaction of the native and thermally aggregated BSA with gelatin in coil conformation.

19.5   CONCLUSION

We have combined dynamic light scattering measurements with circular dichroism, and fluorescence spectroscopy to study the effect of the limited thermo aggregation of BSA on the structure development in the complex forming water-gelatin-BSA system above the temperature of the conformation transition of gelatin. We clearly establish that the structure of the complexes formed and the origin of the forces stabilizing interprotein complexes are different in the case of the native, and the thermally aggregated BSA. Intermacromolecular interaction of the native BSA with gelatin sample leads to collapse gelatin macromolecules, formation compact (30 nm in radius) BSA-gelatin complexes (~6:1 mole/mole), stabilization of the secondary structure of BSA, and the local rearrangements of the tryptophan surroundings in BSA. The complexes formed and stabilized by Coulomb forces.

In the contrast to native BSA sample, intermacromolecular interaction of the BSA^{TA} sample with gelatin leads to formation of the large complex particles of the BSA associates with gelatin, the partial unfolding globular protein and formation of the charged complex particles.

KEYWORDS

- Bacterial flagella
- Dynamic light scattering
- Interbiopolymer complex
- Irreversible precipitation
- Thermal denaturation

REFERENCES

1. Cantor, C. R.; Schimmel, P. R. Part III The Behavior of Biological Macromolecules. In: *Biophysical Chemistry*. Freeman WH & Co: San Francisco, 849–886.

2. Dusenbery, D. B. *Sensory Ecology: How Organisms Acquire and Respond to Information*. Freeman WH & Co, New York, 1992.

3. Xia, J.; Dubin, P. L. Chapter 15 Portein-polyelectrolyte Complexes. Dubin, P. L.; Davis, R. M.; Schultz, D.; Thies, C., Eds., In *Macromolecular Complexes in Chemistry and Biology*. Springer-Verlag, Berlin, 1994; pp. 247–271.

4. Dubin, P. L.; Gao, J.; Mattison, K. *Sep Purif Methods*, **1994**, *23*(1), 1–16.

5. Tolstoguzov, V. B. Functional Properties of Protein-Polysaccharide Mixtures. Hill, S. E.; Ledward, D. A.; Mitchel, J. R., Eds. In *Functional Properties of Food Macromolecules*. Aspen publishers Inc: Gaithersburg, MD, 1998; pp. 253–277.

6. Antonov, Y. A.; Moldenaers, P. *Biomacromolecules*, **2009**, *10*(12), 3235–3245.

7. Record, M. T. Jr.; Anderson, C. F.; Lohman, T. M. *Q. Rev. Biophys.*, **1978**, *11*,103–178.

8. Von Hippel, P. H.; Bear, D. G.; Morgan, W. D.; McSwiggen, J. A. *Annu. Rev. Biochem.*, **1984**, *53*, 389–446.

9. Antonov, Y. A.; Lefebvre, J.; Doublier, J. L. *J. Appl. Polym. Sci.*, **1999**, *71*, 471–482.

10. Antonov, Y. A.; Soshinsky, A. A. *J. Biol. Macromol.*, **2000**, *27*(4), 279–285.

11. Antonov, Y. A.; Dmitrochenko, A. P.; Leontiev, A. L. *Int. J. Biol. Macromol.*, **2006**, *38*(1), 18–24.

12. Serov, A. V.; Antonov, Y. A.; Tolstoguzov, V. B. *Nahr*, **1985**, *29*(1), 19–30.

13. Kiknadze, E. V.; Antonov, Y. A. *Appl. Biochem. Microbiol. (Russia, Engl. Transl)*, **1998**, *34*(5), 462–465.

14. Ottenbrite, R. M.; Kaplan, A. M. *Ann. NY Acad. Sci.*, **1985**, *446*, 160–168.

15. Magdassi, S.; Vinetsky, Y. Novel Cosmetic Delivery Systems, Marcel Dekker Inc: New York, 1997, pp 21–33.

16. Regelson, W. *Interferon*, **1970**, *6*, 353–359.

17. Denning, D.; Patel, S.; Uversky, V.; Fink, A.; Rexach, M. *Proc. Natl. Acad. Sci. USA*, **2003**, *100*, 2450–2455.

18. St Stout, T. J.; Foster, P. G.; Matthews, D. J. *Curr. Pharm. Des.*, **2004**, *10*, 1069–1082.

19. Kayitmazer, B.; Shaw, D.; Dubin, P. Characterization of protein-polyelectrolyte complexes. Zetasizer Nano application note. http://www.malvern.com/common/downloads/campaign/MRK513-02.pdf.

20. Polyakov, V. I.; Popello, I. A.; Grinberg, V. Y.; Tolstoguzov, V. B. *Nahr*, **1986**, *30*(1), 81–88.

21. Polyakov, V. I.; Popello, I. A.; Grinberg, V. Y.; Tolstoguzov, V. B.; *Food Hydrocoll.*, **1997**, *11*(2), 171–180.

22. Peters, T., Jr. *Albumin: An Overview and Bibliography*, 2nd edn. Miles Inc Diagnostics Division: Kankakee IL, 1992.

23. Djabourov, M.; Leblond, J.; Papon, P. *J. Phys. Fr.*, **1988**, *49*, 319–332.

24. Ward, A. G.; Courts, A. *The Science and Technology of Gelatin;* Academic Press:, London-New York-San Francisco, 1977.

25. Kaibara, K.; Okazaki, T.; Bohidar, H. B.; Dubin, P. *Biomacromol.*, **2000**, *1*, 100–107.

26. Bowman, W. A.; Rubinstein, M.; Tan, J. S. *Macromol.*, **1997**, *30*(11), 3262–3270.

27. Migneault, I.; Dartiguenave, C.; Bertrand, M.; Karen, J.; Waldron, C. *BioTechniques*, **2004**, *37*, 790–802.

28. Kirschenbaum, D. M. *Anal. Biochem.*, **1977**, *81*(2), 220–246.

29. Antonov, Y. A.; Van Puyvelde, P.; Moldenaers, P. *Biomacromol.*, **2004**, *5*, 276–283.

30. Middaugh, C. R.; Tisel, W. A.; Haive, R. N.; Rosenberg A. *J. Biol. Chem.*, **1979,** *254,* 367–370.
31. Polyakov, V. I.; Popello, I. A.; Grinberg, V. Y.; Tolstoguzov, V. B. *Nahr,* **1985,** *29,* 323–333.
32. Antonov, Y. A.; Wolf, B. A. *Biomacromol.,* **2005,** *6*(6), 2980–2989.
33. Zhang, Z.; Li, G.; Shi, B. *J. Soc. Leather Technol. Chem.,* **2006,** *90,* 23–28.
34. Woody, R. W. *Methods Enzymol.*, **1995,** *246,* 34–71.
35. Creighton, T. E. *Protein Structure*, 2nd edn; Oxford University Press: Oxford, 1997.
36. Campbell, I. D.; Dwek, R. A. *Biol. Specty;* The Benjamin Cummings Publishing Co; Menlo Park, 1984.
37. D'Alfonso, L.; Collini, M.; Baldini, G. *Biochem.,* **2002,** *241,* 326–333.
38. Lefebvre, J. *Rheol. Acta.,* **1982,** *21,* 620–625.

CHAPTER 20

WILD ORCHIDS OF COLCHIS FORESTS AND SAVE THEM AS OBJECTS OF ECOEDUCATION, AND PRODUCERS OF MEDICINAL SUBSTANCES

E. A. AVERJANOVA, L. G. KHARUTA, A. E. RYBALKO, and K. P. SKIPINA

CONTENTS

20.1 Introduction 374
20.2 Overview 374
20.3 Features of the Biology and Ecology 377
20.4 Orchids as an Object of Ecoeducation 379
20.5 Medicinal Properties 381
20.6 Conclusion 384
Keywords 384
References 384

20.1 INTRODUCTION

The requirements set priorities of our research activities. And the current state of the biosphere could be better. The living conditions of people in urban areas are particularly grim. The result is a dramatic and widespread decline in public health. Meantime, ways of maintenance and restoration of vital forces of person exist of yore. By this methods are forgotten in modern time, how unnecessary rejected. Because natural sources of curative substances are already too not enough. And biotechnology comes here to the aid.

It is, in particular, the so-called "salep," known for centuries. The source of it is the underground storage organs of plants of orchid family, terrestrial species distributed in temperate climate zone. On the properties of "salep" we will later, and now just note that the world human population grows and natural populations of terrestrial orchids are thinning, melting away before our eyes, of course, is not in any way providing the needs of people. It's time to replace the usual gathering of funds in the healing nature to the intensive cultivation of the active substance through biotechnology.

20.2 OVERVIEW

Orchidaceae is one of the largest families among angiosperms. According to one estimate the family includes 800 genera and 25,000 species [1]. Orchids – a family of evolutionarily the most perfect of plants pollinated by insects. Orchids are distributed in all climatic zones, on all continents, not only grow in Antarctica. Most rich family in the tropics, subtropics and then go, but also in the temperate zone, it presents much (about 900 species in the temperate latitudes of the northern hemisphere, about 120 in Europe). List of Flora Orchids Sochi Black Sea region has 46 species [2] distributed from the coast to the upper boundary of the alpine meadows. Species of most large size, long-term healing in the sense of obtaining raw materials, and simultaneously the most ornamental found mostly in a narrow strip of seawater's edge on the slopes of the foothills up to 250–400 m above sea level. This is primarily Orchis mascula, O. purpurea, O. provincialis, Platanthera bifolia, Ophris oestrifera, etc. (Figs. 1–4).

FIGURE 1   Orchis mascula.

FIGURE 2   Orchis purpurea.

FIGURE 3   Orchis provincialis.

FIGURE 4   Ophris oestrifera.

The abundance of species, unfortunately, is not accompanied by a high number of plants in natural populations. Orchids are usually rare and endangered species throughout all areas [3]. That is why the work on cultivation of this species group is so relevant. This is possible without causing harm to populations only by using methods of cultivation in vitro.

20.3 FEATURES OF THE BIOLOGY AND ECOLOGY

Representatives of the orchid family Colchis flora – terrestrial perennial herbaceous plants in height from 20 to 100 cm, with underground storage organs, often very decorative.

Propagation by seeds, often vegetative. The cycle of development is not easy and spread over time – from seed to flowering new plants can take 5–7 (15) years [4]. The seeds have no nutrient reserves, due to this are very light and can travel through the air for long distances. But the effectiveness of seed reproduction is low, because seedling is formed only in suitable soil conditions and using a special fungus-symbiont.

Flowering of many species (respectively, and fruit set) does not happen every year. In unfavorable conditions, plants are able to fall into a state of secondary dormancy. They are not making the leaves and stalk, are not showing up on the soil surface during the year, and even a few years, subsisting by the fungus-symbiont. Pollination of flowers is only by insects of specific group of species. The structure of the flower are usually very closely matches the size and morphology of insect pollinators [5]. Frequent situations where seed production populations of orchids are very low due to insufficient number of suitable insect in the vicinities.

Orchids are not capable of strong competition, they just die, if their habitat is overgrown with tall grasses and shrubs. However, many species, which we call the forest, withstand a very weak light in the tall woods with a solid crown canopy. Other orchids do not occur in shaded areas, they can be found in open meadows and forest edges, among the bushes.

Requirements for soils are different for different species, for orchids of Colchis botanical province are usually neutral or alkaline reaction, good aeration. Withstands poor, dry soil, but most species reach their maximum development on rich, moist soils.

In the Fig. 5 we can see what it looks like the main source of the healing qualities of Orchids – tuberoid. It is an underground organ for storage of nutrients. In gathering tuberoids as medicinal raw plant is completely destroyed. Here is a mature plant Steveniella satyrioides in a phase of beginning of the growing season, the snapshot taken in February. It is clearly visible size of the tuberoid, it does not exceed 2.5 cm in length. Does it make sense to dig tuberoids from natural populations, destroying them completely? After all, medicinal raw materials we receive very little.

FIGURE 5   Tuberoid of orchidea.

The features of biology and environmental preferences cause the low number of Orchids everywhere. In addition, Colchis forest area of Sochi is a powerful influence of civilization, growing from year to year.

The rapid development of the resort, particularly heightened in connection with the Olympic construction site leads to a tremendous reduction in habitat Orchids. It is now impossible to determine exactly how many tens and hundreds of plants irretrievably lost, where now new roads and other infrastructure in Sochi are. But we can confidently say that this process is not stopped. Construction continues, the city is growing strongly, more low-mountain belt of land occupied by the cottage, chalet and other settlements. The band of suitable habitat, and so narrow in many species (from the coast to the heights of 200–300 m above sea level) is rapidly decreasing. At the moment we are witnessing, for example, the loss of populations of Orchis picta and Serapias vomeracea around the holiday village Naval-ishino. Such examples could be cited.

Local populations disappear, lost some part of the gene pool and it's irreparable. Viability of the species as a whole is reduced. Leaves the planet's ecosystem biodiversity.

Prevention of such events – a long and laborious work, for which already has all the scientific, technological and organizational conditions.

Sochi branch of the Russian Geographical Society and the Department of Physiology of the Sochi Institute of People's Friendship University in the SEC "Biotechnology" developed a project to restore the Sochi Orchids. For solution to this problem is planned emergency assistance – Transfer orchids from the destroyable habitats to the special mini-reserves with further development of technologies for conservation by modern methods of biotechnology. Such an approach to make an impact on natural populations of orchids, in most cases, almost imperceptible.

We have started a cycle of research which include breeding in aseptic culture, adapting the plants in greenhouses, then in the open field and repatriation, in other words transport into specially organized reserves in order to restore extinguished local populations and creating new ones.

In the process of reproduction of native forms will inevitably arise promising clones to create new ornamental crops. A lot of attention to the implementation of the project will receive callus cultures, as a source of medical funds. Based on extensive scientific base and technological development of domestic and foreign scholars and practitioners, we have already started work on the introduction of in vitro culture of several species of terrestrial orchids Colchis flora. The main work will be deployed as early as 2012 under a contract with the Olympic Committee Sochi-2014 for the conservation of biodiversity.

20.4 ORCHIDS AS AN OBJECT OF ECOEDUCATION

We aim to preserve orchids Colchian forests and restoration of natural populations is not only in the environmental aspect, but also plan to use them as objects of environmental education, and ecoeducation and upbringing of the youth people. "Wild Orchid" – it sounds quite fresh and new for travelers and Sochi residents. On the basis of laboratory, green-house, hothouse complex, the pilot sites open land, as well as protected areas, which are grown under natural conditions different types of Orchids, planned to create a scientific and practical tour route in a single complex with the environmental path. Visitors will be able to directively learn modern scientific and technological methods of conservation of flora, in the best of their ability and skills to participate in the processes of the laboratory for further breeding and cultivation of orchids, the study of biological singularities and, if desired, to create media products that call preserve native nature (photo albums, posters, videos, etc.). Visitors can also contribute to the arrangement of ecological trails and nursery for plants. This route has a fascination for most of the year, as determined by differences in the timing of flowering of different species. Green leaves of orchids can be seen in Sochi, all year round, the earliest flowering observed in March (Steveniella satyrioides, Fig. 6), a mass flowering begins in April, which continues with the change of mid-July, then in early September, autumn orchid blossoms (Spiranthes spiralis, Fig. 7), and blooms until the end of October. In terms of greenhouse blooms will be more.

FIGURE 6   Steveniella satyrioides.

FIGURE 7   Spiranthes spiralis.

With the participation of other interesting plants we will create a guided tour, an unusual and attractive in all seasons.

20.5   MEDICINAL PROPERTIES

Medicinal properties of Orchids are not well understood. Scattered literature data [6–20] summarized in the table (Table 1).

TABLE 1   Medicinal properties of Orchids.

№	Genus	Phytochemistry	Actions, therapeutics
1	Anacamptis Rich.	Starch, mannan, dextrin, sucrose, albumins	Enveloping, restorative. Cystitis, gastritis, colitis, impotence
2	Corallorhiza Chatel.	Alkaloids, slimy substance	Hypotensive, antipyretic, sedative. Rheumatism
3	Cypripedium L.	Alkaloids, saponins, sugars, binder, resinous substances, essential oils, phenolic compound, tsipripedin, glucoside of o-hydroxycinnamic acid, tannins, anthocyanins hrizantemin, lipids, coumarins, luteolin, apigenin, chrysin, ascorbic acid, calcium oxalate, hydroquinone, arbutin	Hypnotic, anticonvulsant, sedative, vasodilator, laxative, antipyretic, sedative, diuretic. Gastritis, gynecological bleeding, headaches, mental disorders, skin diseases.
4	Dactylorhiza Neck. ex Nevski	Mucilage, starch, glucoside: loroglossin, albumen, volatile oil, ash, dactylorhins, coumarin lo-roglossin, flavonoids, anthocyanins: cyanine, seranin, ofrisanin, serapianin, orhitsianin	Aphrodisiac, expectorant and nervine tonic Diabetes, diarrhea, dysentery, paralysis, convalescence, impotence and malnutrition. coating, anti-inflammatory, emollient, ditoksikatsionnoe tool antitumor, restorative and tonic, sedative. Gastric disorders, flatulence, hemorrhoids, diarrhea, emaciation
5	Epipactis Zinn.	Alkaloids, coumarins	Laxative. The noise in the head
6	Epipogium J. S. Gmel. ex Borkh.	carotenoids: neoksantin, lutein, violaxanthin, α-carotene, β-carotene	Restorative, tonic, pain reliever
7	Goodyera R.Br.	Alkaloids, loroglossin. rutin, kaempferol-3-0-rutinozid, izoramnetin-3-0-rutinozid, gudayerin	Emollient, detoxification. In scurvy in diseases of the kidneys, eyes, scrofula, to improve the appetite, as a remedy for snake bites, diseases of the stomach, bladder, female diseases and diseases of the eye

TABLE 1   *(Continued)*

№	Genus	Phytochemistry	Actions, therapeutics
8	Gymnadenia R.Br.	Flavonoids, coumarin loroglossin, mannan, glucosides, urea, glucoside of o-hydroxycinnamic acids, flavonoids: quercetin, kaempferol, astragalin, izokvertsitrin, 3-β-glucoside-7-β-glucoside, kaempferol 3-β-glucoside-7-β-glucoside, quercetin, anthocyanins: hrizantemin, cyanine, seranin, ofrisanin, orhitsianin I, orhitsianin II, serapianin	Enveloping, emollient, restorative, toning, nourishing for debilitated patients, an expectorant, wound healing, increases the potency, emollient, expectorant, abortifacient, an aphrodisiac. Epilepsy, neuro-psychiatric disorders, toothache, abscesses, neuroses, dysentery, infertility, gastritis, enteritis, enterocolitis
9	Herminium Guett.	Alkaloids, slimy substance	Painkiller for toothache, enveloping
10	Orhis L.	Blennogenic matter – Mann, pentosans, starch, metilpentozany, proteins, dextrin, sucrose, bitter, volatile oil, carotene, kvertsitrin, pectins, coumarin glycoside loroglossin.	Nutritious, enveloping, restorative, stimulant, contraceptive. Enterocolitis, metiorizm, hemorrhoids, atonic constipation, cystitis, impotence, cancer, colds, snake bites, externally for fungal infections of nails
11	Platanthera Rich.	Starch, saponins, mannan	Enveloping, restorative, febrifuge, diuretic, antispasmodic, hypotensive, and uterine contraceptive. Impotence, menstrual disorders, cystitis, enterocolitis, atonic constipation
12	Spiranthes Rich.	Flavonoids, alkaloids, phenolic compounds: n-hydroxybenzaldehyde, n-gidroksi-benzilovy alcohol salt of ferulic acid, polynuclear aromatic compounds: spirantol A, spirantol B orhinol, spirasineol A, spirantezol	Restorative. Kidney disease, generalized weakness, tuberculosis, exhaustion

In generalizing it should be noted that the main active principle tuberoids almost all kinds of substances are blennogenic providing enveloping effect, in combination with other medicinal qualities most often used as a tonic, nourishing way to restore people's health, weakened by a variety of ailments of infectious and traumatic nature. Such a wide range of applications suggests that there is a strong positive effect on the immune system. The universality of the properties allows the use of drugs from Orchids to restore the health of people regardless of their age and condition, without fear of side effects.

20.6 CONCLUSION

The relevance of research and practical measures for introduction to culture of this group of plants is evident. The aseptic culture will give the necessary high multiplication factor, which will, in addition to receiving a sufficient amount of material to isolate the curative substance, and realize research of the reintroduction and repatriation, with the goal of biodiversity conservation of the unique ecosystem of the region in many ways.

KEYWORDS

- **Colchis flora**
- **Fungus-symbiont**
- **Salep**
- **Tuberoids**

REFERENCES

1. Stewart, A.; Griffiths, J. M. *Manual of Orchids;* Timber Press: Portland, Oregon, 1995.
2. Solodko, A. S.; Makarova, E. L. *Orchids Sochi Black Sea.* Sochi, 2011.
3. The Red Data Book of the Russian Federation. *Plants and Fungi;* Moscow, 2008.
4. Vakhrameyeva, M. G.; Denisova L. V.; Nikitina, S. V.; Samsonov, S. K. *Orchids of our Country.* Moscow: Nauka, 1991.
5. Ivanov, S. P.; Kholodov, V. V.; Fateryga, A, V. *Proceedings of Tauric National University. A series of "Biology and Chemistry.* **2009**, *22*(61), 1. S. 24–34.
6. Luning, B. Alkaloid content of Orchidaceae. In: Withner, C. L., ed., *The Orchids Scientific Studies.* John Wiley and Sons: New York, London, 1974.

7. Slaytor, M. B.: The Distribution and Chemistry of Alkaloids in the Orchidaceae. In *Orchid Biology Reviews and Perspectives* Arditti, J. ed., Cornell University Press: Ithaca, London, 1977, pp 96–115.
8. Hew, C. S.; Arditti, J.; Lin, W. S. Three Orchids used as Herbal Medicines in China: an Attempt to Reconcile Chinese and Western Pharmacology. In *Orchid Biology: Reviews and perspectives VII*. J Arditti et al., eds., Kluwer Academic Publishers: Dordrecht, Boston, London, 1997, 213–283.
9. Hory, N. *Chem. Pharm.* Stavropol Publishing House: London, 1982.
10. Kizu, H.; Kaneko, E. I.; Tomimori. T.: XXVI. *Chem. Pharm. Bull.* **1999**, *47*(11), 1618–1625.
11. Jiachen, Z. I., et al. *J. Nat. Prod.* **2008**, *71*, 799–805.
12. Schroeter, A. I.: Moscow*: Meditsina*, **1975**, 328.
13. Hegnauer, R.: *Birkhauser Verlag*, **1963**, 540 s.
14. Gorovoy, P. G.; Salokhin, A. V.; Doudkin, R. V.; Gavrilenko, I. G.: *Turczaninowia*, **2010**, *13*(4), 32–44.
15. Budantsev, A. L., et al., ed. *Wild Useful Plants of Russia*. St. Univ. SPHFA, 2001, 663p.
16. Fruentov, N. *Khabarovsk*, 1987; 352.
17. Yakovlev, G. P. et al., ed. *Dictionary of Medicinal Plants and Animal Products: Textbook. Manual.* St. Petersburg: Special Literature, 1999; 407.
18. *Plant Resources of Russia and Adjacent States: Flowering Plants, Their Chemical Composition and Use.* The family Orchidaceae – Yatryshnikovye. St. Petersburg: Science, 1994. T. 8., 84–99.
19. Singh, A.; Duggal, S. *Ethnobotanical Leaflets*, **2009**, *13*, 351–363.
20. Seredin, R. M.; Sokolov, S. D.: *Chem. Pharm.* Stavropol Publishing House: London, 1969.

CHAPTER 21

EXPRESS ASSESSMENT OF CELL VIABILITY IN BIOLOGICAL PREPARATIONS

L. P. BLINKOVA, Y. D.PAKHOMOV, O. V. NIKIFOROVA, V. F. EVLASHKINA, and M. L. ALTSHULER

CONTENTS

21.1 Introduction...388
21.2 Materials and Methods...388
21.3 Results and Discussion 389
21.4 Conclusion 391
Keywords 391
References..391

21.1 INTRODUCTION

Lyophilized cells in biological preparations exist in a form of artificial anabiosis that in our opinion may be paralleled with natural forms of dormancy caused by environmental stresses (physical, chemical, biological). Activity of bacterial cultures, including probiotic preparations is commonly evaluated as a number of colony forming units (CFU/ml) after plating serial dilutions of bacterial cultures in normal saline. Microbiologists believe that a colony is not always formed from a single cell and real number of cells is somewhat greater. Plating and other culture-based methods also fail to reveal cells that are not readily culturable at the moment of sampling. Such cells may be dormant, sublethally injured or exist in viable but nonculturable state [1–4]. We tested the commercially available kit for differentiating between viable and dead for accuracy when viability of probiotics is assessed.

In this chapter we discuss suitability of a commercially available kit that visually discriminates between viable and dead cells (Live/DeadBaclight™) for assessment of viability rates of lyophilized biological preparations. We have shown that this kit allows rapid and accurate evaluation of viability of studied samples, which may be several orders of magnitude higher than may be revealed by traditional culture-based methods.

21.2 MATERIALS AND METHODS

We studied five samples of commercially available colibacterin (*Escherichia coli* M17) produced by different manufacturers and with different storage period after expiration date. These include:

(1) Perm'RIVS, batch # 572-2 and expiration date 11.1982.
(2) ImBiO, batch #170-3and expiration date06.2001.
(3) Mechnikov Biomed, batch #40-3and expiration date03.2008.
(4) Mechnikov Biomed, batch #240-3and expiration date05.2009.
(5) Mechnikov Biomed, batch #270-3and expiration date07.2009.

Samples were resuspended in 1ml per dose of normal saline. Aliquot was taken from the sample, diluted 10-fold and immediately stained with Live/Dead staining kit (Baclight™). For our study we used a kit with separate dyes. Equal volume of each dye was mixed and diluted 10-fold. From this mixture 3 µl was added to 100 µl of a sample. Viable and dead cells emit different colors, therefore viable/dead cells ratio can easily be evaluated as well as in terms of numbers in given volume. As described in the instructions, cells that emitted green fluorescence were considered viable and red fluorescing cells – dead. Cells that emitted yellow or orange fluorescence (where present) were counted as viable, but damaged during the processes of sample preparation and visualization, by the DNA binding dyes or the exciting light.

Total cell counts were made in a counting chamber after further 10-fold or 20-fold dilution depending on the sample.

CFU/ml value was assessed by plating on nutrient agar immediately after rehydration and after 48–72 h.

21.3 RESULTS AND DISCUSSION

We propose an approach to visual assessment of viable bacteria in probiotics and other biological preparations using commercially available fluorescence staining kit Live\Dead (Baclight™). Using this method we assessed viability of a model strain of E. coli M17 in several samples of colibacterin probiotics preparations produced by different manufacturers (Mechnikov Biomed; ImBiO; Perm' RIVS) with different storage period after their expiration date (1982–2009), kinds of vessel (ampoules and vials). According to the results of fluorescent staining viability of studied preparations varied from 52.2%, for the oldest sample, to 91.3% for the one that expired in July 2009 (see Table). Results of direct counting of viable and dead cells were compared with CFU/ml values. Using Live/ Dead it is possible to assess at least two sample preparations per hour. We showed that 4–99% of lyophilized E. coli M17 cells didn't form colonies immediately after rehydration while remaining viable in normal saline. Further incubation of serial decimal dilutions of the sample in normal saline for 48–72 h resulted in increase of number of colony forming units by 3–4 orders of magnitude. Such increase occurred in tubes where initially only some individual colonies have formed. Such increase is clearly impossible if only due to division of initially culturable cells. Apparently processes similar to delayed multiplication of some rehydrated cells may occur in the intestines of warm-blooded animals. These cells might exist in some form of anabiosis, or be sublethally injured which requires more extended recovery period before first division is possible. It should be noted that presence of yellow fluorescing cells increased with time of storage (data not shown). It is probably due to the increase of internal damage in dry cells with time. In our study we might have slightly overestimated total numbers of viable cells, because some of them could be structurally intact, but too severely damaged internally. In contrast some cells considered as dead might be able to repair their membranes and become able to form colonies. However, we have shown that real viability of even long expired preparations may be far greater then revealed by traditional plating procedures. In fact it can be very close to the initial values.

Colibacterin features			Biological parameters					
Manufacturer	Batch number, expiration date	Vessel, capacity	Total cell counts		% of viable cells (Live/Dead) Total viable counts		CFU/ml (0 h)	% of viable cells unable to form colonies
			0 h	72 h	0 h	72 h		
Perm'RIVS	572-2 11.1982	Ampoules, 3 doses	2.11×10^{10}	3.72×10^{10}	52.2%	95.91%	2.89×10^{7}	99.73%
					1.1×10^{10}	3.56×10^{10}		
ImBiO	170-3 06.2001	Ampoules, 3 doses	3.11×10^{10}	4×10^{10}	86.7%	96.99%	5.55×10^{9}	79.5%
					2.7×10^{10}	3.88×10^{10}		
Mechnikov Biomed	40-3 03.2008	Vials, 5 doses	1.21×10^{10}	3.6×10^{10}	90.3%	97.66%	1.58×10^{9}	85.5%
					1.09×10^{10}	3.52×10^{10}		
Mechnikov Biomed	240-3 05.2009	Vials, 5 doses	1.44×10^{10}	2.58×10^{10}	88.2%	94.98%	1×10^{10}	21.3%
					1.27×10^{10}	2.45×10^{10}		
Mechnikov Biomed	270-3 07.2009	Vials, 5 doses	1.32×10^{10}	3.56×10^{10} (48 h)	91.3%	99.05% (48 ч)	1.16×10^{10}	4.13%
					1.2×10^{10}	3.53×10^{10}		

21.4 CONCLUSION

Therefore the described express method in comparison to traditional plating allowed rapid and accurate assessment of cell viability in commercial batches for biological control and reveal higher potential for retention of activity in probiotic preparations.

KEYWORDS

- **Colibacterin**
- **Live/Dead staining kit**
- **Lyophilized cells**
- **Rehydration**

REFERENCES

1. Aersten, A.; Michielis, C. W. *Rev. Microbiol.*, **2004**, *30*, 263.
2. D'Oliver, J. *Microbiol. J.* **2005**, *43(S)*, 93.
3. Ganesan, B.; Mark, R.; Stuart, M. R.; Weimer, B. C. *Appl. Environ. Microbiol.*, **2007**, *73(8)*, 2498.
4. Kaprelyants, A. S.; Kell, D. B. *Appl. Environ. Microbiol.*, **1993**, *59(10)*, 3187.

CHAPTER 22

ACTIVITY OF LIPOSOMAL ANTIMICROBIC PREPARATIONS CONCERNING *STAPHYLOCOCCUS AUREUS*

N. N. IVANOVA, G. I. MAVROV, S. A. DERKACH, and E. V. KOTSAR

CONTENTS

22.1 Introduction 394
22.2 Background 394
22.3 Materials and Methods 395
22.4 Results and Discussion 396
22.5 Conclusion 397
Keywords 398
References 398

22.1   INTRODUCTION

The skin of patients with atopic dermatitis planted on different microorganisms, whose number is much bigger than the skin of healthy people. For example, Staphylococcus aureus (S. aureus) sow from the skin of patients with atopic dermatitis in 80–100% of cases. Skin diseases of microbial etiology in most cases basic drug treatment and prevention are antibiotics. Widespread use of antibiotics has negative consequences, one of which is the emergence of pathogens with resistance to penicillin, gentamicin, tetracycline, methicillin, lincomycin, sulfonamides, as well as a new generation of antibiotics: quinolones, cephalosporins. Consequently, the problem of prevention and treatment of infectious diseases is urgent. One of the ways of its solution is the introduction of new chemotherapeutic agents into medical practice.

It is known that nanoparticles and liposomal forms of medicines allow significantly improving the efficacy, reducing toxicity, therapeutic dose and qualitatively changing the nature of their actions. Thus, in the work [4] it was shown that, despite the low concentrations, the efficiency of liposomal benzyl penicillin to inhibit growth of bacterial biofilms of S. aureus was higher than the intact benzyl penicillin. According to the above-mentioned issues, the purpose of the study was to investigate the efficiency of antimicrobial agents in liposomal form relatively to Staphylococcus aureus.

22.2   BACKGROUND

It has been found the antimicrobic activity of liposomal lincomycin depends on the composition and charge of liposomes. The antimicrobic activity of liposomal lincomycin, obtained on the basis of egg lecithin is higher than antimicrobic activity of the solution lincomycin in 3 times concerning planktonic cells of Staphylococcus aureus. Negatively charged liposomes received on the basis of polar lipids and lincomycin were more effective, than neutral liposomes. Minimum inhibitory concentration of them less of minimum inhibitory concentration of lincomycin in 7 times concerning planktonic cells of Staphylococcus aureus. The antimicrobic activity of liposomal benzoyl peroxide obtained on the basis of egg lecithin was in 14 times higher than antimicrobic activity of the solution benzoyl peroxide concerning of Staphylococcus aureus.

22.3 MATERIALS AND METHODS

The strain ATCC 25923 Staphylococcus aureus was taken from SE "Mechnicov Institute of Microbiology and Immunology AMSU." We have also used the following items: egg lecithin (Ukraine, "Biolek"), DMSO (Russia), the mixture of negatively charged lipids that were obtained by the original technology of Dr. Nina Ivanova, substance of benzoyl peroxide ("Aldrich" USA), lincomycin (JSC "Darnitsa", Ukraine), Mueller-Hinton agar (HiMedia Laboratories Pvt. Limited (Индия)), meat-peptone broth (MPB).

The Receiving of Liposomes. The substance of benzoyl peroxide (BP) dissolved in chloroform due to its poor solubility in aqueous solutions and added to an alcohol or chloroform solution of lipids in the ratio of BP: Lipids 1:10, 1:20. The liposomes were obtained by evaporating the lipids and antibiotics on a rotary vacuum evaporator (Switzerland). Next mixture was suspended in sterile buffered saline. Liposomes prepared in the extruder EmulsiFlex–C5 (Canada "Avestin"), punching with compressed air (10 cycles) to achieve a constant optical density on spectrophotometer (DU–7Spectrophotometer Beckman, USA) at temperatures that above the phase transition temperature of any of the lipid components was present. The average size of liposomes was 160–180 nm, concentration of lipids in the liposomes was 2%, ratio LN : lipids and BP : lipids was 1:20 [2].

Cleaning of Switched Antibiotics in Liposomes (Ls) from those that are not involved in making liposomes using ultracentrifuge (MSE-Superspeed Centrifuge 65, England) for an hour at 105,000 g. The output of liposomes was determined spectrophotometrically at 450 nm.

The determination of minimum inhibitory concentration (MIC) of antibiotics was performed by microtiter method. Antimicrobial agents were diluted by serial dilutions of meat-peptone broth (MPB) in flat-bottomed plates, they were also added to the culture of Staphylococcus aureus, and were incubated for 24 hours at 34°C. The control was culture

Staphylococcus aureus without antimicrobial agents. After that mixture sowed on solid nutrient medium Mueller-Hinton agar for calculation of amount of colony forming particles and MIC definitions. MIC was considered to be the lowest concentration, which retards the growth of Staphylococcus aureus during the incubation period.

22.4 RESULTS AND DISCUSSION

Among the antibiotics we stopped on lincomycin. It has a bacteriostatic effect on a wide range of microorganisms, with increasing doses of lincomycin it has a bactericidal effect. Antimicrobial mechanism of action of lincomycin is the inhibition of protein synthesis in the cells of microorganisms. The drug is active with respect to Gram-positive aerobic and anaerobic microorganisms, including Staphylococcus spp

BP has a wide spectrum of antimicrobial activity. It is active against the bacteria, in the case it is also resistant to antibiotics [3, 4].

In the first phase of our study the minimum inhibitory concentrations (MIC) of lincomycin and BP in Ls on the basis of egg lecithin were used, which is a soft lipid and traditionally used in the creation of liposomal forms of drugs. As a result, the definition MIC of lincomycin and its liposomal preparations were found and also Ls received on the basis of egg lecithin and lincomycin are more effective than epy solution of lincomycin under the action of Staphylococcus aureus planktonic cells (Fig. 1, №2). MIC of Ls this composition decreased in 3 times in comparison with the MIC of the lincomycin solution.

Negatively charged liposomes received on the basis of polar lipids and lincomycin were the most effective. The using of negatively charged liposomes that contained lincomycin reduced the MIC of the lincomycin solution in 7 times concerning planktonic cells of Staphylococcus aureus. (Fig.1, №3).

Liposomal BP on the basis of egg lecithin was also more effective against Staphylococcus aureus in comparison with the MIC of intact BP. At definition MIC ofliposomal forms BP on the basis of egg lecithin it has been found that its MIC made 31 µg/ml that was in 14 times less MIC BP dissolved in DMSO.

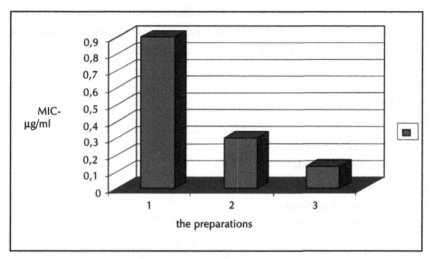

FIGURE 1   The minimum inhibitory concentration of the liposomal lincomycin. (a) the control (lincomycin solution), (b) liposomal lincomycin on the basis of egg lecithin, (c) negatively charged liposomal lincomycin.

22.5   CONCLUSION

1. It is found that minimally inhibitory concentration of liposomal antibiotic solution with the neutral charge (lecithin liposomes) decreased in 3 times in comparison with minimally inhibitory concentration of the lincomycin solution concerning Staphylococcus aureus.

2. The using of negatively charged liposomes on the basis of polar lipids with the antibiotic, strengthens efficiency of antibiotic action in greater degree than using lecithin liposomes. Negatively charged liposomes containing lincomycin reduced its minimally inhibitory concentration in 7 times concerning Staphylococcus aureus.

3. The minimally inhibitory concentration of the liposomal antimicrobic preparation benzoyl peroxide with the neutral charge (lecithin liposomes) decreased in 14 times in comparison with minimally inhibitory concentration of benzoyl peroxide concerning Staphylococcus aureus.

4. The received results enable to predict the using of liposomal forms of antimicrobic substances for increase of pharmacological efficiency in treatment of Staphylococcus infections.

KEYWORDS

- **Bactericidal effect**
- **Benzoyl peroxide**
- **Liposomes**
- **Microtiter method**

REFERENCES

1. Kim, H. J.; Jones, M. N. *J. Liposome. Res.*, **2004**, *14*-C, 123–139.
2. Sorokoumova, G. M.; Selishcheva, A. A.; Kaplun, A. P. *Educational Tools in Bioorganic Chemistry*, 2000; 105.
3. Tanghetti, E. A.; Popp, K. F.: *Dermatol. Clin.*, **2009**, *27*, 17–24.
4. Tanghetti, E. A.: *Cutis.*, **2008**, *82*, 5–11.

CHAPTER 23

POLYELECTROLYTE MICROSENSORS AS A NEW TOOL FOR METABOLITES' DETECTION

L. I. KAZAKOVA, G. B. SUKHORUKOV, and L. I. SHABARCHINA

CONTENTS

23.1 Introduction... 400
23.2 Experimental Details.. 401
    23.2.1 Materials 401
    23.2.2 Preparation of SNARF-1 Dextran and SNARF-1
        Dextran/Urease Containing CaCO3 Microparticles 402
    23.2.3 Fabrication of SNARF-1 Dextran Loaded
        Microcapsules.. 402
    23.2.4 Spectroscopic Study 403
    23.2.5 Spectrofluorimetric Study.. 404
    23.2.6 Confocal Laser Scanning Microscopy........................ 404
23.3 Results and Discussions.. 414
23.4 Conclusion 415
Keywords 416
References... 416

23.1   INTRODUCTION

Design of micro and nanostructured systems for in-situ and in-vivo sensing become an interesting subject nowadays for biological and medical-oriented research [1–3]. Miniaturization of sensing elements opens a possibility for non-invasive detection and monitoring of various analytes exploiting cell and tissues residing reporters. Typical design of such sensor is nano- and microparticles loaded with sensing substances enable to report on presence of analyte by optical means [4–9]. The particles containing fluorescent dye can be use as a sensor for relevant analytes such as H^+, Na^+, K^+, and Cl^- et al. [10–13]. The fluorescent methods are most simple and handy among the possible ways of registration. They provide high sensitivity and relative simplicity of data read-out. For analysis of various metabolites it is necessary to use the enzymatic reactions to convert analyte to optically detectable compound [14, 15]. In order to proper functioning all components of sensing elements (fluorescence dyes, peptides, enzymes) are to be immobilized in close proximity of each other. That "tailoring" of several components in one sensing entity represents a challenge in developing of a generic tool for sensor construct. One approach to circumvent problem has been introduced by the PEBBLE (Photonic Explorers for Bioanalyse with Biologically Localized Embedding) system [5, 16]. PEBBLE is a generic term to describe use co-immobilization of sensitive components in inert polymers, substantially polyacrylamide, by the microemulsion polymerization technique [17]. This technique is useful for fluorescent probe, but to our mind, is too harsh for peptides and enzymes capsulation due to organic solvents involved in particle processing. Multilayer polyelectrolyte microcapsules have not this shortcoming as they are operated fully in aqueous solution at mild condition. These capsules are fabricated using the Layer-by-Layer (LbL) technique based on the alternating adsorption of oppositely charged polyelectrolytes onto sacrificial colloidal templates [18, 19]. Immobilization of one or more enzymes within polyelectrolyte microcapsules can be accomplished by the coprecipitation of these enzymes into the calcium carbonate particles, followed by particle dissolution in mild condition leaving a set of protein retained in capsule [15, 20, 21]. A fluorescence dye can be included in polyelectrolyte capsules as well. Thus, the multilayer polyelectrolyte encapsulation technique microcapsules allows in principle combining enzyme activity for selected metabolite and registration ability of dyes in one capsule. In this work we demonstrate urea detection using capsules containing urease and pH sensitive dye.

The concentration of urea in biological solutions (blood, urine) is a major characteristic of the condition of a human organism. Its value may suggest a number of acute and chronic diseases: myocardial infarction, kidney and liver dysfunction. The measurement of urea concentration is a routine procedure in clinical practice. Urease based enzymatic methods are most widely used for urea detection and use urease enzyme as a reactant. There are multiple urease-based methods, which differ from each other by the manner of monitoring the enzymatic reaction. Despite of high specificity, reproducibility and extremely sensitivity for urea as urea is the only physiological substrate for urease, these methods are all laborious and time-consuming since freshly prepared chemical solutions and calibration are required daily. They all are lacking in-situ live monitoring what makes them inappropriate for analysis in vivo as residing sensors. Embedding of urease into polyelectrolytes microcapsules can help to solve these problems [22, 23]. The encapsulated urease completely preserves its activity at least 5 days at the fridge storage [22].

Aim of this work was to demonstrate a particular example of a sensor system, which combines catalytic activity for urea and at the same time, enabling monitoring enzymatic reaction by optical recording. The proposed sensor system is based on multilayer polyelectrolyte microcapsules containing urease and a pH-sensitive fluorescent dye, which translates the enzymatic reaction into a fluorescently registered signal.

23.2   EXPERIMENTAL DETAILS

23.2.1   MATERIALS

Sodium poly(styrene sulfonate) (PSS, MW = 70,000) and poly(allylamine hydrochloride) (PAH, MW = 70,000), calcium chloride dihydrate, sodium carbonate, sodium chloride, ethylenediaminetetraacetic acid (EDTA), TRIS, maleic anhydride, sodium hydroxide (NaOH) and Bromocresol purple were purchased from Sigma-Aldrich (Munich, Germany). Urease (Jack bean, Canavalia ensiformis) was purchased from Fluka. SNARF-1 dextran (MW = 70,000) was obtained from Invitrogen GmbH (Molecular Probes #D3304, Karlsruhe, Germany). All chemicals were used as received. The bidistilleted water was used in all experiments.

23.2.2 PREPARATION OF SNARF-1 DEXTRAN AND SNARF-1 DEXTRAN/UREASE CONTAINING CACO₃ MICROPARTICLES

The preparation of loaded $CaCO_3$ microspheres was carried out according to the coprecipitation-method [20, 21]. To prepare the $CaCO_3$ microspheres loaded with SNARF-1 dextran were used: 1.6 ml H_2O, 0.5 ml 1M $CaCl_2$, 0.5 ml 1M Na_2CO_3 and 0.4 ml SNARF-1 dextran solution (1 mg/ml).

To prepare the $CaCO_3$ microspheres contained different ratio of SNARF-1 dextran and urease were used:

Sample I: 0.6 ml H_2O, 0.5 ml 1M $CaCl_2$, 0.5 ml 1M Na_2CO_3, 0.4 ml SNARF-1 dextran solution (1 mg/ml) and 1 ml urease (3 mg/ml);

Sample II: 0.8 ml H_2O, 0.5 ml 1M $CaCl_2$, 0.5 ml 1M Na_2CO_3, 0.2 ml SNARF-1 dextran solution (1 mg/ml) and 1 ml urease (3 mg/ml).

The solutions were rapidly mixed and thoroughly agitated on a magnetic stirrer for 30 s at 4°C. After the agitation, the precipitate was separated from the supernatant by centrifugation (250x g, 30 s) and washed three times with water. The procedure resulted in highly spherical microparticles containing SNARF-1 dextran or SNARF-1 dextran and urease with an average diameter ranging from 3.5–4 µm.

23.2.3 FABRICATION OF SNARF-1 DEXTRAN LOADED MICROCAPSULES

Microcapsules were prepared by alternate layer-by-layer (LbL) deposition of oppositely charged polyelectrolytes poly(allylamine hydrochloride) (PAH, MW = 70,000) and poly(styrene sulfonate) (PSS, MW = 70,000) onto CaCO3 particles containing SNARF-1 dextran or SNARF-1 dextran and urease to give the following shell architecture: (PSS/PAH)4PSS. Short ultrasound pulses were applied to the sample prior to the addition of each polyelectrolyte in order to prevent particle aggregation. The decomposition of the $CaCO_3$ core was achieved by treatment with EDTA (0.2 M, pH 7.0) followed by triple washing with water. The microcapsules were immediately subjected to further analysis or stored as suspension in water at 4°C.

23.2.4 SPECTROSCOPIC STUDY

All spectroscopic studies were carried out with UV–vis spectrophotometer *Varian Cary 100* at constant agitation and thermostatic control at 20°C.

The SNARF-1 dextran concentrations in different capsules samples were estimated by matching absorption intensity of supernatant after co-precipitation of the dye with $CaCO_3$ to intensity of calibrated of SNARF-1 dextran concentrations in free solution. The average content of SNARF-1 dextran per capsule was calculated to be: (1) for SNARF-1 dextran $CaCO_3$ microparticales – 1 pg; (2) for SNARF-1 dextran/urease $CaCO_3$ microparticles: *Sample I* – 0.6 pg, *Sample II* – 0.2 pg.

The amount of active urease immobilized into the polyelectrolyte microcapsules was determined under assumption that the enzyme retain its activity while encapsulated. Free urease had 100 U/mg according to the data sheets. The activity of free and encapsulated enzyme were determined from the decomposition of urea into two ammonia molecules and CO_2 using a pH-sensitive dye Bromocresol purple [24]. The urease aliquot solutions were added to a reaction mixture contained a necessary amount of urea and 0.015 mM Bromocresol, whose pH was apriory brought up to 6.2. The reaction kinetics was recorded as a change in the optical absorption of the dye at 588 nm to obtain the linear calibration plot. Then, the known number of microcapsules containing urease and SNARF-1 dextran was added to the reaction solution. The revealed activity of enzyme was compared with amount of free urease.

23.2.5 SPECTROFLUORIMETRIC STUDY

All spectrofluorimetric studies of SNARF-1 dextran and SNARF-1 dextran/urease were carried out with the spectrofluorimeter *Varian Cary Eclipse*, at constant agitation, thermostatic control at 20°C, λ_{exc}=540 nm, slit width: excitation at 10 nm and emission at 20 nm. The microcapsule suspensions were used at concentration 2×10^6 capsules/ml, which was estimated with the cytometer chamber. All solutions were prepared on bidistilleted water.

TRIS-maleate buffer solutions for pH setting were prepared by adding appropriate quantity of 0.2 NaOH to 0.2 M TRIS and maleic anhydride mixture and diluted to 0.05 M concentration.

23.2.6 CONFOCAL LASER SCANNING MICROSCOPY

Confocal images were obtained by Leica Confocal Laser Scanning Microscope TCS SP. For capsules visualization 100× oil immersion objective was used throughout. 10 μl of the SNARF-1 dextran/urease capsules suspension was placed on a coverslip. To this suspension 10 μl of 0.1 mol/l urea is added. After about 20 min confocal images were obtained. The red fluorescence emission was accumulated at 600–680 nm after excitation by the FITC-TRIC-TRANS laser at 543 nm.

23.3 RESULTS AND DISCUSSIONS

Degradation of urea ($CO(NH_2)_2$) is catalyzed by urease and results in the shift of the medium pH into the alkaline range.

$$CO(NH_2)_2 + H_2O = CO_2 + 2\,NH_3$$

Monitoring of the urea degradation can be done by using SNARF-1 as pH-sensitive dye to follow changes of the pH in the enzyme driven reaction. In order to fabricate the sensing microcapsule the urease and SNARF-1 bearing dextran were simultaneously co-precipitated to form $CaCO_3$ spherical particles 3.5–4 μm in size, containing both components urease and fluorescent dye SNARF-1 coupled dextran (MW=70000) [15]. Then the particles were coated by standard layer-by-layer protocol with nine alternating layers of oppositely charged polyelectrolytes PSS and PAH. The formed shell had a $(PSS/PAH)_4PSS$ architecture. After the dissolution of $CaCO_3$ with the EDTA solution the obtained capsule samples contain an enzyme and a fluorescent dye in its cavity.

The dye and enzyme concentrations inside the polyelectrolyte capsule are predetermined essentially at the stage of formation of the $CaCO_3/$ SNARF-1 dextran/urease conjugate microparticles. Obviously, the amount of both components of urease and SNARF-1 dextran in the capsules and their ration should play an important role while functioning of entire sensing system is concerned. However, the final composition of co-precipitated particles and later capsules in fabricated samples may be different, though the same initial concentration of components used while preparing co-precipitating particles. It depends on a number of factors: adsorption, capturing and distribution of the components among the $CaCO_3$ particles, the size and the number of the particles yielded.[25] These parameters might vary from one experiments to another and therefore, it makes problematic to obtain two capsule samples with exactly the same content of encapsulated substances while relying on single capsule detection. Yet, it is imperative to observe this condition to reproduce the efficiency of any sensor. Thus, we always run experiments with at least two samples of capsules in parallel produced independent and having the same parameters at preparation. One of major problem on single particle/capsule detecting is deviation of fluorescent intensity from one particle to another due to uneven fluorescent distribution over population of capsules. To avoid this bottleneck and to obtain a sensor whose reliability and efficiency would not depend on the concentration of the reacting and registering substances in single capsule we opted the SNARF-1 fluorescent dye for the present study (Fig. 1).

The emission spectrum of SNARF-1 undergoes a pH-dependent wavelength shift from 580 nm in the acidic medium to 640 nm in alkaline environment. The ratio $R = I_{580nm}/I_{640nm}$ of the fluorescence intensities from the dye at two emission wavelengths allows to determine the pH value according to the ratiometric method. Dual emission wavelength monitoring is well established method eliminating a number of fluorescence measurement artifacts, including photobleaching, sample's size thickness variation, measuring instrument stability and non-uniform loading of the indicator [26]. It becomes very important particularly if one is using not an ensemble of capsules but only single or few capsules in the analysis, e.g., in the experiments with cells.

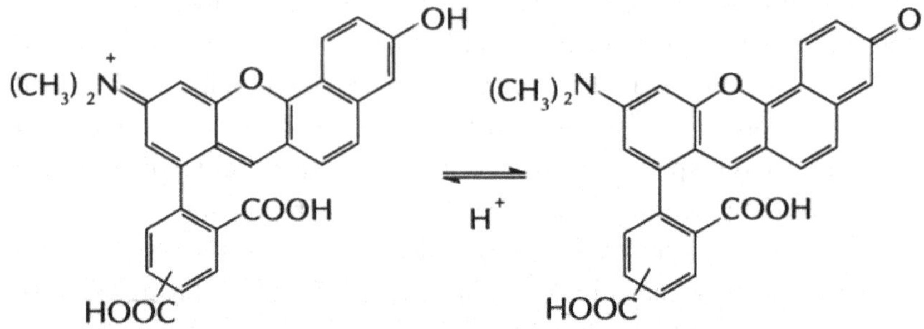

FIGURE 1   Structure of the protonated and deprotonated forms of the SNARF-1 dye.

In order to verify a feasibility of fluorescence based urea sensing on two component co-encapsulation we fabricated two polyelectrolyte capsule samples of the (PSS/PAH)$_4$PSS shell architecture with different content of dye and urease. The first sample (*Sample I*) contained in average 0.6 pg SNARF-1 dextran per capsule, while the content of the SNARF-1 dextran in the other sample was 0.2 pg per capsule (*Sample II*). The concentration of active urease in samples was opposite 0.2 and 0.6 pg/capsule respectively what gives an average SNARF-1 dextran/urease ratio of 3:1 and 1:3 in these investigated samples.

Spectrofluoremetric studies were carried out to determine the correlation between the fluorescence intensity of the SNARF-1 dextran/urease capsules and the pH of the medium. Both the capsule samples were stored for 10 min in the 0.05 M TRIS-maleate buffer at pH in the range 5.5–9. The excitation wavelength was 540 nm. The capsules fluorescence spectra of the first sample are shown in Fig. 2. The spectra obtained for both samples were similar. The encapsulated dye is capable to provide information of the medium acidity in a reasonably wide range of pH. It is seen that fluorescence spectra are characteristic for every pH value. This fact can be used for calibration regardless amount of dye per capsules in studied samples.

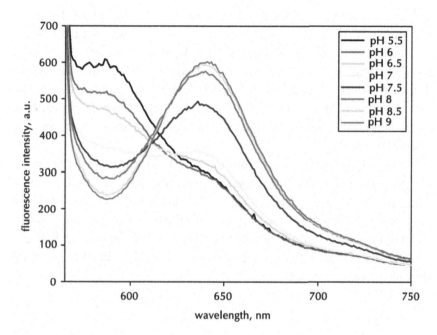

FIGURE 2   Fluorescence spectra of the SNARF-1 dextran/urease capsules in the 0.05 M TRIS-maleate buffer at pH in the range 5.5–9.

The ratio between fluorescence intensity and pH can be described by the following equation according to [26]:

$$pH = pK_a - \log\left(\frac{R - R_{min}}{R_{max} - R} * \frac{I_{640nm}(B)}{I_{640nm}(A)} \right) \qquad (1)$$

where $R = I_{580nm}/ I_{640nm}$; R_{min} and R_{max} – are the minimal and maximal R values in the titration curve (Fig. 3, curves 3,4); $I_{640nm}(A)$ and $I_{640nm}(B)$ – fluorescence intensities at 640 nm for the protonated and deprotonated forms of the dye, i.e. in the acidic and alkaline media, respectively. The R_{min} and R_{max} meanings depend on the experimental conditions. In this study they were determined to be:

Sample 1: $R_{min} = 0.41$, $R_{max} = 1.96$, $I_{640nm}(B)/ I_{640nm}(A)=2$;
Sample 2: $R_{min} = 1.06$, $R_{max} = 2.14$, $I_{640nm}(B)/I_{640nm}(A)= 1.69$.

To yield of the pK_a value the data were plotted as the log of the $[H^+]$ versus the $\log\{(R-R_{min})/(R_{max}-R)*(I_{640nm}B)/I_{640nm}A\}$. In this form, the data gave a linear plot with an intercept equal to the pK_a (Fig. 4, curves 3,4). As follows from this data pK_a for sample 1 is equal 7.15, for the sample 2 – 7.25. Thus, the pK_a value differed on 0.1 for different samples whereas the concentration of dye in them differed in 3 times. However, dependence of fluorescence intensity for the two samples was linear in the smaller interval of values.

In Fig. 3, the curves 1, 2 shows the ratio R dependences on pH value for free fluorescent dye in comparison with SNARF-1 dextran capsules. The calculation of pK_a values (Fig. 4, the curves 1, 2) has demonstrated that for encapsulated dye it is less, than for the free dye solution. It is reasonable to assume that this effect is the result of the interaction between SNARF-1 dextran and polyelectrolyte shell, notably with PAA, because they have opposite charges.

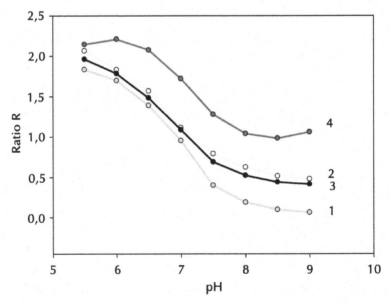

FIGURE 3 Change of fluorescence intensity ratio –R at 580 and 640 nm for: curve 1-SNARF-1 dextran water solution; curve 2 – containing SNARF-1 dextran capsules; curve 3 – SNARF-1 dextran/urease capsules (sample I, 0.6 pg dye/capsule; curve 4 – SNARF-1 dextran/urease capsules (sample II, 0.2 pg dye/capsule) in 0.05M TRIS-maleate buffer at pH in range 5.5–9.

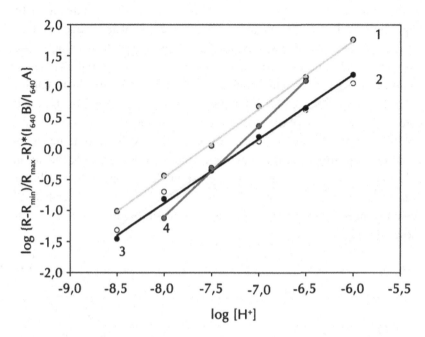

FIGURE 4 Experimental points and theoretical curves generated by the ratiometric method to determine the pK_a values: curve 1 – for SNARF-1 dextran water solution, pK_a=7.58; curve 2 – for containing SNARF-1 dextran capsules, pK_a=7.15; curve 3 – for SNARF-1 dextran/urease capsules (sample I, 0.6 pg dye/capsule), pK_a=7.15; curve 4 – for SNARF-1 dextran/urease capsules (sample II, 0.2 pg dye/capsule), pK_a=7.25.

Figure 5 illustrates the effect the urea in concentration from 10^{-6} to 10^{-2} mol/l produces on fluorescence spectra of capsules containing SNARF-1 dextran and urease. Urea concentration dependence is reflected spectro-scopically by apparent pH change in the course of enzymatic reaction in-side the capsules. The ammonium ions generated via enzymatic reaction in capsule interior effect on pH shift what is recorded by SNARF-1 on the plot (Fig. 5). The fluorescent spectra were measured at the 30-min time point after adding urea solutions at these concentrations to the SNARF-1 dextran/urease capsule's samples. Our particular attention was paid to the kinetics of the change at the fluorescence intensity ratio at 580 nm to 640 nm R (Fig. 6) and its relevance to amount of SNARF/urease. Parameter R was plotted versus time and as one can see on curves at high concentration of the urea substrate (10^{-3}M) level off at about 15–20 minutes after the beginning of the of the enzymatic reaction, while at low concentrations of urea (10^{-5} M) the time needed for flattening out spectral characteristics reaches 25–30 minutes. Remarkably, there are no substantial changes for samples at variable concentrations of the dye and urease inside the capsule at least at studied range of 0.2–0.6 pg per capsule. SNARF-1 dextran indicates only the course of the enzymatic reaction, therefore the time needed for the R parameter curve to level off correlates with the time needed to reach the equilibrium in the enzymatic reaction urea/urease occurring inside the capsule. Presumably, it takes about few minutes to equilibrate concentration of urea and its access to urease. We assume rather fast diffusion of urea through the multilayers due to small molecular size of the molecules [27].

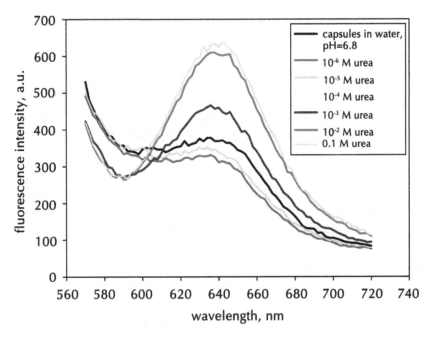

FIGURE 5   SNARF-1 dextran/urease capsule fluorescence spectra in water in the presence of urea from 10^{-6} to 0.1 M.

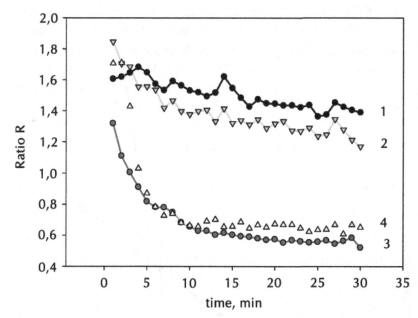

FIGURE 6 Kinetics of the change of the SNARF-1 dextran/urease capsule fluorescence intensity expressed through the fluorescence intensity ratio R at 580 and 640 nm in the presence of 10^{-5} M (curve 1 – sample I, curve 2 – sample II) and 10^{-3} M urea (curve 3 – sample I, curve 4 – sample II).

To calculate the apparent pH caused by urea concentration the values of R_{min}, R_{max}, pK_a and $I_{640nm}(B)/I_{640nm}(A)$ were used. R_{max} was determined as a fluorescence intensity ratio at 580 nm and 640 nm in the capsules stored in bidistilled water (pH = 6.4). R_{min} is the ratio of the fluorescence intensity spectrum related to the minimal R value at 580 nm and 640 nm in the SNARF-1 dextran/urease capsules when "high" concentration of urea (0.1 M) was added and 30 min after the enzymatic reaction begun. The pK_a value were assumed to be =7.15 and 7.25 for sample 1 and 2 respectively, that was in accord with the experiment with buffer solutions (Figs. 3, and 4). From the values obtained a calibration curve of the pH dependence inside the capsules on the urea concentration present in the solution was plotted. The calibration curve is presented in Fig. 7.

Thus, to determine the urea concentration in the solution it is necessary to obtain the following three spectra: (1) of the capsules without substrate (urea); (2) of the capsules at "high" concentration of urea in the solution,

e.g. 0.1 M as used for our calibration plot; (3) of the capsules in the investigated sample studied. These data will suffice to calculate the values of
R_{min}, R_{max} and $I_{640nm}(B)/I_{640nm}(A)$, which are characteristic of each particular sample of the sensing capsules. Using Eq. (1) one then can calculate the
pH as apparently reached in the capsules in the course of urea degradation
and to compare its value with the calibration curve in Fig. 7.

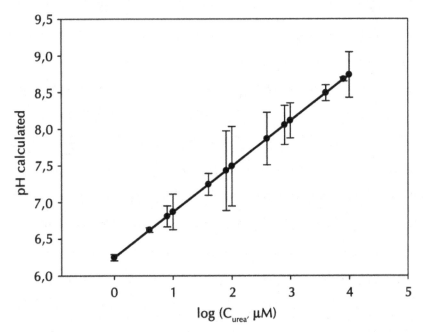

FIGURE 7   Calibration curve for detecting urea using the SNARF-1 dextran/urease
capsules in water solutions.

It is worth to notice, that this calibration curve is obtained for SNARF-1
dextran/urease capsule in pure water without substantial contamination of
any salt, which could buffer the systems and spoil truly picture for urea
detection. We carried out experiments to build a similar calibration curve
in the presence of the 0.001 M TRIS-maleate buffer (were used solutions
with the pH 6.5 and 7.5) but it resulted in overwhelming effect of pH buffering. Buffering the solution eliminates the pH change caused in a course
of enzymatic reactions. Thus, it sets a limit for detection of urea concentration

using SNARF-1-dextran/urease capsules. However the calibration in conditions of particular experimental system is reasonable at salt free solution assumption. Summarizing, one can state the presented in Fig. 7 calibration curve as suitable for estimation of urea concentrations *in situ* in water solutions.

Feasibility studies on single capsule detection of urea presence were carried out using Confocal Fluorescent Microscopy. CLSM image of SNARF-1 dextran/urease capsules in absence and at 0.1M urea added to the same capsules are presented on Fig. 8. Small distinctions in the form and the sizes of capsule population are connected with non-uniformity of SNARF-1 dextran/urease $CaCO_3$ particles received in co-precipitation process what is rather often observed for calcium carbonate templated capsules containing proteins [20].

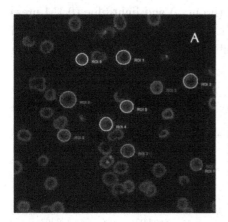
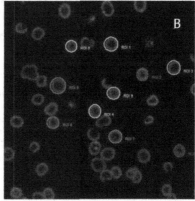

FIGURE 8 Confocal fluorescence microscopy images of (PSS/PAH)4PSS capsules loaded with SNARF-1 dextran (MW = 70kDa) and urease enzyme in water (image A) and 0.1 M urea concentrations (image B). The Table 1 shows the increase of Mean energy of individual capsules before (Flow) and after the addition 0.1 M urea solution (Fhigh) to water capsule suspension. The red fluorescence emission was accumulated at 600–680 nm after excitation by the FITC-TRIC-TRANS laser at 543 nm.

The images have been processed with Leica Confocal Laser Scanning Microscope TCS SP software to quantify the effect. The image areas corresponding to location of 10 selected capsules are set off in different colors – ROI_{1-10} (Region Of Interest). The value of average intensity of a luminescence is defined as parameter Mean Energy by formula:

$$I^2_{Mean} = \frac{1}{N_{Pixel}} \sum_{Pixel} I_i^2$$

were I_{mean} – the average image energy of ROI areas;
N_{pixel} – the total number of pixels that are included in the calculation;
I_i – the energy correspond for particular pixel.

The energy meanings of individual capsules are presented in the Table 1.

Although, value of fluorescence intensity is seen to be different for each capsule the more than double increase of integrated intensity is well pronounced in all monitored capsules upon addition of urea solution. The distribution of energy relation therefore remains almost constant for each capsule. These data on single capsules are in good agreement with data presented on Fig. 5 obtained on entire capsule population. Indeed, integrated area under spectra for black (no urea) and light blue (0.1M urea) with spectral range of 600–680 nm is about twice in difference. This fact demonstrates the principal applicability to use single capsule for carrying out analysis of urea presence.

TABLE 1   The change of mean fluorescence intensity of 10 selected capsules in presence of 0.1 M urea.

ROI	Mean Energy, I_{low}	Mean Energy, I_{high}	I_{high}/I_{low}
ROI 1	2947.95	6467.42	2.1939
ROI 2	4635.87	10316.34	2.2253
ROI 3	2424.13	5694.92	2.3493
ROI 4	4021.83	11161.52	2.7752
ROI 5	1666.76	3727.05	2.2361
ROI 6	3177.11	7237.68	2.2781
ROI 7	2237.61	5932.76	2.6514
ROI 8	3192.70	7158.54	2.2422
ROI 9	4100.76	8944.83	2.1813
ROI 10	2758.57	6127.32	2.2212

23.4 CONCLUSION

In this study we have demonstrated a particular example of a sensor system, which combines catalytic activity for the substrate (urea) and at the same time enabling to monitor the enzymatic reaction by co-encapsulated pH sensitive dye. Substrate sensitive enzyme urease was co-encapsulated together with SNARF-1 coupled to dextran in multilayer microcapsules. Enzymatic activity was recorded by fluorescent changes caused by increasing of pH in course of enzymatic cleavage of urea as measured on population of capsules and on single capsule imaging by confocal fluorescent microscope. Suggested method can be used to measure the concentration of urea in solutions where the content of urea is fairly high (blood, urine) and also able to detect urea at concentration down to 10^{-5} M at non-buffering solution. Spectroscopic parameters of microencapsulated sensors were found stable in regardless ratio between urease and SNARF-dextran in concentration range of 0.2–0.6 pg per capsules what encourages such as microencapsulated sensing system as robust. Although, that pH sensitivity of dye has a limitation to function in buffers the concept of co-encapsulation of metabolite active enzymes and dyes sensitive to product of enzymatic reactions is illustrated to be workable in reasonable concentration range and applicable for single capsule based detecting.

The presented results prove the concept of feasibility of microencapsulated enzyme/dye systems for local metabolite sensing and optical online recording. These co-encapsulating sensors have advantages over well known PEBBLE systems as they could sense the substances via extra enzymatic reaction what is much more prospective in term of analytes to be monitored, especially in biological systems as cells and tissue. The micron size of the sensors will pave the way for producing and applicability of injectable and implantable sensing systems like 'a smart tattoo' or delivered to the cell or tissue and serving for on-line monitoring of various biological processes. Aspects of, considered here, urea sensing has a particular challenge to measure concentration of urea in vivo (in cells and tissue, e.g., in skin epithelium), what remains subject of further research.

KEYWORDS

- A smart tattoo
- Bromocresol
- Layer-by-Layer
- PEBBLE
- Spectrofluorimeter
- Spectrophotometer
- Tailoring

REFERENCES

1. Arregui, F. J. *Sensors Based on Nanostructured Materials*, 2009.
2. Fehr, M.; Okumoto, S.; Deuschle, K.; Lager, I.; Looger, L.L.; Persson, J.; Kozhukh, L.; Lalonde, S. Frommer, W. B. *Biochem. Soc. Trans.*, **2005**, *33*(1), 287–290.
3. Vo-Dinh, T.; Griffin, G. D.; Alarie, J. P.; Cullum, B.; Sumpter, B.; Noid, D. *Summary Nanomedicine*, **2009**, *4*(8), 967–979.
4. Sukhorukov, G. B.; Rogach, A. L.; Garstka, M.; Springer, S.; Parak, W. J.; Muñoz-Javier, A.; Kreft, O.; Skirtach, A. G.; Susha, A. S.; Ramaye, Y.; Palankar, R.; Winterhalter, M. *Small,* *3*(6), 944–955.
5. Lee, Y. E.; Smith, R.; Kopelman, R. *Annu. Rev. Anal. Chem. (Palo Alto Calif)*, **2009**, *2*, 57–76.
6. Sukhorukov, G. B.; Rogach, A. L.; Zebli, B.; Liedl, T.; Skirtach, A. G.; Köhler, K.; Antipov, A. A.; Gaponik, N.; Susha, A . S.; Winterhalter, M.; Parak, W. J. *Small*, **2005**, *1*(2), 194–200.
7. De Geest, B. G.; De Koker, S.; Sukhorukov, G. B.; Kreft, O.; Parak, W. J.; Skirtach, A. G.; Demeester, J.; De Smedt S. C.; Hennink, W. E. *Soft Matter*, **2009**, *5*, 282.
8. Peteiro-Cartelle, J.; Rodríguez-Pedreira, M.; Zhang, F.; Rivera, P.; Gil, L.; del Mercato, L.; Parak, W. J. *Nanomedicine*, **2009**, *4*(8), 967–979.
9. Sailor, M. J.; Wu, E. C. *Adv. Funct. Mater.*, **2009**, *19*(20), 3195–3208.
10. Nayak, S.; McShane, M. J. *Sensor Lett.*, **2006**, *4*, 433–439.
11. Kreft, O.; Muñoz Javier, A.; Sukhorukov, G. B.; Parak, W. J. *J. Mater. Chem.*, *42*, 4471–4476.
12. Brown, J. Q.; McShane, M. J. *IEEE Sensors J.*, **2005**, *5*, 1197–1205.
13. del Mercato, L, L.; Abbasi, A. Z.; Parak, W. *J. Small*, **2011**, doi:10.1002/smll.201001144.
14. Brown, J. Q.; McShane, M. J. *Biosens. Bioelectron.*, **2005**, *21*, 1760–1769.
15. Stein, E. W.; Volodkin, D. V.; McShane, M. J.; Sukhorukov, G. B. *Biomacromolecules*, **2006**, *7*, 710–719.
16. Brasuel, M.; Aylott, J. W.; Clark, H.; Xu, H.; Kopelman, R.; Hoyer, M.; Miller, T. J.; Tjalkens, R.; Philbert, M. *Sens. Mater.*, **2002**, *14*, 309–338.

17. Xu, H.; Aylott, J. W.; Kopelman, R.; *Analyst*, **2002**, *127*, 1471–1477.
18. Donath, E.; Sukhorukov, G. B.; Caruso, F.; Davis, S. A.; Möhwald, H. *Angew. Chem., Int. Ed.*, **1998**, *37*, 2202–2205.
19. Sukhorukov, G. B.; Donath, E.; Davis, S.; Lichtenfeld, H.; Caruso, F.; Popov, V. I.; Möhwald, H. *Polym. Adv. Technol.*, **1998**, *9*, 759–767.
20. Petrov, A. I.; Volodkin, D. V.; Sukhorukov, G. B. *Biotechnol. Prog.* **2005**, *21*, 918–925.
21. Sukhorukov, G. B.; Volodkin, D. V.; Gunther, A. M.; Petrov, A. I.; Shenoy, D. B. Möhwald, H. *J. Mater. Chem.*, **2004**, *14*, 2073–2081.
22. Lvov, Y.; Antipov, A. A.; Mamedov, A.; Möhwald, H.; Sukhorukov, G. B. *Nano Lett.*, **2001**, *1*, 125.
23. Lvov, Y.; Caruso, F. *Anal. Chem.*, **2001**, *73*, 4212.
24. Paddeu, S.; Fanigliulo, A.; Lanzin, M.; Dubrovsky, T.; Nicolini, C. *Sens. Actuators*, **1995**, *25*, 876–882.
25. Halozana, D.; Riebentanz, U.; Brumen, M.; Donath, E.; *Colloids and Surfaces A: Aspects*, **2009**, *342*, 115–121.
26. Whitaker, J. E.; Haugland, R. P.; Prendergast, F. G. *Anal. Biochem.* **1991**, *194*(2), 330–44.
27. Antipov, A. A.; Sukhorukov, G. B.; Leporatti, S.; Radtchenko, I. L.; Donath, E.; Möhwald, H.; *Colloids Surfaces A: Physicochem. Eng. Aspects*, **2002**, *198–200*, 535–541.

CHAPTER 24

SELECTION OF MEDICAL PREPARATIONS FOR TREATING LOWER PARTS OF THE URINARY SYSTEM

Z. G. KOZLOVA

CONTENTS

24.1 Introduction.. 420
24.2 Characteristics of Preparations 420
24.3 Method of Experiment 421
24.4 Results and Discussion 423
Keywords 426
References.. 426

24.1   INTRODUCTION

Stones – a metabolism illness due to various endogenous or exogenous causes and often of a hereditary nature characterized by the urino-formation of stones in the urinary system.

There are people in all age groups who suffer from irretention of urine. Thirty percent of women suffer from this in one form or another, i.e., unable to regulate the functioning of the urinary bladder. This problem may be solved by strengthening the wall of the urinary bladder, decreasing the inflammatory process in the urinary tract and strengthening the connective tissue.

Antioxidants (AO) are nutrient substances (vitamins, microelements etc.), which human organisms require constantly. They serve to maintain a balance between free radicals and AO forces.

Modern medicine uses AO to improve people's health and as a prophylaxis. Therefore, in recent times, the AO properties of various compounds are being widely studied. The most prevalent source of AO is considered to be vegetative objects on the basis of which medicinal preparations and BAAs are prepared.

As a criterion for evaluating the quality of a medicinal preparation, we took its Antioxidant activity (AOA), i.e., concentration of natural AO in it, which was measured on a model chain reaction of liquid-phase oxidation of hydrogen by molecular oxygen.

Set task: Quantitatively measure AO content in investigated preparations since they constitute a vegetation composition and evaluate their effectiveness in improving the quality of life.

The following are medicinal preparations and BAAs investigated for treating illnesses of the urethra (cystitis, enuresis and urine stones): Cyston (India), Urotractin (Italy), Promena (USA), Spasmex (Germany), Urolyzin (Russia), Contrinol (USA), Tonurol (Anti-Enuresis) (Russia), Blemaren (Germany).

24.2   CHARACTERISTICS OF PREPARATIONS

CYSTON (India) – Complex therapy for stones in the bladder, crystallization, infection of the urethra, podagra.

Each tablet contains: Didymocarpus pedicellata R. Br., Saxifraga Ligulata Wall, Rubia cordifolia L.; Cyperus scariosus R.Br.; Achyranthes aspera L.; Onosma bracteatum Wall.; Vernonia cinerea (L.) Less. BAA.

CONTRINOL (USA) – A Mixture of eastern and western medicinal plants used for normal functioning of the urethra. It strengthens the connective tissue.

Each capsule contains: Horsetail Herb, White Poplar Bark, Dogwood Berry and Schizsandra Berry. BAA.

PROMENA (USA) – provides important support for the prostate, possesses anti-inflammation properties, normalizes functioning of the urethra and strengthens the immune system.

Each capsule contains: Vitamin E, Vitamin C, Zinc, Vitamin A, Parsley Leaf, Echinacea, Pumpkin Seed, Gravel Root, Corn silk, Bee Pollen. BAA.

SPASMEX (Germany), TROSPIYA CHLORED (PRO. MED. CS PRAHA a.s.) – quarter amine is safer to use because of its unique chemical structure. Lowers tone of the smooth muscles of the urinary bladder, reduces detrusion of the bladder.

Each capsule contains: Dry birch bark extract, Irish Moss, Origanum vulgare L.

UROLYZIN (Russia) – source of Arbutin and Flavonoids used as a diuretic and antiseptic.

Content: Extracts of Folium Betula, Polygonum avicular L., Orthosiphon stamineus Benth, Sprout vaccinium myrtillus L., Fructus Aronia melanocarpa Elliot, Fructus Sorbus aucuparia L., Burdock Root. BAA.

TONUROL (ANTI-ENURESIS) (Russia) – supports urethra organs, strengthens bladder wall, helping in case of incontinence and has antiseptic and anti-inflammation properties.

Capsule content: Equisetum arvense L., White Poplar Bark, Hypericum perforatum L. flowers, Dogwood Berry, Schisandra chinensis Baill (Turez), Matricaria recutita (L.) extract.

BLEMAREN (Germany) – for prophylactic and treatment of stones in the bladder. Burbling tablets.

24.3   METHOD OF EXPERIMENT

Chain reactions of oxidation can be used for quantitative characterization of the properties of inhibitors (antioxidants). The investigated samples were analyzed by means of a model chain reaction of initiating oxidation of cumene [1–4]. Initiated oxidation of cumene in the presence of studied AO proceeds in accordance with the following scheme:

Initiation of chain   Origination of $RO_2^{\cdot}$ radicals, initiation rate W_i
Continuation of chain $RH+RO_2^{\cdot}$ $(_3+O_2ROOH+RO_2^{\cdot}$
Break of chain $2RO_2^{\cdot}$ molecular products
$InH+RO_2^{\cdot}$ $RO_2H+In^{\cdot}$
$RO_2^{\cdot}+In^{\cdot}$ molecular products
(We use the widely accepted numeration of rate constants of elementary reaction of inhibited oxidation.)

In accordance with this scheme, each independent inhibiting group of AO breaks two chains of oxidation.

Cumene (isopropyl benzene) was used as oxidizing hydrocarbon and azo-bis-isobutyronitrile as initiator, which forms free-radicals upon thermal decay. Initiating rate was determined from the following formula:

$W_i = 6.8 \times 10^{-8}$ [AIBN] mole / l·s,

where [AIBN] (AZO-bis-ISOBUTYRONITRILE) is the initiator concentration in mg per ml of cumene.

The period of induction τ is determined by plotting the dependence of the quantity of absorbed oxygen ΔO_2 in the reaction against time t. The end of the kinetic curve is a linear portion representing non-inhibited reaction, i.e., the portion after expending AO. The AO expenditure time τ is determined graphically on the kinetic dependence of oxygen absorption by the point of intersection of two straight lines: one of the lines is the line that the kinetic curve assumes after AO is consumed and the other is a tangent to the kinetic curve, the tangent of the angle of inclination of which is one half the tangent of the angle of the first. The greater the amount of AO in the sample the greater the period of induction τ.

The method is direct and based on the use of the chain reaction of liquid-phase oxidation of hydrocarbon by molecular oxygen.

The method is functional, i.e., the braking of the oxidizing reaction is determined only by the presence of AO in the system being analyzed. Other possible components of the system (not AO) do not exert a significant effect on the oxidizing process, which enables to analyze AO in complex systems, avoiding a stage of separation.

The method is very sensitive, exact and informative.

The method is absolute, i.e., does not require calibration, is simple to apply and does not require complex equipment [1–5].

24.4 RESULTS AND DISCUSSION

The following medicinal preparations were taken for investigation: Cyston, Spasmex, Urotractin, Urolyzin, Contrinol, Promena, Tonurol (Anti-Enuresis), Blemaren. The AOA of these preparations was determined.

As an example, the Fig. 1 shows the kinetic dependences of oxygen absorption in a model reaction of initiated cumene oxidation in the absence of antioxidant (straight line 1) and in the presence of Urotractin (curve 2), Spasmex (curve 3), Promena (curve 4), Contrinol (curve 5), Cyston (curve 6), Urolyzin (curve 7).

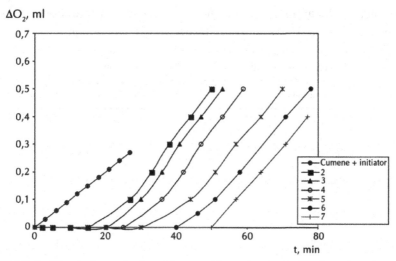

FIGURE 1   Kinetic Dependences of Oxygen Absorption 1 ml of hydrocarbon, 1 mg of initiator, t = 60°C.

1 – hydrocarbon (cumene) + initiator AZO-bis-IZOBUTYRONI-TRILE, 1 mg),

2 – with Urotractin added (6.6 mg), $\tau = 20$ min,

3 – with Spasmex added (6.0 mg), $\tau = 25$ min,

4 – with Promena added (7.5 mg), τ = 30 min,

5 – with Contrinol added (20 mg), τ = 38 min,

6 – with Cyston added (9.6 mg), τ = 45 min,

7 – with Urolyzin added (14.6 mg), τ = 50 min.

It can be seen from the figure that, in the absence of the additive, hydrocarbon oxidation proceeds at constant rate (straight line 1). When the preparation is added, the oxidation rate at the beginning is strongly retarded but begins to increase after a certain period of time. This is indicative of the presence of antioxidant in the additive.

The rise in reaction rate is due to the expenditure of antioxidant. When it is used up, the reaction proceeds at the constant rate of an uninhibited reaction. The time of antioxidant (τ) is determined graphically by the intersection of two straight lines on the kinetic curve. One of these is the straight-line portion of the kinetic curve after AO has been used up. The other is the tangent to the kinetic curve whose inclination angle is one- half the tangent angle of the first.

Data on the antioxidant content of investigated preparations are presented in Table 1. These data are illustrated by the diagram.

TABLE 1   Concentration of Antioxidants in Studied Preparations.

№	Preparations	Antioxidant Concentration (M/kg)
1	Cyston	9.6×10^{-3}
2	Spasmex	8.5×10^{-3}
3	Promena	8.2×10^{-3}
4	Urolyzin	7.0×10^{-3}
5	Urotractin	6.2×10^{-3}
6	Tonurol (Anti-Enuresis)	5.4×10^{-3}
7	Contrinol	3.9×10^{-3}
8	Blemaren	–

Diagram Illustrating Results in Table

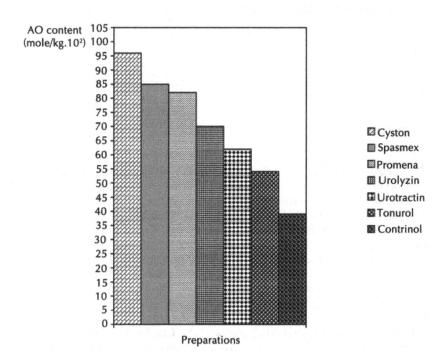

Results of analysis show that AOA is in the interval of values (3.9–9.6) × 10⁻³ M/kg, which correlates with the values determined earlier for AO in plants: Basilicum, Hypericum perforatum L. – 8.8 × 10⁻³ M/kg, Coriandrum sativum L. – 3.9 × 10⁻³ M/kg, Valeriana officinalis L. – 7.7 × 10⁻³ M/kg, Eleutherococcus Senticosus maxim. – 9.4 × 10⁻³ M/kg. Thus, the effectiveness of certain BAAs may be due to their AO content (more than 10⁻³ M/kg of dry substance) and their inclusion in complex therapy of corresponding illnesses may be justified. The close values of AO concentration between the studied preparations and medicinal plants gives basis for assuming their having a positive effect on the human organism [6].

KEYWORDS

- Antioxidant content
- Connective tissue
- Prophylaxis
- Urinary bladder

REFERENCES

1. Tsepalov, V. F. "A Method of Quantitative Analysis of Antioxidants by means of a Model Reaction of Initiated Oxidation," In *Investigation of Synthetic and Natural Antioxidants in vitro and in vivo*. Moscow, Russian, 1992.
2. Kharitonova, A. A.; Kozlova, Z. G.; Tsepalov, V. F.; Gladyshev, G. P. *Kinetika i Kataliz. J.*, Russian, **1979**, *20*(3), 593–599.
3. Tsepalov, A V. F.; Kharitonova, A.; Kozlova, Z. G. "Physicochemical Characteristics of Natural Food Products" In *Bioantioxidant*. Moscow, Russian, 1998; 94.
4. Tsepalov, V. F.; Kharitonova, A. A.; Kozlova, Z. G.; Bulgakov: V. G. "Natural Antioxidants: An Express Method of Analysis and Prospects of Use as Food Additives." In *Pishchevye Ingredienty*. Moscow, Russian, 2000; 7–8.
5. Tsepalov, V. F. *Zavodskaya Lab.*, Russian, **1964**, *1*, 111.
6. Kozlova, Z. G.; Eliseyeva, L. G.; Nevolina, O. A.; Tsepalov, V. F. *Content of Natural Antioxidants in Spice-aromatic and Medicinal Plants*. Theses of Report at International Scientific Conference: Populations Quality of Life-Basis and Goal of Economic Stabilization and Growth, Oryol, 23.09–24.09.1999, Russian, 184–185.

CHAPTER 25

IMPROVEMENT OF THE FUNCTIONAL PROPERTIES OF LYSOZYME BY INTERACTION WITH 5-METHYLRESORCINOL

E. I. MARTIROSOVA and I. G. PLASHCHINA

CONTENTS

25.1 Introduction.. 428

25.2 Background... 428

25.3 Materials and Methods... 429

    25.3.1 Surface Tension Measurement... 429

    25.3.2 Surface Dilatational Properties Measurement 430

25.4 Results And Discussion 431

    25.4.1 Adsorption of Methylresorcinol in the Presence of

           Lysozyme... 431

    25.4.2 Surface Dilatational Properties 433

25.5 Conclusion 436

Keywords 437

References... 437

25.1 INTRODUCTION

Wide using of lysozyme in medicine for the treatment of various infections and in food and cosmetic industry to prevent bacterial contamination of the stuffs leads to production of the resistance of microorganisms to lysozyme action. In recent years, a great deal of attention has been devoted to the investigations aimed at studying lysozyme modifications, affecting lysozyme properties while retaining its enzymatic activity. One of the ways to solve this problem is modification of lysozyme structure with using weak nonspecific interaction with 5-methylresorcinol. It is known numeral effects of MR on the structure and functions of lysozyme. In particularity, in concentration range of 10^{-7}–10^{-3}M MR the specific and non-specific enzymatic activities raise, its substratum specificity extends and temperature range of lysozyme catalysis expands [1].

Earlier the interrelation between the influence of MR concentration on nonspecific activity of lysozyme and the destabilizing effect of MR on its native conformation were established. It is shown, that activity and thermostability of lysozyme depend on the concentration of MR and the incubation time of their mixed solutions [2]. The MR ability to self-organization in a solution with micelle like structure formation was established via methods of mixing microcalorimetry and dynamic light scattering [3].

Due to diphylic character of MR and its ability of self-organization in a solution it can behave similarly to nonionic surfactants. Adsorption of MR was shown to be mixed-diffusion barrier controlled (pH 7.4). MR was also shown to change thermodynamic affinity of lysozyme to the solvent with the resulting the protein being slightly less or more surface active depending on MR concentration. Under used conditions MR and lysozyme can compete in adsorption process at air/water interface [4].

25.2 BACKGROUND

The effect of 5-methylresorcinol (MR) on the surface activity of lysozyme (LYS) and rheological properties of its adsorption layers at the air/solution (0.05 M PBS, pH 6.0) interface at 25°C was researched using a dynamic drop tensiometry and dilatometry methods. The influence of methylresorcinol on adsorption of protein was studied at fixed protein concentration 5.1×10^{-6} M and varying (0.16–67.2 mM) MR concentration. MR changes the adsorption behavior of LYS. LYS-MR mixtures are more surfaces active than the pure LYS in all MR concentration range. Complex modulus of elasticity (E) increases in all range of frequencies used (0.007–0.625 rad/s) with MR concentration from 0.16 mM to CMC_{MR}. The viscosity component (E_{ip}) and phase angle increase with MR concentration up to 3.21 mM and then decreases up to CMC_{MR}. The rheological behavior of the adsorption layers as well of intact lysozyme as modified ones is solid-like (E=E_{rp}, $E_{rp} \gg E_{ip}$, where E_{ip} is elastic component, and low frequency dependence of complex modulus E).

The aim of this work is to study the effect of MR on the surface activity of lysozyme and rheological properties of its adsorption layers at the air/solution (0.05M phosphate buffer, pH 6.0) interface.

25.3 MATERIALS AND METHODS

A sample of hen egg white lysozyme (Sigma, USA) with activity 20,000 U/mg and molecular mass 14,445 Da was used.

Alkyl-substituted hydroxybenzenes, 5-Methylresorcinol (5-Methylbenzene-1,3-diol) (Sigma, USA) with molecular mass 124.14 g/mol (anhydr) was taken.

All reagents for phosphate buffer preparation in Milli-Q water were of analytical grade.

25.3.1 SURFACE TENSION MEASUREMENT

Dynamic surface tension was measured with an automatic drop Tracker tensiometer (ITC Concept, France), connected to thermostatic bath to maintain the temperature constant at 25°C during the measurements. The principle of tensiometer is to determine the surface tension of the studied solution from the axisymmetric shape of a rising bubble analysis [5]. Due to the active control loop, the instrument allows long-time experiments with a constant drop/bubble volume or surface area.

Surface tension, σ (mN/m), was measured in 7 ml samples at constant lysozyme (0.075 mg/ml) and varying 5-methylresorcinol (0.16–67.2 mM) concentrations and their mixed solutions in 0.05M phosphate buffer, pH 6.0 at 25°C.

The mixed solutions LYS-MR were prepared with using of lysozyme and MR after 24 h incubated separately at 25°C in darkness. After mixing of these solutions 1:1 they were stored 3 h at 25°C away from the light. The dynamic surface tension was measured over 60,000 s to guarantee steady-state of the adsorption layer. A t→∞ asymptotic extrapolation was used to find the steady-state surface tension values. Standard deviations were always less, than 0.5 mN/m, and duplicate measurements were made for each MR concentration. Finally, phosphate buffer were confirmed not to present surface activity by measuring separately the surface tension of a phosphate buffer solution, obtaining values practically equal to those of pure water.

From the kinetic curves (surface tension versus time) for the solutions of different composition, the steady-state surface tension isotherms (surface tension versus concentration) were obtained. As the concentration of the protein in the mixture was fixed, the abscissa in all graphs represents the concentration of the surfactant in the mixed solution. From surface tension data the MR critical concentration of self-organization (micelle-formation) in presence of lysozyme were estimated. For this purpose surface tension values were fitted with the logarithm of the MR concentration.

25.3.2 SURFACE DILATATIONAL PROPERTIES MEASUREMENT

The surface rheological parameters of adsorbed LYS-MR films at the air-water interface, the surface dilatational modulus (E) and the phase angle (θ) were measured as a function of time with constant amplitude ($\Delta A/A$) of 3% and angular frequency (ω). The range of angular frequencies used was 0.007–0.625 rad/s. Measurements of rheological properties were performed after formation of enough stable adsorption layer during of 60,000–70,000 s. The sinusoidal oscillation for surface dilatational measurement was made with 3–5 oscillation cycles followed by a time of 3–5 cycles without any oscillation. The average standard accuracy of the surface pressure is roughly 0.1 mN/m. The reproducibility of the results (for at least two measurements) was better than 0.5%.

25.4 RESULTS AND DISCUSSION

25.4.1 ADSORPTION OF METHYLRESORCINOL IN THE PRESENCE OF LYSOZYME

MR is weak non-ionic surfactant. In the case of protein/non-ionic surfactant mixtures, there is a competitive adsorption phenomenon, which is combined with weak hydrophobic interactions between protein and surfactant, resulting in modification of the protein molecule.

Earlier we have demonstrated the ability of MR to increase surface activity of lysozyme [4]. We used phosphate buffer with pH 7.4, I = 0.05 M as a solvent, that is corresponds to the conditions of lysozyme catalytic activity determination in the presence of MR [1]. However, using dynamic light scattering method we have shown that there were associates (or aggregates) of some lysozyme molecules in solution at this pH value. To avoid this effect we changed the pH value from 7.4 to 6.0.

The dynamic surface tension measurements performed in phosphate buffer solutions (pH 6.0, $I = 0.05$ M) at constant lysozyme concentration, 5.1×10^{-6} M, and varying concentrations of MR, show that the LYS-MR mixture is characterized by more high rate of adsorption and more low quasi-equilibrium surface tension, than pure LYS (Fig. 1). As concerning the effect of MR on the surface activity of lysozyme, it is seen from Fig. 2 that MR enhances it at all using concentration-making lysozyme more hydrophobic. MR critical micelle concentration in lysozyme presence is 32.7 mM.

In literature the surface activity of phenols were investigated in [6]. The effect mainly depended on the nature of the phenolic compound and its physicochemical properties in relation to their affinity for the air/water interfaces [6]. For example, catechin was proven to be able to accumulate at the air/water interface, decreasing the surface tension values with increasing its concentration.

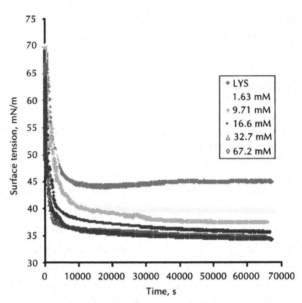

FIGURE 1   Dynamic curves of adsorption of MR in mixture with lysozyme at air/water interface at 25°C; MR concentrations 1.63–67.2 mM and constant protein concentration 5.1×10^{-6} M.

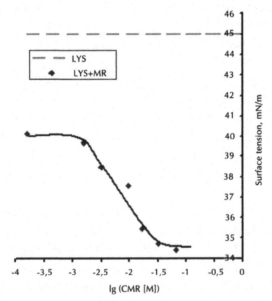

FIGURE 2   Surface tension isotherms for MR in mixed solution with lysozyme at constant protein concentration (5.1×10^{-6} M) at 25°C.

25.4.2 SURFACE DILATATIONAL PROPERTIES

It is known that interfacial rheology of protein–surfactant mixed layers depends on the protein (random or globular), the surfactant (water-soluble or oil-soluble surfactant, ionic or non-ionic), the interface (air–water or oil–water), the interfacial (protein/surfactant ratio) and bulk (i.e., pH, ionic strength, etc.) compositions, the method of formation of the interfacial layer (by spreading or adsorption, either sequentially or simultaneously), the interactions (hydrophobic and/or electrostatic), and the displacement of protein by surfactant [7].

The quasi-equilibrium adsorption layers (the formation time of 60,000–70,000 sec) were subjected to compressive/tensile deformation sinusoidally in the field of linear viscoelasticity. The dependences of the complex viscoelastic modulus of adsorption layers (E), as well as its elastic (real part, E_{rp}) and viscous (imaginary part, E_{ip}) components from the frequency of the applied deformation were obtained (Fig. 3).

It was found that the value of the complex viscoelastic modulus is almost equal to its elastic component in all range of frequencies used (0.007–0.625 rad/s) (Fig. 3). With increasing the frequency, the rise of the viscoelastic modulus was established at practically unchanged value of the imaginary (viscous) component. The predominance of the elastic component above of viscous component in several times was seen. The practical coincidence of complex modulus and its elastic component values, weak dependence of complex modulus from applied frequency of loading and the predominance of elastic component value compared with viscous ones indicate a "solid-like" rheological behavior of modified lysozyme adsorption layers. This behavior is typical for all globular proteins and LYS among them.

In Figs. 4 and 5 the results of complex modulus and phase angle measuring as functions of MR concentration are presented. The drop lines indicate the interval of change corresponding parameters for pure LYS. As seen, MR can both increase and decrease the viscoelastic parameters of LYS adsorption layers depending on concentration. The effect is more expressed at low frequencies.

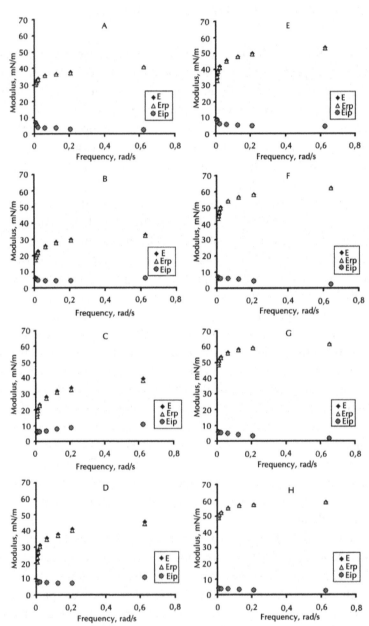

FIGURE 3 Dependence of elasticity modulus on frequency in mixtures of lysozyme and MR different concentration (A – native lysozyme, B – 0.16 mM, C – 1.63 mM, D – 3.21 mM, E – 9.71 mM, F – 16.6 mM, G – 32.7 mM, H – 67.2 mM).

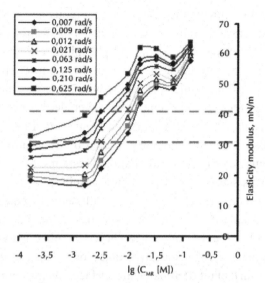

FIGURE 4 Dependence of Lysozyme complex elasticity modulus on MR concentration at frequency 0.007 and 0.625 rad/s. The drop lines indicate the interval of change corresponding parameters for pure LYS.

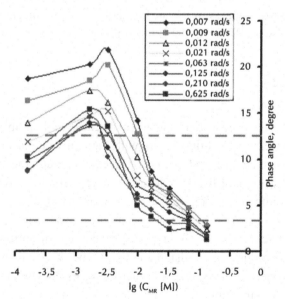

FIGURE 5 Dependence of Lysozyme phase angle on MR concentration at frequency 0.007 and 0.625 rad/s. The drop lines indicate the interval of change corresponding parameters for pure LYS.

There are some information about monotonous decrease of the equilibrium surface tension, dilatational elasticity, and adsorption of lysozyme for non-ionic surfactant decyl dimethyl phosphine oxide (C_{10}DMPO) as the concentration of surfactant increases in the mixture. However, in the case of mixtures of non-ionic surfactants with more flexible proteins like β-casein, the elasticity of the interfacial layer decreases before passing through a maximum as the concentration of surfactant increases [7]. Possibly, the weaker interfacial network formed by β-casein as compared to globular proteins determines the dilatational response of the mixtures. The same picture was shown for the system β-casein mixed with dodecyl dimethyl phosphine oxide (C_{12}DMPO). For all studied frequencies (0.005–0.1 Hz) the elasticities for adsorption layers have a maximum about 4×10^{-5} mol/l C_{12}DMPO concentration. It was shown the obtained values are very close to those measured for the surfactant alone. Thus, in this concentration region the surfactant dominates the surface layer. In our case we have the close picture with maximum on the complex viscoelastic modulus line (Fig. 4). But MR has a smaller molecular size and its concentration of maximum is bigger than C_{12}DMPO ($1.7–3.3 \times 10^{-2}$ mol/l). We can assume the surfactant dominates in the surface layer in this concentration region in agreement with [8] for C_{12}DMPO results.

25.5 CONCLUSION

Proteins are often used as foaming agents because of their ability to unfold at the interface, thus creating layers with high surface elasticity and also steric resistance against coalescence of layers. Although they decrease the interfacial tension and hence reduce the driving force for disproportionation, often quite high protein concentrations are required for the formation of stable foams [9]. For this reason, mixing lysozyme with low-molecular weight surfactants MR can be used for production of antibacterial pharmacological products of emulsion and foam types with enhanced physical stability and period of using while increasing its enzymatic activity and wider spectrum of action.

Methylresorcinol belongs to the class of phenols. It opens more attractive prospects of its possible industrial application. Outstanding representatives of phenolic compounds – antioxidants such as gallic acid, catechin, quercetin can combine both the surface-active properties and act as stabilizers of emulsions. Quercetin improved the dispersion state of the emulsions with the increasing of its concentration. Gallic acid, despite its partitioning in the water phase due to its polarity, delayed the formation of both the hydroperoxides and thiobarbituric acid reactive substances (TBARs) and limited their accumulation. Catechin did not affect the formation of oxidation products whilst quercetin, among the tested antioxidants, caused the lowest formation of both hydroperoxides and TBARs through 33 days of storage [6].

Based on previous studies, we can assume that the MR has antioxidant properties [10]. In this case MR behavior will allows a stabilization mechanism that is based on interactions with protein molecules which lead the formation of a stiff viscoelastic surface film that can also work as barrier towards pro-oxidant species as well as exert some antioxidant properties. This is a subject of further investigation.

KEYWORDS

- **Catechin**
- **Dynamic light scattering**
- **Lysozyme**
- **Non-ionic surfactant**
- **Solid-like**

REFERENCES

1. Petrovskii A. S.; Deryabin D. G.; Loiko N. G.; Mikhailenko N. A.; Kobzeva T. G.; Kanaev P. A.; Nikolaev Yu. A.; Krupyanskii Yu. F.; Kozlova A. N. El'-Registan G.I. Regulation of the functional activity of lysozyme by alkylhydroxybenzenes. *Microbiology* **2009**, *78(2)*, 146–155.
2. Plashchina I. G.; Zhuravleva I. L.; Martirosova E. I.; Petrovskii A. S.; Loiko N. G.; Nikolaev Yu. A. El'-Registan G.I. Effect of Methylresorcinol on the Catalytic Activity and Thermostability of Hen Egg White Lyzozyme. In Biotechnology, Biodegradation, Water and Foodstuffs. Nova Science Publishers: N.Y. 2009; 45–57.

3. Martirosova E. I. Regulation of hydrolase catalytic activity by alkylhydroxybenzenes: thermodynamics of C$_7$-AHB and hen egg white lysozyme interaction. In Biotechnology and the Ecology of Big Cities Nova Science Publishers: N.Y. 2011; 105–113.

4. Martirosova E. I.; Plashchina I. G. Adsorption Behavior of 5-Methylresorcinol and its Mixtures with Lysozyme at Air/Water. In Biochemistry and Biotechnology: Research and Development. Nova Science Publishers: N. Y. 2012.

5. Loglio G.; Pandolfini P.; Miller R.; Makievski A. V.; Ravera F.; Ferrari M.; Liggieri L. Drop and bubble shape analysis as tool for dilational rheology studies of interfacial layers, In *Novel Methods to Study Interfacial Layers, Studies in Interface Science.* Möbius, D.; Miller, R., Eds.; **2001**, *11*, 439–484.

6. Di Mattia C.D., Sacchetti G., Mastrocola D., Sarker D.K., Pittia P. Surface properties of phenolic compounds and their influence on the dispersion degree and oxidative stability of olive oil O/W emulsions. *Food Hydrocolloids* **2010**, *24*, 652–658.

7. Maldonado-Valderrama J., Patino J.M.R. Interfacial rheology of protein–surfactant mixtures. *Current Opinion in Colloid and Interface Science.* **2010**, *15*, 271–282.

8. Kotsmar Cs., Pradines V., Alahverdjieva V.S., Aksenenko E.V., Fainerman V.B., Kovalchuk V.I., Krägel J., Leser M.E., Noskov B.A., Miller R. Thermodynamics, adsorption kinetics and rheology of mixed protein–surfactant interfacial layers. *Adv. Colloid Interface Sci.* **2009**. *150*, 41–54.

9. Alahverdjieva V.S., Khristov Khr., Exerowab D., Miller R. Correlation between adsorption isotherms, thin liquid films and foam properties of protein/surfactant mixtures: Lysozyme/C10DMPO and lysozyme/SDS Colloids and Surfaces A: *Physicochem. Eng. Aspects* **2008**, *323*, 132–138.

10. Revina A.A., Larionov O.G., Kotchetova M.V., Zimina G.M., Zolotarevski V.I., El-Registan G.I. Spectrophotometric and chromatographic investigation of the radiation-induced oxidation products of 3,5-hydroxytoluene aqueous solutions. *Chem. High Energ.* **2004**, *38*, 176–182.

CHAPTER 26

INTRODUCTIONS IN CULTURE IN VITRO RARE BULBOUS PLANTS OF THE SOCHI BLACK SEA COAST (SCILLA, MUSCARI, GALANTHUS)

A. O. MATSKIV, A. A. RYBALKO, and A. RYBALKO

CONTENTS

26.1 Introduction.. 440
26.2 Material and Method... 440
26.3 Results... 441
26.4 Discussion... 447
26.5 Conclusion 448
Keywords 448
References.. 448

26.1 INTRODUCTION

Liliaceae plants have the important meaning for cultivation of the cut off flowers, potting of culture and creation of landscape compositions. Analyzed the literature on problems of micropropagation, receptions of the new forms and virus free plants. The directions of researches on perfection, technology of cultivation of a landing material, improvement from viruses are planned. Many of these plants are medicinal.

In the Sochi Black Sea Coast the different kinds of order Liliales having prospect of use in a pharmaceutical industry grow. They are representatives of family Amaryllidaceae – a snowdrop (Galanthus alpinus Sosn., G. platiphyllus Traub et Moldenke, G. risechense Stern, G. woronowii Losinsk., Leucojum aestivum L.); Hyacinthaceae – species of muscari (Muscari coeruleum Losinsk., M. dolichanthum Woronow et Tron, M. neglectum Guss., M. pallens M. Bieb.), a Scilla bifolia L., S. siberica Haw.); Colchidaceae – (Colchicum speciosum Steven, C. laetum Steven), etc. They have the status vulnerable and require protection. Along with it, they can be used in a national economy as primroses for early (February–March) gardening.

We have developed regimes for micropropagation in the culture of cultivars have been developed for lily bulbs and rare plants of the North Caucasus (muscari, scilla, galanthus) on various types of explants scales of bulbs collected in the wild Sochi suburban forests, and micropropagation of plants grown in culture in vitro. Various plant growth regulators in MS basal medium were used for initiation of organogenesis and the formation of bulblets tested, including cytokinins-benzyladenine (BA) and auxin-naphthaleneacetic acid. Within 8 weeks of the multiplication factor was a lily – 1:10, muscari – 1:21 and scilla – 3:5.

26.2 MATERIAL AND METHOD

Due to the limited quantity of an initial material we carried out methodical development on lily plants. Initial material of Muscari coeruleum Losinsk in the form of a bulb it was received from the Sochi office of the Russian Geographical society. Scilla bifolia L. and Galanthus woronowi Losinsk bulbs are selected in the suburban Sochi woods. Before introduction in culture of in vitro of a bulb previously cleared of pollution under flowing water within an hour. Further with a scalpel deleted the infected sites and sterilized in 25% whiteness solution (during 20 min) and washed out in the 3rd portions of the sterile distilled water. For introduction in culture of in vitro we used MS nutrient medium [1].

26.3 RESULTS

The studied plants – a scilla, galanthus and muskari are vulnerable at the expense of gathering flowers that reduce the reproduction of seeds providing genetic variability. One of preservation factors population is ensuring formation of seeds. We carried out collecting seed boxes and counted up quantity of seeds. It is established by us that in places of collecting flowers there are no daughter plants of these species (Fig. 1). In a protected environment it is formed seed-boxes and daughter plants (Figs. 2–4). We defined the seed productivity of scilla and galanthus. The seed-cases of scilla contain 3.03 ± 1.99 pieces of seed, galanthus – 20.59 ± 11.46.

FIGURE 1   Galanthus woronowi – in a protected environment Seed-cases (1), daughter plants (2).

FIGURE 2   Galanthus woronowi – influence of anthropo-genous factor. No seed-cases (1), daughter plants.

FIGURE 3   Flowering of Scilla bifolia.

FIGURE 4   Seed-cases (1), daughter plants (2) and bulbs of Scilla bifolia.

Thus, we established that one of the factors causing degradation of populations of these cultures is mass collecting flowers. In controllable conditions there is a formation of seeds and germination of affiliated plants. For preservation of these important components of a biodiversity of the Sochi Black Sea Coast creation dwelling place for these and other the small bulb plants is necessary. In our researches when using MS nutrient medium with a various combination of plant growth regulator, the following results are received.

Considering importance of these plants for working out of the program of new generation of preservation of disappearing flora in our region, we have spent work on introduction to culture in vitro for micropropagation of muscari, scilla and galanthus. Studies are before undertaken on micropropagation of lilies [2] and muscari [3]. Preliminary materials on this research are reported on International scientific conference "Pharmaceutical and Medical Biotechnology" (Moscow, on March, 20–22th, 2012) [4]. For realization of researches used the sterile plants of lily and muscari (Figs. 5, and 6), and also scales of bulbs of scilla and galanthus (Figs. 7, and 8) selected the suburban forest Sochi. With that end in view on Murashige and Skoog medium [1] added benzyladenine (1 mg/l) and naphthaleneacetic acid (0.2 mg/l) landed explants of bulb scale. Within 8 weeks on them were formed affiliated bulblets. At their change on the fresh environment again accrued bulblets. At the subsequent division after 8 – week cultivation the reproduction factor has averaged lily – 1:10, scilla – 3:6 and muscari – 1:21 piece (Table 1). Such procedure we spend regularly, that gives the chance will receive a considerable quantity of homogeneous landing units. Similar results are received for galanthus. By the same technique we spend micropropagation various industrial cultivars of lilies. The applied standard technique can be used and for other plants of order Liliales.

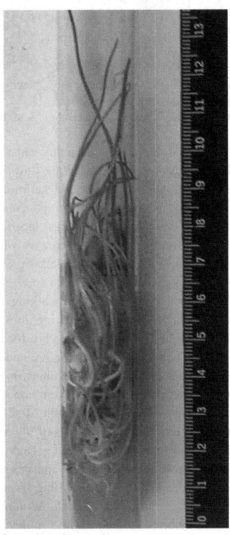

FIGURE 5 Muscari coeruleum in test tubes.

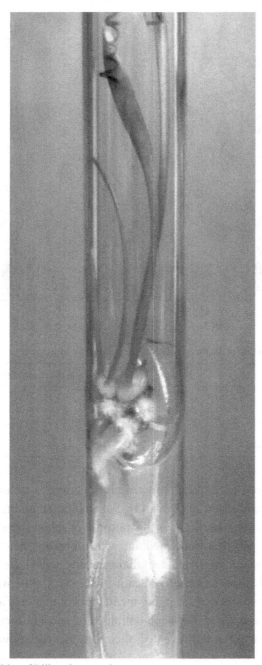

FIGURE 6 Bulblet of Lilium in test tubes.

FIGURE 7   Bulb of Scilla bifolia.

TABLE 1   Definition of factor of reproduction in culture of in vitro.

N	Plant	Number of tubes	Total microcutting	Mean
1.	Scilla	13	46	3,5
2.	Muscari	3	64	21,3
3.	Lilium	18	181	10,1

We also used an embriogeniya. Embriogeniya is a new direction in vegetative reproduction of plants [5]. We studied an embriogeniya on an example muscari and scilla. For this purpose eksplant from sterile culture landed on the MS nutritious medium with gibberelic acid (2 mg/l), iso-pentenil adenine (1 mg/l) and naphthaleneacetic acid (0.2 mg/l). Within 4 weeks of cultivation it is received callus (Fig. 8). Callus will be used for studying of possibility of receiving medicinal components and new forms of plants of these types. Between options of a nutrient medium of an es-sential difference it is not revealed, therefore this technique is applied to different species.

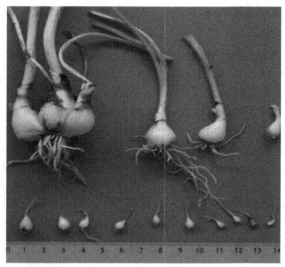

FIGURE 8   Bulb of Galanthus woronowi.

26.4   DISCUSSION

The region the Sochi Black Sea Coast is the most capacious enclave of Russia by quantity of species. More than 2000 species of wild-growing vascular plants here grow. Among them a large number endemic, relicts, rare types demanding protection. Seventy-eight from them are included in the Red book of Russia [6]. There are considerable stocks of effective natural medicines [7]. Needs for medicinal raw materials, uncontrollable collecting these valuable resources put threat of disappearance of many versions. It causes of expansion of researches on application of methods of biotechnology for biodiversity preservation. It will provide possibility of use of new types of raw materials in a national economy. Usual methods of reproduction don't correspond to cope with this problem. Large-scale production in culture of in vitro will provide advance on the market at lower prices. Besides, equipment quite simple, also works can be executed by people with average qualification. The micro multiplied plants can be introduced in natural conditions in wild populations, and as to provide material delivery for herbariums [8]. At the same time, legislation improvement is required stopped illegible collecting plants [9].

26.5 CONCLUSION

1. Introduced into the culture in vitro geophytes Sochi Black Sea Coast (Scilla bifolia, Muscari coeruleum Losinsk., Galanthus woronowii Losinsk).
2. It was confirmed the dependence of the nature of the concentration of cytokinins morphogenesis.
3. Obtaining sterile culture, which makes it possible to carry out research on the use of these crops in the economy, and specifically in the restoration of natural habitats in the ornamental purposes in the project "wild flora from Russia," in developing the program receive a new generation of drug substances on the basis of domestic raw materials.

KEYWORDS

- Bulb
- Cytokinins morphogenesis
- Liliales
- Test tubes

REFERENCES

1. Murashige, T.; Skoog, F. A revised medium for rapid growth and bioassays with tobacco cultures. *Physiol. Plant.* **1962**, *15(3)*, 473–497.
2. Sibileva V. N.; Titova S. M.; Rybalko A. E. Studi of micropropagation of lilies. In Student scientific researches in the field of tourism and resort business. Materials of the 2th International research and practice conference. 20–23 May, 2011; pp. 218-219
3. Pavlenko P. V.; Rybalko A. E. Influence of cytocinins on micropropagation of muscari. Ibid. pp. 175–177.
4. Matskiv A. O. Arakeljan M. A.; Rybalko A. E. Introductions in culture in vitro rare bulbous plants of the Sochi Black Sea Coast (Scilla, Muscari, Galanthus). Proceeding of the Moscow international scientific and practical conference "Pharmaceutical and Medical Biotechnology" (Moscow, on March, 20–22th, 2012) M: Joint-Stock Company "Expo-biohim-tehnologies", D.I. Mendeleyev University of Chemistry and

Technology of Russia, D.I. Mendeleyev University of Chemistry and Technology of Russia, 2012; pp. 455.

5. Batygina T. B. Reproduction, propagation and renewal of plants. Embriologiya of floral plants. Terminology and concepts. Reproduction systems. SPb.: World and family, **2000**, *3*, 35–39.

6. Solodko, A .S.; Kirii, P. V. In *Atlas of the healing flora of Sochi region of the Black Sea Coast.* Moskow, Sochi, 2010; 321 p.

7. Solodko, A. S.; Kirii, P. V. The Red Book of Sochi (rare and vaniching plants). V. 1. Plants and fungi. Beskovs: Sochi 2002; p. 148.

8. Afolayan A. J.; Adebola P. O. In vitro propagation: A biotechnological tool capable of solving the problem of medicinal plants decimation in South Africa. *African J. Biotech.* **2004**, *3(12)*, 683–687.

9. Rybalko, A. A. Features of Teaching of the Ecological Right at Saving Biodiversity of Sochi Black Sea Coast. In: *Biochemistry and Biotechnology: Research and Development.* Nova Science Publishers, Inc.: N.Y. **2012**; pp. 173–176.

CHAPTER 27

CHANGE OF SOME PHYSICO-
CHEMICAL PROPERTIES
OF ASCORBIC ACID AND
PARACETAMOL HIGH-DILUTED
SOLUTIONS AT THEIR JOINT
PRESENCE

F. F. NIYAZY, N. V. KUVARDIN, E. A. FATIANOVA, and S. KUBICA

CONTENTS

27.1 Introduction and Background 452
27.2 Experimental 542
27.3 Conclusion 459
Keywords 460
References 460

27.1   INTRODUCTION AND BACKGROUND

During last decades there is a tendency of the growing interest to the study of high-diluted solutions of bioactive substances. Besides, concentration ranges under study are related to the category of supersmall or, in other words, 'illusory' concentrations. Such solutions, unlike more saturated ones, but with pretherapeutic content of active substance, may possess high biological activity.

Use of bioobjects to reveal supersmall doses effect (SSD) in the substances is the most exact method today allowing not only to define the effect existence and to find out how it shows itself, but also to determine concentration ranges of its action. However, the use of this method will entail great difficulties. In this connection it is necessary to search for other methods, including physico-chemical ones, allowing defining presence of SSD effect in the compounds vif only at the stage of preliminary tests. Study of physico-chemical bases of SSD effect display is one of the most interesting questions in the given sphere of research and attracts attention of many scientists [1–4].

Antineoplastic and antitumorous agents, radioprotectors, neutropic preparations, neupeptides, hormones, adaptogenes, immunomodulators, antioxidants, detoxicants, stimulants and inhibitors of plants growth and so on are included into the group of bioactive substances possessing SSD effect. Study of high-diluted solutions of bioactive compounds was carried out on one-component solutions, that are ones containing only one solute. But now, mainly multicomponent medical preparations, possessing several therapeutic actions, are used in medicine. So, preparations of analgesic-antipyretic action are possibly used at sharp respiratory illnesses, accompanied by muscular pain and rise of temperature. It is possible that effects of multicomponent medical preparation in supersmall concentrations and its separate components will differ.

27.2   EXPERIMENTAL

We studied some physico-chemical properties of high-diluted aqueous solutions of paracetamol and ascorbic acid with the purpose of finding out peculiarities of their change at solutions dilution and also of definition of possible concentration ranges of SSD effect action. Paracetamol and ascorbic acid are the components of combined analgesic-antipyretic and antiinflammatory preparations [5]. Ascorbic acid is used as fortifying and stimulating remedy for immune system. Paracetamol (acetominophene) has anaesthetic and febrifugal effect.

There have been prepared one-component solutions of ascorbic acid and paracetamol, two-component solutions of paracetamol with ascorbic acid (relation of dissoluted compounds in solutions is 1:1), in the following concentrations of dissolved substances (mol/l): 10^{-1}, 10^{-3}, 10^{-5}, 10^{-7}, 10^{-9}, 10^{-11}, 10^{-13}, 10^{-15}, 10^{-17}, 10^{-19}, 10^{-21}, 10^{-23}. Water cleansed by reverse osmosic was used as solvent. Solutions were prepared by successive dilution by 100 times using classical methods. Initial solution was 0,1M one. Before choosing of solution portion for the following dilution the sample was subjected to taking antilogs.

Prepared solutions were studied by cathetometric method of substances screening, the ones acting in supersmall concentrations, and also by method of electronic spestroscopy.

Cathetometric method of substances screening is based on the study of the change of solution meniscus height in capillary [6]. Results of measuring meniscus height of ascorbic acid, paracetamol, paracetamol with ascorbic acid solutions are given in Figs. 1, 2, and 3, accordingly.

Meniscus height of dilution with ascorbic acid concentration of 10^{-3} mol/l was 0.8 mm (Fig. 1). During further dilution value of meniscus height in the capillary reduces, but changes are not uniquely defined. The most lowering of meniscus height is observed in samples in which content of ascorbic acid is 10^{-9}, 10^{-13}, 10^{-15}, 10^{-17} mol/l.

Meniscus height reduces on an average by 13.75%. Lowering of meniscus height by 23.7%, in comparison with more concentrated solution, is also observed for the sample with ascorbic acid content of 10^{-23} mol/l.

Equivalent lowering of meniscus in the capillary has been stated for solutions with ascorbic acid concentration of 10^{-9} mol/l and 10^{-13}–10^{-17} mol/l. Between these concentration ranges there is concentration range in which there are no essential changes. Being based on literature data and also on cathetometric method for screening of substances activity in supersmall doses, we can assume that dilutions of ascorbic acid in concentrations of 10^{-9} mol/l and 10^{-13}–10^{-17} mol/l show biological activity regarding bioobjects.

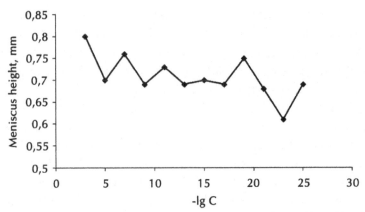

FIGURE 1   Values of meniscus height in the capillary of ascorbic acid solutions (concentration, mol/l).

While studying solutions of paracetamol by cathetometric method it can be observed that in first dilutions by 100 times (paracetamol concentrations being 10^{-3}–10^{-7} mol/l) height of meniscus reduces slightly, maximum 5.7%, regarding meniscus height of the first dilution. This slight lowering of meniscus height in the capillary is caused by rather large dose of active substance in these dilutions. However, sudden lowering of meniscus height in the capillary up to 0.71 mm, which is by 19.3% lower than meniscus height of the first dilution, is observed at diluting paracetamol solution up to the concentration of 10^{-9} mol/l (Fig. 2).

The same dependence is observed for dilution of paracetamol solution with concentration of 10^{-15} mol/l. So, at this concentration meniscus height is 0.7 mm.

Growth of meniscus height in the capillary is observed further for solutions with the following dilution. This process is motivated by very high dilution, which is the solution, according to its composition and properties, tries to attain the state of pure solvent.

These changes are polymodal dependence effect – concentration, described for different substances and different properties in domestic and world scientific literature. From the figure and its description it is clearly seen that there are concentrations of paracetamol solution of 10^{-9} mol/l and 10^{-15} mol/l, for which there has been stated essential lowering of meniscus in the capillary regarding other concentrations. Between these concentration ranges there is a concentration range in which there are no essential changes, this range being the so-called "dead zone."

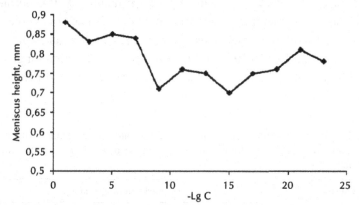

FIGURE 2 Values of meniscus height in the capillary of paracetamol solution (concentration, mol/l).

We have studied solutions of ascorbic acid and paracetamol mixture. It is necessary to note that not uniquely defined change of meniscus height is observed in solutions containing simultaneously two active substances. Meniscus of one-component solutions is more narrow than that of water, but in two-component solutions both reduction and increase of meniscus height values are possible in comparison with water (Fig. 3).

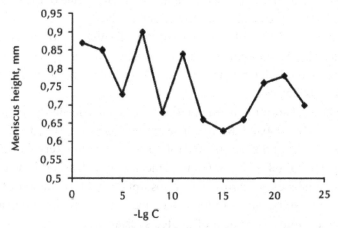

FIGURE 3 Values of meniscus height in the capillary of paracetamol solutions with ascorbic acid.

Lowering of meniscus height is observed in solutions with concentration of paracetamol with ascorbic acid 10^{-5}, 10^{-9}, 10^{-13}, 10^{-15}, 10^{-17} mol/l on 11.5 %, 17.5%, 20%, 23.6% and 20% correspondingly. It is possible to distinguish two concentration ranges in the field of supersmall concentrations where reduction of meniscus height in the capillary takes place: this is – 10^{-9} mol/l and 10^{-13}–10^{-17} mol/l. Received concentration ranges coincide with data of cathetometric studies of ascorbic acid and paracetamol solutions (Table 1).

TABLE 1 Results of study of ascorbic acid, paracetamol, paracetamol with ascorbic acid solutions of wide concentration range by method of electronic spectroscopy and cathetometric method.

Method of study	Concentration ranges, mol/l		
	Ascorbic acid solutions	Paracetamol solutions	Solutions of paracetamol with ascorbic acid
Cathetometric method of screening	10^{-9}, 10^{-13} – 10^{-17}, 10^{-23}	10^{-9}, 10^{-15}	10^{-9}, 10^{-13}–10^{-17}
Electronic spectroscopy	10^{-9}, 10^{-15}, 10^{-19}	10^{-9}, 10^{-13}, 10^{-15}, 10^{-21}	10^{-9}, 10^{-13}, 10^{-15}, 10^{-21}

Insrease of meniscus height in comparison with the solvent is observed in solutions with concentrations of diluted compounds of 10^{-1}, 10^{-3}, 10^{-7}, 10^{-11} mol/l. Growth of meniscus height above such value of the solvent can be explained by the change in SSD effect display.

We have analyzed ultra-violet spectra of solutions of ascorbic acid, paracetamol, paracetamol with ascorbic acid. All the spectra were read from spectrophotometer Cary 100, UV-Visible Specrtophotometer in the range of 200–350 nm.

The most absorption in all solutions takes place in short waves part of the effective wave band. More concentrated solutions (10^{-1}, 10^{-3} mol/l) of ascorbic acid have maximum absorption in the length interval of 220–280 nm, of paracetamol – 220–310 nm, paracetamol with ascorbic acid – 200–320 nm. Tendency to narrowing of absorption field, reduction of optical density value and quantity of tops up to one or two is registered in all solutions, under study, as content of diluted substance reduces.

Solutions spectra with ascorbic acid concentration of 10^{-15} and 10^{-19} mol/l are equal in form and differ from spectra of other dilutions. Lowering of ascorbic acid concentration is not accompanied by uninterrupted reduction of optical density value. Irregular growth of absorption in comparison with more concentrated solutions is observed in solutions with ascorbic acid concentrations of 10^{-9}, 10^{-15}, 10^{-19} mol/l (Table 2).

TABLE 2 Maximum values of absorption of ascorbic acid solutions in ultra-violet field.

Concentration, mol/l	Absorption maximum	
	Wave length, nm	A
10^{-3}	230–270	0.79
10^{-5}	265	0.04
10^{-7}	220–230	0.012
10^{-9}	225	0.03
10^{-11}	235	0.01
10^{-13}	220	0.015
10^{-15}	220	0.04
10^{-17}	220–240	0.005
10^{-19}	220	0.03
10^{-21}	220	0.005
10^{-23}	220	0

Community in given structures may be assumed taking into account coincidence of waves length and values of optical density for solutions with concentrations 10^{-9}, 10^{-15}, 10^{-19} mol/l.

Gradual reduction of paracetamol concentration in solutions is not accompanied by the same reduction of optical density value. Growth of absorption in comparison with more concentrated solutions is observed for solutions with concentrations 10^{-9}, 10^{-13}, 10^{-15}, 10^{-21} mol/l (Table 3). All this allows assuming the rise of structural changes in these solutions.

TABLE 3 Maximum values of absorption of paracetamol solutions in ultra-violet field.

Concentration, mol/l	Absorption maximum	
	Wave length, nm	A
10^{-1}	230	0.58
10^{-3}	270	0.53
10^{-5}	245	0.15
10^{-7}	240	0.02
10^{-9}	235	0.015
10^{-11}	240	0.007
10^{-13}	245	0.007
10^{-15}	220-245	0.01
10^{-17}	235	0.003
10^{-19}	290	0.001
10^{-21}	225	0.02
10^{-23}	230	0.005

Spectrum of ascorbic acid and paracetamol solution in concentrations of 10^{-1} mol/l is characterized by wide absorption band in the field of 200–320nm. Dilution of 0,1M solution by 100 times is accompanied by reduction of absorption field width up to 200–280nm without changing of spectrum form and intensity of absorption on peaks.

Changing of spectrum form accompanied by essential reduction of absorption from 4 to 0.095 with the maximum on the length 243nm takes place while diluting solution of ascorbic acid with paracetamol up to the concentration 10^{-5} mol/l (Table 4).

TABLE 4 Maximum values of absorption of ascorbic acid with paracetamol solutions (1:1) in ultra-violet field.

Concentration, mol/l	Absorption maximum	
	Wave length, nm	A
10^{-1}	237	4.651
10^{-3}	235	4.208

TABLE 4 *(Continued)*

Concentration, mol/l	Absorption maximum	
	Wave length, nm	A
10^{-5}	243	0.095
10^{-7}	205	0.242
10^{-9}	205	0.293
10^{-11}	207	0.084
10^{-13}	206	0.367
10^{-15}	205	0.186
10^{-17}	207	0.086
10^{-19}	207	0.089
10^{-21}	206	0.261
10^{-23}	207	0.105

Two peaks on lengths 205–206 nm and 270–273nm are shown on spectra of solutions with paracetamol and ascorbic acid concentrations 10^{-7}, 10^{-9}, 10^{-13}, 10^{-15}, 10^{-21} mol/l. One peak on the length 207 nm is shown on spectra of solutions with paracetamol and ascorbic acid concentrations 10^{-11}, 10^{-17}, 10^{-19}, 10^{-23} mol/l. Display of maximum on the length 270 nm occurred under conditions that absorption on maximum 205 nm was not less than 0.1.

Increase of optical density is observed in spectra of solutions with paracetamol and ascorbic acid concentrations 10^{-7}, 10^{-9}, 10^{-13}, 10^{-19}, 10^{-21} mol/l.

27.3 CONCLUSION

As a result of this work we, by cathetometric method, have defined concentration ranges of possible display of medical compounds biological activity that forms 10^{-9} mol/l and 10^{-13}–10^{-17} mol/l for ascorbic acid and 10^{-9}, 10^{-15} of paracetamol, and also 10^{-9}, 10^{-13}–10^{-17} for the mixture of ascorbic acid – paracetamol. This allows assuming compatibility of components data in given concentration ranges.

While studying solutions in super-small doses of ascorbic acid, paracetamol and also at their joint presence by method of electronic spectroscopy it has been found out that irregular growth of absorption is observed for solutions with ascorbic acid concentrations 10^{-9}, 10^{-15}, 10^{-19} mol/l in comparison with more concentrated solutions and for paracetamol solutions with concentrations 10^{-9}, 10^{-13}, 10^{-15}, 10^{-21} mol/l. But for the mixture ascorbic acid – paracetamol such interval is 10^{-9}, 10^{-13}, 10^{-15}, 10^{-21} mol/l.

This fact allows speaking about appearance of medical compounds structures with water in this solution, which differ by relatively high absorption.

Information received by method of ultra-violet spectroscopy, agrees with data of cathetometric screening both for solutions of ascorbic acid and paracetamol.

KEYWORDS

- Acetominophene
- Cathetometric method
- Dead zone
- Illusory

REFERENCES

1. Konovalov A. I. Physico-chemical mystery of super-small doses. *Chemistry and Life.* **2009**, *2*, 5–9.
2. Kuznetsov P. E.; Zlobin V. A.; Nazarov G. V. On the question about physical nature of superlow concentrations action. *Heads of reports at III International Symposium "Mechanisms of super-small doses action"*, Moscow, December, 3–6, 2002; p. 229.
3. Chernikov F. R. Method for evaluation of homoeopathic preparations and its physico-chemical foundations. *Matherials of Congress of homoeopathists of Russia.* Novosibirsk, 1999; p. 73.
4. Pal'mina N. P. Mechanisms of super-small doses action. *Chemistry and life.* **2009**, *2*, 10.
5. Maslikovsky M. D. Combined analgesic-febrifugal and anti-inflammatory preparations. M., 1995; p. 208.
6. Niyazi F. F.; Kuvardin N. V. Method for determining abilities of bioactive substances to display "super-small doses" effect. Patent N 2346260 of 10.02.2009.

CHAPTER 28

THE METHODS OF THE STUDY THE PROCESSES OF THE ISSUE TO OPTICAL INFORMATION BIOLOGICAL OBJECT

U. A. PLESHKOVA and A. M. LIKHTER

CONTENTS

28.1 Introduction 462
28.2 Mathematical Model 462
28.3 Conclusion 469
Keywords 470
References 471

28.1 INTRODUCTION

This chapter is devoted to the mathematical modeling of the process of the optical information transmission to the insects of various classes, which is performed in the biocybernetical system (BCS), containing the monitor signal source, the information transmission channel (environmental) and the monitoring object. In the observed model we have considered noises from natural and artificial optical range electromagnetic emission, and the natural luminance mode influence on the energetic and information process characteristics of the information transmission process in different seasons and time of the day. We have investigated the "relation" dependence functions from BCS geometric parameters and landscape peculiarities and we have received BCS elements optimum parameters values, which allow their effectiveness increasing.

28.2 MATHEMATICAL MODEL

According to the rapid development of the technical cybernetics sphere, based on the biophysical processes implementing for the biological objects behavior management [1, 2, 4, 12], there is a sharp necessity for actual problem solving which is in biocybernetical systems effectiveness increasing (BCS) [8]. One of the main stages in this problem solving is mathematical model construction, which adequately describes the information transmission process from the signal source to the monitoring object [11].

This task becomes more sophisticated if we take insects for the monitoring object, which are notable for the high level of their behavior uncertainty. As the main information amount, necessary for the insects vital activity, they take with their visual analyzer, so the construction of the model of optical information transmission to the insects with different vision types is the basic during the insects behavior biocybernetical monitoring system construction. Conversely, BCS effectiveness increasing for insects behavior monitoring (Fig. 1) with optical range electromagnetic emission sources can be reached as the result of the implementing the optoelectronic system optimum designing approach and as the result of the calculating of their elements operating parameters on the information quality criteria basis [9], in analytical expressions of which all BCS elements parameters are included: signal/noise function $C/Ш$ and informational bandwidth $П$ correspondingly.

$$S/N, \tag{1}$$

$$П = \Delta f . \log_2 (1+S/N) \tag{2}$$

where Δf is a frequency band, sensing by the insect's visual organ.

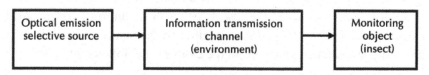

FIGURE 1   Insect behavior biocybernetical monitoring system scheme.

Taking into account the fact that all noises form natural and artificial emission sources can be additively summed [4], we get the following general noise expression:

$$N = N_S + N_{S.E} + N_{L.E} + N_{S.O} + N_{L.O} \tag{3}$$

where, $N_S, N_{S.E}, N_{L.E}, N_{S.O}, N_{L.O}$ are noises, caused by the direct sun flash, and by the reflecting sun and artificial emission from the earth surface and from the underage cloudiness.

Mathematical modeling of the optical information transmission process into BCS needs to take into account the environmental influence, where artificial and natural emission is spread. The main natural optical clutters, in this case, are celestial bodies, The Earth and its surface, atmosphere.

It is certain that the optical signal sensed by the insects eye [10] can be written as follows:

$$S = \frac{m}{l^2} \cdot \int_{\lambda_1}^{\lambda_2} r(\lambda) \cdot \tau(\lambda) \cdot \exp\left(-q(\lambda)\cdot l\right) d\lambda, \tag{4}$$

where $l = \sqrt{x^2 + (h_2 - h_1)^2}$, x is the distance between the selective emission source and the monitoring object horizontally, h_1 is the distance between the earth surface and the monitoring object, h_2 is the source height over the earth surface, λ is the wave length, $r(\lambda)$ is the function of the selective light source spectral emissivity, $\tau(\lambda)$ is the function of the insect relative visibility, m is the coefficient registrating the difference between human and insect visibility functions, $q(\lambda) = k(\lambda) + \sigma(\lambda)$, $k(\lambda)$ is the atmospheric transmission spectral coefficient in UV and in the visual spectrum; $\sigma(\lambda) = 0.83 N A^3 \cdot \lambda^{-4}$ is a Rayleigh dispersion spectral coefficient [5], N is the molecule number in $1 m^3$, A is the molecule cross-section square, m^2.

Taking into account the mathematical model of the information transmission optical channel in the insects behavior monitoring systems [11] expressions for noises $III_{S.O}$, $III_{L.O}$, specified by the Sun emission reflection and artificial selective source from the clouds is the following:

$$N_{S.O} = \frac{m}{\pi} \cdot \left(\frac{R_c}{R_{cf}}\right)^2 \int_{\lambda_1}^{\lambda_2} \xi(\lambda, T) \cdot k(\lambda) \cdot \tau(\lambda) \cdot Noise(\lambda) \cdot \exp(-q(\lambda)) \cdot (1 - \rho(\lambda)) d\lambda \quad (5)$$

$$N_{L.O} = \frac{m}{h_2^2 \cdot \pi} \int_{\lambda_1}^{\lambda_2} r(\lambda) \cdot \tau(\lambda) \cdot Noise(\lambda) \cdot \exp(-q(\lambda) \cdot h_2) \cdot (1 - \rho(\lambda)) d\lambda, \quad (6)$$

where $\xi(\lambda, T)$ is a Sun spectral radiation distribution, R_S, $R_{E.O}$ Sun and Orbit radius respectively, $Noise(\lambda) = (\mu(\lambda) S_1 + \upsilon(\lambda) S_2 + \psi(\lambda) S_3)$, $\mu(\lambda), \upsilon(\lambda), \psi(\lambda)$ is the spectral characteristics of the soil, water and vegetation reflection respectively, S_1, S_2, S_3 are their weight coefficients, which are fixed with the random numbers generator, making the sequence with the proportional distribution in the given value range (7), $\rho(\lambda)$ is the spectral absorption coefficient of the underedges cloudiness [6, 13].

Random numbers generator is used for imitation of the monitoring systems functioning real conditions [3]:

$$\begin{aligned} S_1 &= rnd(1) \\ S_2 &= rnd(1 - S_1) \\ S_3 &= 1 - S_1 - S_2 \end{aligned} \quad (7)$$

where $rnd(1)$ is a function, allowing to take the equally distributed random number in the given $[0,1]$ range.

As an example for the following calculations we will use the some variations N_1, N_2, N_3 of the S_1, S_2, S_3 weight number sets (Fig. 2).

	N_1	N_2	N_3
S_1	0,696	0,543	0,211
S_2	0,133	0,199	0,437
S_3	0,171	0,257	0,353

FIGURE 2 Sets of the weight numbers of the natural surfaces reflective characteristics.

As a result of the calculations $N_{S.O}$ and $N_{L.O}$ at (5), (6) we defined that the given noises have a weak influence at the "signal/noise" function value so it is possible to neglect them in the Eq. (3).

For describing noises N_S, $N_{S.E}$ let's observe the season and the daytime influence on the intensity of the optical range sun emission, reaching the Earth surface. The Earth surface irradiance at the given latitude φ when the weather is fine at the given time t equals [7]:

$$E(n,t) = \begin{cases} Q\cos\theta(n,t), & \cos\theta(n,t) > 0 \\ 0, & \cos\theta(n,t) < 0 \end{cases}, \qquad (8)$$

where Q is the constant insolation, equal to the sun constant, n is the full earthday number from the beginning of the year, the time t is in the $(0 < t < \tau_0)$ range, where τ_0 is the sun earthday ($\tau_0 = 24$ hours).

The dependence on the sunrays dip θ cosine is the following:

$$\cos\theta(n,t) = \cos\delta(n)\cdot\cos\left[\frac{2\pi}{\tau_0}\left(t + \frac{\tau_0}{2}\right)\right] + \sin\delta(n). \qquad (9)$$

Here δ is a declination, which adds the angle between the earth axis and direction to the sun disk center to $\pi/2$. Solar declination sine δ as a function of n daytime from the beginning of the year is expressed by the following formula:

$$\sin \delta(n) = \sin \eta \cdot \cos \varepsilon(n), \qquad (10)$$

where η is an angle between the earth axis and the vertical to the earth orbit plane ($\eta = 23°27'$), $\varepsilon(n)$ is an earth axis azimuth angle, the dependence of which on the season, i.e., on the day number n, is expressed by the formula:

$$\varepsilon(n) \approx \frac{2\pi\tau_0}{\tau_1}(n-172). \qquad (11)$$

In accordance with the taken above assumptions the noise N_c from the direct sun illumination can be the following:

$$N_S = m\left(\frac{R_S}{R_{E.O}}\right)^2 \cos\theta(n,t) \cdot \int_{\lambda_1}^{\lambda_2} \xi(\lambda,T)k(\lambda)\tau(\lambda)d\lambda, \qquad (12)$$

and the noise at the sun emission reflection from the geological substrate $N_{S.E}$ will be the following:

$$N_{S.E} = \frac{m}{\pi}\left(\frac{R_c}{R_{ç.i}}\right) \cos\theta(n,t) \cdot \int_{\lambda_1}^{\lambda_2} \xi(\lambda,T)k(\lambda)\tau(\lambda) \cdot Noise(\lambda)d\lambda \qquad (13)$$

Then, substituting (3), (12), (13) in (2), finally we will take:

$$\frac{S}{N} = \frac{m\int_{\lambda_1}^{\lambda_2} r(\lambda)\tau(\lambda)\exp(-q(\lambda)\cdot l)d\lambda}{l^2(N_S + N_{S.E} + N_{L.E})} \qquad (14)$$

Let's observe the energy S and informational S/N, Π characteristics dependence from the different BCS parameters.

At the Fig. 3 there are the graphics of the "signal/noise" function dependence from the earth surface energy illumination in different seasons and at a different daytime (in this case winter is not observed as it doesn't correspond with the insects vital active phase).

From the curve analyses we can notice that the "signal/noise" function changes greatly depending on the season and on the daytime and reaches its maximum at night and then decays quickly and at $t = 12 \div$ period reaches its minimum.

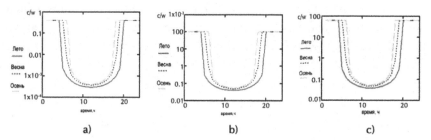

a)   b)   c)

FIGURE 3   The graphics of the "signal/noise" function dependence from the earth surface energy illumination for insects with different vision types: (a) monochrome; (b) dichrome; (c) trichromatic.

While modeling the optical information transmission process to the insects it is necessary to define the dependence features of the useful signal C from the altitude h_2 (Fig. 4) and from the distance x (Fig. 5).

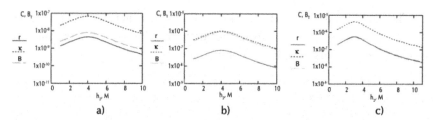

a)   b)   c)

FIGURE 4   The dependence graphics from the altitude h_2, of the useful signal producing by the optical emission selective sources (H–halogen tube, X–xenon lamp, T–tungsten lamp at T = 1500 K, x = 4м) in the insect's visual organ with the different vision types: (a) monochrome; (b) dichrome; (c) trichromatic.

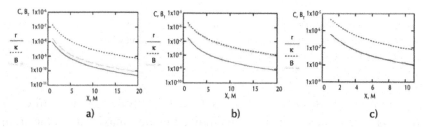

a)   b)   c)

FIGURE 5   The dependence graphics from the distance of the x useful signal, produced by the optical emission selective sources (H–halogen tube, X–xenon lamp, T–tungsten lamp at T = 1500 K, h_2 = 4м) in the insect's visual organwith different vision types: (a) monochrome; (b) dichrome; (c) trichromatic.

As a calculation result there are "signal/noise" function values, which are within 0.003 and 1, which can be seemed as insufficient for effective providing of the insects behavior monitoring process (Fig. 6).

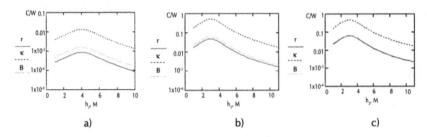

a)   b)   c)

FIGURE 6   The dependence graphics from the altitude h_2 of the "signal/noise" function for the insects with different vision types: (a) trichromatic; (b) monochrome; (c) dichrome and for the electromagnetic emission selective sources (H–halogen tube, X–xenon lamp, T–tungsten lamp at T = 1500K, x = 4м, N_2).

The graphics analysis at the Fig. 7 shows that there is a dependence of the signal/noise function from the distance x between the monitoring signal source and the monitoring object.

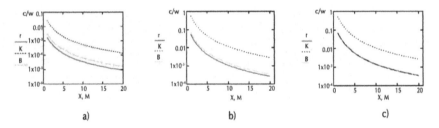

a)   b)   c)

FIGURE 7   The dependence graphics from the altitude x of the signal/noise function for the insects with different vision types: (a) trichromatic; (b) monochrome; (c) dichrome and for the electromagnetic emission selective sources (H–halogen tube, X–xenon lamp, T–tungsten lamp at T = 1500K, x = 4м, N_2).

From the graphics analysis (Figs. 6, and 7) it follows that in all cases there is a dependence of the signal/noise function maximum value from the altitude of the electromagnetic emission selective source above the Earth surface what defines its optimal value for the insects with different vision types. For the insects with monochrome vision type there is the signal/noise function maximum value at $h_2 = 4,2i$ and $x = 1,7i$, and for the insects with di-, and also with trichromatic vision types it is at $h_2 = 3i$ and $x = 1i$. It is also noticed that the signal/noise maximum value is reached while using a xenon lamp for the insects with all vision types, and the signal/noise values for a halogen tube and a tungsten lamp are practically equal for the insects with mono- and dichrome vision types.

While evaluating the dependence of the signal/noise function (Fig. 8) created with natural and artificial optical emission source (*halogen tube*), we observed the different surfaces weight coefficients combinations in (Figure 2).

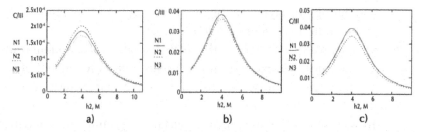

FIGURE 8   The dependence graphics of the signal/noise function from the source altitude above the Earth surface at the random set of the natural surfaces weight coefficients (W–water surface, S–soil, F–flora) for the insects with different vision types: (a) monochrome; (b) dichrome; (c) trichromatic.

From the graphics analysis at the Fig. 8 it follows that for all natural surfaces weight coefficients sets there is the dependence of the signal/noise function maximum value from the electromagnetic emission selective source altitude above the Earth surface. Signal/noise function reaches its maximum at the altitude $h_2 = 4i$, than it has a steady decreasing character. Also for all vision types there is a vivid dependence from the random set of the natural surfaces weight coefficients.

28.3   CONCLUSION

1. The mathematical model of the process of the optical information transmission in the insects behavior monitoring systems was made;

2. Analytical expressions for energy S and information S/N, Π characteristics of the optical information transmission process to the insects were received;

3. The dependence of the energy and information characteristics of the optical information transmission channel to the insects with different vision types from different BCS parameters were investigated;

4. The dependence of the energy and information characteristics from the different BCS parameters of the optical information transmission channel to the insects with different vision types were investigated;

5. It was defined that the natural illumination mode during the year and during the daytime strongly influences onto the optical information transmission process to the insects;

6. The optimal parameters of the related position of the monitoring signal source and receiver of the insects with different and of the electromagnetic emission artificial selective source were calculated;

7. It was stated that the natural surfaces spectral characteristics weight coefficients changes lead to the signal/noise function values change;

8. It was shown that the use of one standard source of the monitoring signal in BCS can be insufficient for providing the effective process of the insects behavior monitoring.

KEYWORDS

- **Earthday**
- **Halogen tube**
- **Orbit**
- **Signal/noise**
- **Sun**

REFERENCES

1. Andreevsky, A. S. Certification to the useful model 66889 – Insecticide device.
2. Belenov, V. N. Gazalov V. S. *Zelenograd*, **2003**, 3, 30–33.
3. Bobnev, M. P. *Random Number Generation.* **1997**, 230.
4. Gazalov, V. S. *Agriculture Rational Motorization*, **1984**, 6–9.
5. Van der Hulst, G. *Light Reradiation by Small Particles.* Foreign literature publishing house, 1962; 537.
6. Zuev, V. E.; Krekov, G. M. *Gidrometeoizdat*, **1986**, 256.
7. Kladieva, A. S.; Dzhalmukhambetov A. U. *Calculations in the MATLAB Environment of the Latitude-Temporal Distribution of the Sun Energy on the Earth Surface.* Projecting of the Engineering and Scientific Applications in the MATLAB Environment: Materials of the IVth Russian Scientific Conference. Astrakhan: Publishing House Astrakhan University, 2009, 146–155.
8. Lihter, A. M. *Optimal Projecting of the Optic-Valved Systems.* Monograph. Astrakhan: Publishing House, Astrakhan State University, 2004, 241.
9. Lihter, A. M. *Modeling of the Fishing Processes Management System.* Monograph. Astrakhan: Publishing House, Astrakhan University, 2007; 290.
10. Mazohin–Porshnyakov, G. A. *Insects Sensors Physiology Manual.* Publishing House of the Moscow University, 1977; 456.
11. Pleshkova, U. A.; Lihter A. M. *Ecol. Sys. Dev.*, **2010**, *12*, 24–27.
12. Ryibkin, A. P.; Kazakov, V. P. *Patent for Invention* 2001100190 – *Protective System from the Sanguivorous Flies.*
13. Timofeev, U. M.; Vasyil'ev A. V. *Atmospheric Optics Theoretical Basis. Science Publishing House*, 2003; 474.

CHAPTER 29

RESEARCH UPDATE ON CONJUGATED POLYMERS

A. JIMENEZ and G. E. ZAIKOV

CONTENTS

29.1 Introduction 474

29.2 Experimental 475

    29.2.1 Material and Reagents 475

    29.2.2 Instrumental Setup 475

    29.2.3 Reaction of 3-chloropropylthiol and Ferric Chloride with
    PANI Deposited on the GC Electrode 475

    29.2.4 Reaction of 3-chloropropylthiol and Ferric Chloride with
    PANI 476

    29.2.5 Characterization 476

29.3 Results and Discussion 477

29.4 Conclusion 480

29.4.1 Supplementary Data 481

Appendix 483

Transistors 485

Display Devices 491

Biosensors 497

Keywords 500

References 501

29.1 INTRODUCTION

Conducting polymers because of their electronic, magnetic, and optical properties are an attractive class of materials for variety of advanced technologies [1–5]. Among conducting polymers, polyaniline (PANI) has been extensively studied due to its good environmental stability, electrical properties and inexpensive monomer. Some of potential applications of PANI are:

As component of electrochemical batteries [6, 7]; as chemical sensors in solution [8–10]; optical wave guides [11]; microelectronic devices [12, 13]; to produce electrochromic devices [14]; as antioxidant for rubbers and plastics [15, 16]; to produce conductive rubber and plastics composites [17, 18]; catalyst for chemical reaction [19]; a component in gas-sensing devices for the "electronic nose" [10, 20, 21]; as antistatic coating and electromagnetic shielding material [22, 23]; and etc. Substituted PANI is of great interest for variety of applications as above mentioned for PANI. There has been considerable interest in developing new synthetic approaches for their production. Substituted PANI with altered properties can be made by adding functional groups to the backbone of the polymer. Ring substituted of alkyl PANI [24, 25]; and alkoxy (ethoxy, methoxy) ring substituted PANI [26, 27]; strong electron withdrawing groups ($-CF_3$, $-CN$, $-NO_2$) ring substituted PANI [28], Ring substituted polymerization of o-toluidine, m-toluidine and o-chloroaniline was reported by Sazou [29]. Ring substituted PANI was also obtained by coupling of 4-sulfobenzene diazonium ion and poly (N-methylaniline) [30]. A typical synthesis of substituted polymer may be carried out via oxidative polymerization of the corresponding monomer [31]. In many cases the substituted monomer with desired functional group is either too difficult to oxidize or sensitive to oxidative or acidic conditions. A new synthetic strategy was developed by Freund [32] to obtain ring substituted poly(hydroxyaniline) and poly(iodoaniline), using poly (aniline boronic acid) as a precursor for substitution. Herein, we report the synthesis of propylthiol ring substituted PANI by reaction of 3-chloropropylthiol and ferric chloride with dimethylsulfoxide solution of emeraldine base form of PANI. The 3-chloropropylthiol ring substitution of PANI was also carried out on the surface glassy carbon electrode.

29.2 EXPERIMENTAL

29.2.1 MATERIAL AND REAGENTS

3-chloropropylthiol, aniline, ammonium persulfate, acetonitrile, dimethylsulfoide (DMSO), tetrahydrofurane (THF) and ferric chloride were purchased from Aldrich Chemical Inc. Hydrochloric acid was purchased from Fisher Scientific.

29.2.2 INSTRUMENTAL SETUP

The glassy carbon (GC) electrodes (3 mm diameter) were purchased from bioanalytical Science. Cyclic voltammetry was performed with a BAS (Bioanalytical System) potentiostat (Model 100). In the voltammetric experiments, a three-electrode configuration was used, including Ag/AgCl reference electrode and a platinum wire counter electrode. The reflectance FTIR spectra of polymers were obtained using Nicolet NEXUS 870 FTIR instrument. The uv-vis spectra using a Agilent 8453 spectrophotometer.

29.2.3 REACTION OF 3-CHLOROPROPYLTHIOL AND FERRIC CHLORIDE WITH PANI DEPOSITED ON THE GC ELECTRODE

The oxidative polymerization of aniline was performed to produce PANI, when 40 mM of aniline was dissolved in 25 ml 0.5 M aqueous hydrochloric acid solution. The potential of the GC electrode was scanned between -0.4 to 1.1 V vs. Ag/AgCl at a scan rate 100 mV/s, the polymerization was stopped at -0.4 V when the charge passed from the reduction of deposited polymer reached to 0.65. The PANI film had a green color. The film was washed with distilled water, then with tetrahydrofurane (THF). The GC electrode with film was immersed in a large glass tube with THF solution contained 0.01 g $FeCl_3$ and 0.5 ml of 3-chloropropylthiol the top of tube was closed to prevent evaporation of the solvent. The reaction was carried out at 60°C for 24 h in a temperature controlled mineral oil bath. The cyclic voltammogram of the film after reaction is similar to those 1, 2, 4 trisubstituted PANI reported by Freund.

29.2.4 REACTION OF 3-CHLOROPROPYLTHIOL AND FERRIC CHLORIDE WITH PANI

The emeraldine base form of PANI (compound 1) was prepared by chemical oxidation method described by MacDiarmid and co-workers [33]. The PANI was extracted with THF until the extract was colorless. The dried PANI (0.25 g) reacted with 3-chloropropylthiol (0.5 ml) and ferric Chloride (0.01 g) in anhydrous dimethylsulfoide at 81°C for 24 h. The black solution (propylthiol substituted PANI) was precipitated with 1 M HCl, a dark green to black precipitate was formed the filtrate was greenish blue; the precipitate was washed with 1 M HCl, then with acetonitrile. The precipitate (propylthiol substituted PANI) was dried under continuous vacuum for 48 h.

The HCl doped Propylthiolpolyaniline was sent for chemical analysis to Guelph Chemical Laboratories LTD (Guelph, Ontario, Canada). The result of Chemical analysis suggests that substitution occurred on all benzenoid rings in the polymer (compound 2).

29.2.5 CHARACTERIZATION

The propylthiol substituted PANI was studied by reflectance FTIR using Nicolet NEXUS 870. The polymer was treated with 1 M aqueous NaOH solution to remove the doped HCl from the polymer. The polymer was washed with water, and then, it was dried under continuous vacuum for 24 h. The FTIR reflectance spectrum of sodium polyanilinepropylthiolsulfide (NaPAPS) was obtained and compared with the FTIR reflectance spectrum of emeraldine base.

The UV-vis spectrum of dimethylsulfoxide solution of Propylthiolpolyaniline was studied using a Agilent 8453 spectrophotometer.

The cyclic voltammogram of Propylthiolpolyaniline was obtained in 0.5 M HCl solution using BAS (Bioanalytical system) Instrument Model

100, a thin film was smeared on ITO glass slide, the working electrode, a platinum wire counter electrode and a platinum wire counter electrode, and Ag/AgCl reference electrode.

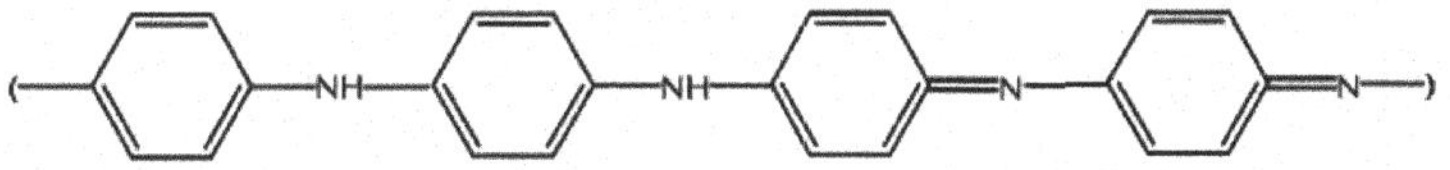

compound 1

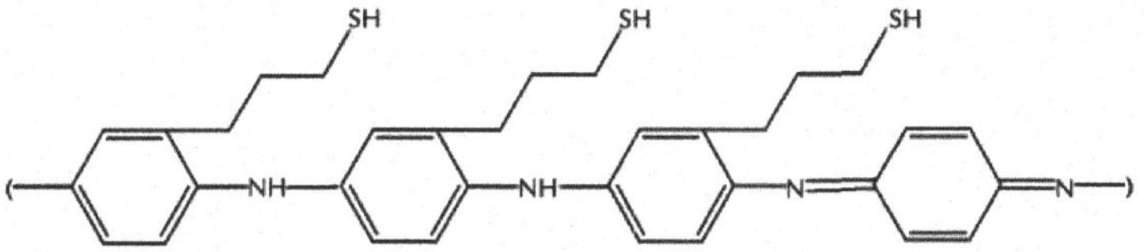

compound 2

29.3 RESULTS AND DISCUSSION

The effect of substituent on solubility, therefore processibility of PANI has been a subject of interest of scientists and engineers for many years. To develop a new synthetic approach to obtain substituted PANI, in present investigation the Friedel-Craft synthetic method was used for production of propylthiol substituted PANI.

PANI was prepared by both direct oxidation of aniline using ammonium persulfate and electrochemical oxidation of aniline on GC electrode.

The FTIR spectrum of sodium salt of propylthiol substituted polyaniline, Fig. 1 shows that the ratio of absorption intensity at 1594 cm^{-1} due to the quinoid ring to that at 1506 cm^{-1} due to that benzenoid ring is 1:3 [34, 35], as it presented in the structure 1. The absence of an absorption peak at 831 cm^{-1}, as it is in emeraldine base, Fig. 2, characteristic of C-H out of plain bending vibrations of 1, 4-disubstituted benzene rings [34] with the presence of splitting of this peak into two peaks at 823 cm^{-1} and 944 cm^{-1} indicative of 1,2,4-trisubstitution of benzene ring in the polyaniline. The C-S stretching band of propylthiol appeared at 775 cm^{-1} [36, 37]. The aliphatic C-H stretch at 2947 cm^{-1}, 2915 cm^{-1} and symmetrical stretching $v_{s.}$ CH$_2$ at 2847 cm^{-1} confirms the propylthiol substitution of polyaniline.

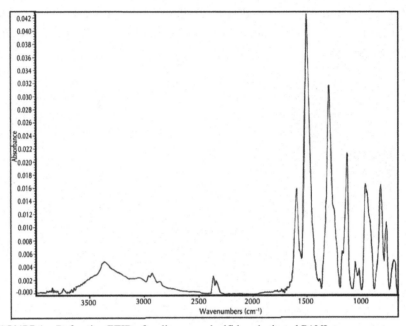

FIGURE 1   Refractive FTIR of sodiumpropylsulfide substituted PANI.

The cyclic voltammograms of electrochemically prepared PANI and after its reaction with 3-chloropropylthiol and ferric chloride are in Fig. 2.

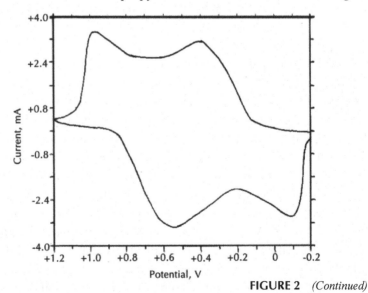

FIGURE 2   *(Continued)*

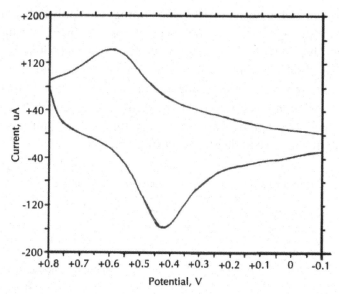

FIGURE 2 Cyclic voltammograms of electrochemically prepared PANI (a) and propylthiol substituted PANI (b).

The changes in voltammogram are a clear indication of 1, 2, 4-trisubstitution of benzene ring in the polyaniline [32]. The cyclic voltammogram of chemically prepared propylthiol substituted PANI is a typical voltammogram obtained with smeared solid polymer on ITO glass slide.

The uv-vis spectrum of dimethylsulfoxide solution of propylthiol substituted PANI demonstrates a partial oxidation as a broad band at 600 nm, Fig. 3.

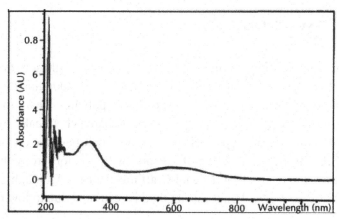

FIGURE 3 UV-VIS spectrum of propylthiol substituted PANI.

The chemical analysis of HCl doped propylthiolpolyaniline Table 1, perfectly suggests that substitution occurred on all benzenoid rings in the polymer. Three out of four rings in polyaniline were substituted, it makes the degree of substitution 75%, while the degree of substitution of electrochemically prepared ring substituted dialkylamine is 25% [34]. Electrophilic substitution of chloroalkylthiole in the presence of ferric chloride makes it possible to prepare any type of alkylthiole, alkyl amine, and other substitution with a good degree of substitution.

TABLE 1 Chemical analysis of o-propylthiol substituted polyaniline.

Sulfur	Nitrogen	Hydrogen	Carbon	
14.6	8.5	5.5	60	Theoretical values
15	8.3	5.4	59.8	Corrected chemical analysis values
13.43	7.38	4.82	54.50	Observed chemical analysis * values of substituted polymer

*The propylthiol substituted PANI was precipitated with 1 M solution of HCl then it was washed with HCl solution then with acetonitrile. The propylthiol substituted polyaniline contained two moles of HCl.

The result of chemical analysis of ring substituted propylthiolpolyaniline (Fig. 2) is presented below:

29.4 CONCLUSION

Production of substituted poly(aniline)s seem to be straight forward and it can be generated via oxidative polymerization of corresponding monomer. However, in many cases the desired substituted polymer is hard to obtain. In present investigation PANI(s) have been prepared chemically by direct oxidation of aniline and electrochemical oxidation of aniline on GC electrode. The substitution reaction of polymer with 3-chloropropylthiol and ferric chloride occurred via Friedel-Craft mechanism. Although for bulk chemically prepared polymer this method is restricted due to low to moderate solubility of polymer in the proper solvent, but this technique works perfectly for polymers prepared on the surface of electrode or prepared on the surface of conducting glass slide. We suggest Friedel-Craft method for synthesis is a good strategy to overcome the above-mentioned problems for preparation of substituted polymers.

29.4.1 SUPPLEMENTARY DATA

The refractive FTIR of sodiumpropylsulfide substituted PANI (the colored, Fig. 4). The FTIR spectrum of PANI (emeraldine, Fig. 5), cyclic voltammograms of chemically prepared PANI (Fig. 6) and propylthiol substituted PANI (Fig. 7) is presented as Appendix.

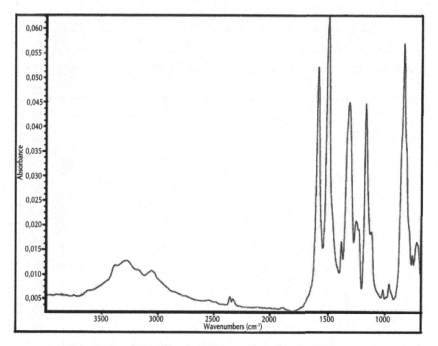

FIGURE 4   Refractive FTIR of sodiumpropylsulfide substituted PANI.

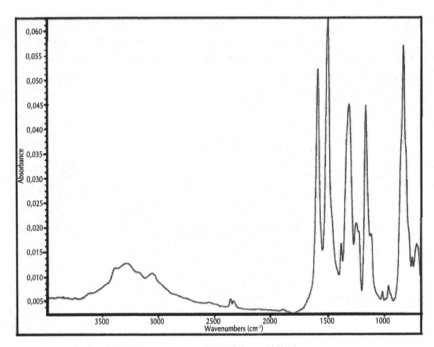

FIGURE 5   Refractive FTIR spectrum of PANI (emeraldine).

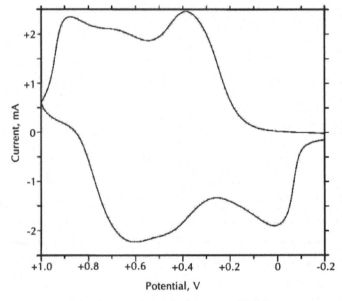

FIGURE 6   Cyclic voltammogram of chemically prepared PANI.

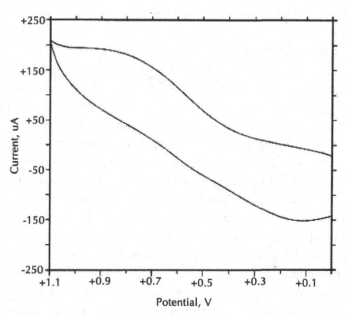

FIGURE 7   Cyclic voltammogram of chemically prepared propylthiol substituted PANI.

APPENDIX

An organic polymer that possesses the electrical, magnetic and optical properties by doping/de-doping process commonly associated with a conventional polymer is termed an Intrinsically Conducting Polymer (ICP) more commonly known as synthetic metals. In this appendix, the application of ICPs in different types of sensors, transistors, and display devices are discussed. Although, there are a lot of applications for ICPs, we are focusing mainly on these more interesting ones. The emphasis is on major development in above-mentioned field over the last 5–10 years. The polymer-based transistors would be expected to be superior to conventional inorganic transistors. Moreover, the difficult process of constructing the inorganic transistors can be replaced by a simple technology involving organic ICPs. ICPs are best suited for electrochromic display devices (ECDs). They show different colors depending on the oxidation-reduction state. The function of a sensor is to provide information on our physical, chemical and biological environment. ICPs can be used for constructing different type of sensors such as biosensors, gas sensors, humidity sensors and ion sensors.

Since their discovery in 1970s ICPs have impacted different fields of industry. One of these emerging fields is microelectronics, which is of great importance in manufacturing different kinds of electrical devices. The discovery of conducting polymers has been regarded as so important that it was recognized with the 2000 Nobel prize in chemistry.

The emphasis in electronic technology is to make better, faster and smaller electronic devices for application in modern life. Almost all electronic devices are fabricated from semiconductor silicon. Organic materials such as proteins, pigments and ICPs have been considered as alternatives for carrying out the same functions that are presently being performed by conventional semiconductors. Among these materials, ICPs have attracted most attention for possible applications in molecular electronics (ME) because of their unique properties and versatility. These ME materials differ from conventional polymers. The presence of an extended π -conjugation in polymer, confers the required mobility to changes that are created on the polymer backbone (by the process of doping) and make them electrically conducting. π electrons are delocalized across the molecule, leading to the semiconducting or conductive characteristics of a conjugated system.

The first conducting polymer was trans-polyacetylene which was doped with bromine and was produced at 1970s. Soon other conjugated polymers such as poly (p-phenylene), polypyrrole (PPy), polyethylene dioxythiophene (PEDOT) and polyaniline (PANi) and their derivatives, which are stable and processable, were synthesized. The molecular structures of a few ICPs are shown in Fig. A.1.

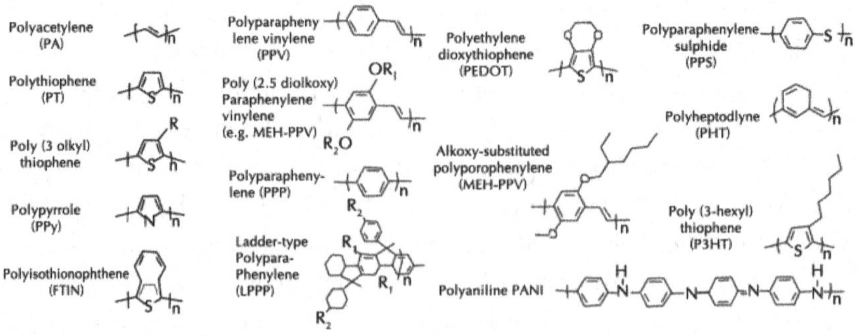

FIGURE A.1   Structures of a few ICPs.

The conductivity of these polymers can be tuned from insulating regime to superconducting regime, by chemical modification, by the degree and nature of doping. Besides these, polymers offer the advantages of lightweight, flexibility, corrosion-resistivity, high chemical inertness, electrical insulation, and the ease of processing. Another advantages lie in the fact that these materials possess specific advantages such as high packing density, possibility of controlling shape and electronic properties by chemical modification. However there are several problems associated with these materials, namely reproducibility and stability. Therefore ICPs also referred to as "synthetic metals" are promising materials for ME devices such as light-emitting diodes (LEDs), thin-film transistors (FETs), sensors and electrochromic display devices (ECDs).

TRANSISTORS

Organic thin film transistors (OTFTs) have a potential as the active matrix driving element for many electronic devices such as flexible paper-like displays, biological plastic chips or low-cost identification tags, whose driving elements may not be achieved from conventional inorganic materials and processing technologies. In general, OTFTs utilize a thin film of organic semiconducting material as the active layer of the transistor. Two electrodes ("source" and "drain") in contact with the organic semiconductor are used to apply a source–drain voltage and measure the source–drain current that flows through the organic semiconductor, while a third electrode ("gate") is used to modulate the magnitude of the source–drain current. The gate can be used to switch the transistor "on" (high source–drain current) and "off" (negligible source–drain current). Depending on the organic semiconducting material used as the active layer, the mobile charge carriers can either be electrons (n-type material) or holes (p-type material).

OTFTs can be roughly classified into two primary categories: organic field-effect transistors (OFETs) and organic electrochemical transistors (OECTs). Figure A.2 shows the schematic cross section of an OEFT and an OECT. In OFETs, the source–drain current is modulated by field-effect doping, where the charge carrier density in the organic semiconductor is controlled by the gate electrode via an electric field applied across an insulating layer (gate dielectric). In OECTs the source–drain current is modulated by electrochemical doping or de-doping, where the change in conductivity of the organic semiconductor is mediated by ions from an adjacent electrolyte. OECTs exhibit much lower operating voltages than OFETs, but due to the movement of ions involved in OECTs, their switching times are considerably slower (on the scale of seconds or longer) than those for OFETs (on the scale of milliseconds or shorter).

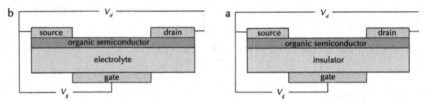

FIGURE A.2   (a) Schematic cross section of an organic field-effect transistor (OFET). (b) Schematic cross section of an organic electrochemical transistor (OECT). The applied source–drain voltage V_d and gate voltage V_g are also shown.

Classical (inorganic) field effect transistors (FETs) operate by the principle that the conductivity of an (inorganic) semiconductor, which is connected to an electrical circuit by a "source" terminal and a "drain" terminal attached to opposite ends of the active semiconductor, can be increased or decreased by the presence of an electric field which is supplied by an electrically conducting "gate" electrode. The gate electrode is not attached directly to the source or drain electrodes. The electric field associated with a given positive or negative potential applied to the gate electrode passes through a nonconducting material "dielectric" that separates the source/drain electrode material from the gate electrode. The change in conductivity of the active semiconducting source/drain material is, in effect, modified by the "through space" electric field effect.

In contrast to the small molecules, polymers can be processed easily from solution and therefore are useful for applications on flexible plastic substrates and in multilayer device configurations. However, solution-processed polymer films are inherently more disordered than in case of molecular FETs. It can be concluded from the above results that conjugated polymer based FETs are much cheaper than Si based device. However, the slow response and limited lifetime of the device restrict them to replace current Si technology. Nevertheless, the possibility of making flexible and flat panel with conjugated polymers open a new area of large-area low-cost plastic electronics.

One of the important parameters deciding the performance of FETs is the mobility. Charge carrier mobilities in ICP based FETs are found to be around 10^{-5} $Cm^2/V.s$ depending on the applied voltage and nature of the gate insulator. These values are significantly lower than that of inorganic semiconductor device in which mobilities are in the range of $0.1-1$ $Cm^2/V.s$. In an FET, the interface of the gate insulator and organic semiconductor layer plays an important role in charge transport. Electrical characteristics of the FETs are estimated by observing the change of the relationship between source-drain current (I_{SD}) and voltage (V_{SD}) upon sweeping gate voltage (V_G). From the $I_{SD}-V_{SD}$ and $I_{SD}-V_G$ relationships of FETs, the turn-off gate voltage, on/off ratio (defined as the current ratio of $I_{on,max}$ to $I_{off,min}$), the trans conductance (g_m, defined as the slope of the plot of I_{SD} versus V_G) and the response rate of the FETs to the electric field can be obtained.

The polymer-based transistor would be expected to be superior to conventional inorganic transistors. Furthermore, the sophisticated technology required for constructing the inorganic transistors can be replaced by a simpler technology involving ICPs. Researchers constructed a Polycarbazole-based electrochemical transistor. Since this polymer has a high redox potential compared to the other polymers, a transistor made of polycarbazole would be expected to have better transfer and saturation characteristics. The transistor was fabricated using Pt-coated glass electrodes, polycarbazole was deposited on both Pt electrodes. One half of the polycarbazole-deposited Pt plate was used as the source and the other half as the drain. The transistor designed for their study is shown in Fig. A.3. Remarkable features of the polycarbazole transistor include the low hysteresis of the device in scanning forward and reverse directions of the drain voltage and the maximum positive voltage needed for the saturation of the drain current.

During the last decade, various types of inorganic–organic hybrid FET devices have been reported. In the hybrid FETs, the electrodes and the dielectric layer were formed using inorganic materials, which are same as those used for conventional inorganic FETs, while the active channel was

formed using organic semiconducting materials such as pentacene, per-
ylene and various oligomers. Nevertheless, the hybrid OFETs are hardly
able to be employed in low-cost, large area and flexible display, because
they require complicated manufacturing processes and high-cost special
equipments to evaporate each component of the device at the elevated
temperatures. The processes for All-polymer FETs are expected to give
rise to many advantages in manufacturing the active matrix-driving element for large area display panel with mechanical flexibility and optical
transparency. Since the All-polymer FETs composed of only polymeric
materials, the devices could be fabricated by using a simple and low-cost
process at room temperature and possessed fairly high optical transmittance and mechanical flexibility. We, therefore, consider the All-polymer
FETs may be applicable to an active matrix-driving element for the flexible and transparent electronic systems.

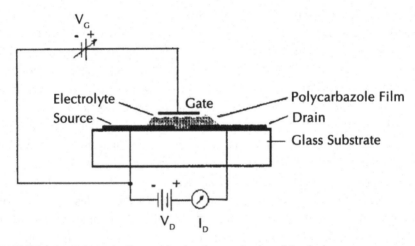

FIGURE A.3 Schematic diagram for the characterization of the device.

Scientists fabricated an all polymer FET type (metallic polymer)–insulator–(metallic polymer) (PIPFET) by line patterning technique which
involves no printing of a ICP and which uses a localized, as distinct from
a global, gate electrode. A hybrid FET device is also described in which a
commercial micrometer is used as the adjustable gate electrode to control-
lably vary the thickness of the dielectric material. Figure A.4 shows their

fabricated FET device. They reported a curious effect whereby an electric field apparently greatly affects the conductivity of an organic polymer, PEDOT, doped to the metallic conducting regime when it is used in an All-polymer field effect transistor configuration. It has been postulated that a doped ICP consists of metallic "islands" surrounded by lowly conducting "beaches" as shown in Fig. A.5. The field changes the conductivity of the semiconducting beaches but not that of the metallic islands; hence, the field changes the extent of electrical percolation between the metallic islands, and therefore changes the bulk conductivity of the material.

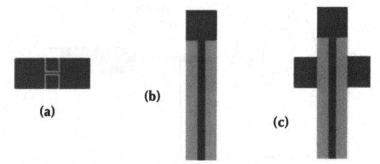

FIGURE A.4 (a) indicates the source-drain electrode (b) the gate electrode coated with PEDOT (c) the fabricated device made by superposition of (b) on (a).

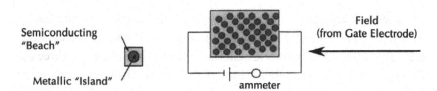

FIGURE A.5 Percolation field effect in doped ICPs. Metallic islands separated by beaches of non- or lowly conducting (semiconducting) polymer.

Scientists have proposed a technique called electrospinning to electrostatically lay down the active semiconducting polymer on substrates. Since electrospinning can be used in the controlled assembly of parallel, periodic fiber arrays, this technology is attractive for fabricating low-cost logic and switching circuits based on ICPs. Figure

A.6 shows the basic elements of the electrospinning apparatus. FET behavior in doped electrospun PANi/PEO nanofibers was observed. Saturation channel currebts were observed at surprisingly low source-drain voltages. Reducing or eliminating the PEO content in the fiber enhanced device parameters.

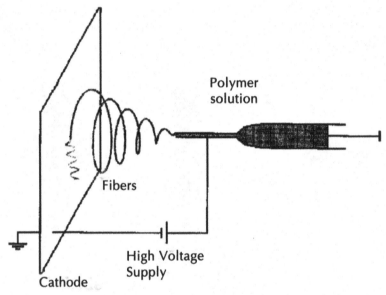

FIGURE A.6   Schematic showing the main components of an electrospinning apparatus.

Scientists reported a novel photolithographic method to fabricate the flexible All-polymer FET with optical transparency, where all components are formed from the polymeric materials. Active channel and all three electrodes were formed on a flexible polymer substrate using the simple photolithographic patterning technique of the electrically conducting PEDOT or PPy. Transparent photocrosslinkable polymers such as PVCN or Epoxy/MAA polymer were used as the dielectric layer. Figure A.7 shows the structure of the All-polymer FET. These FET devices exhibited relatively longer response times (order of minute) than those of the conventional inorganic FETs, The source-drain current of the FETs decreased with increase of the positive gate voltage, implying the p-type FETs worked in the depletion mode.

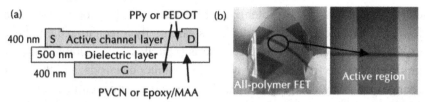

FIGURE A.7 Schematic structure of the flexible All-polymer FET with optical transparency.

Doped and de-doped nanotubes and nanowires of PPy, PANi and PEDOT were synthesized by the electrochemical polymerization method, using Al_2O_3 nanoporous templated. The electrical and optical properties of the nanotubes and nanowires were controlled through various synthetic conditions such as doping level, dopant and template dissolving solvent. It was observed that the nano-systems were transformed from a conducting state to a semiconducting or insulating state through the process of de-doping with the treatment of NaOH as a solvent. From the gate dependence of I-V characteristic curves for the systems, it was clear that charge carriers were p-type. Therefore, these nanotubes and nanowires could be used in the fabrication of polymer-based transistors.

DISPLAY DEVICES

One important application of ICPs is electrochromism. The absorption and emission spectra of certain dyes may be shifted by hundreds of angstroms upon application of a strong electric field. This effect is called "electrochromism." In other words, Electrochromism is the property of a material where its color is changed by an electrochemical redox reaction. An electrochromic material is the one that changes color in a persistent but reversible manner by an electrochemical reaction and the phenomenon is called electrochromism. Electrochromism is the reversible and visible change in transmittance and/or reflectance that is associated with an electrochemically induced oxidation–reduction reaction. The color change is commonly between a transparent ("bleached") state and a colored state, or between two colored states. These color changes are directly related to the conductivity changes of the polymer and so are affected by a small electric current at low dc potentials of the order of a fraction of volts to a few volts. An electrochromic device is essentially a rechargeable battery in which the electrochromic electrode is separated by a suitable solid or liquid electrolyte from a charge balancing counter electrode, and the color changes occur by charging and discharging the electrochemical cell with applied potential of a few volts. Figure A.8 illustrates an electrochromic device configuration. An electrochromic device is composed as follows: glass/transparent conductor (ITO)/electrochromic layer (WO_3)/ion conduction layer (PPy, electrolyte)/ion storage layer (V_2O_5)/transparent conductor (ITO)/glass.

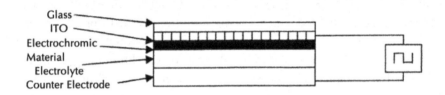

FIGURE A.8   Schematic diagram of the ECD.

In the recent years, ICPs have gained a lot of attention for ECDs. This is due to the fact that all electroactive and ICPs are potentially electrochromic materials, and are more processable than inorganic electrochromic materials and offer the advantage of a high degree of color tailorability. Furthermore, we can tune the optical properties of these materials by controlled doping and/or dedoping. ECD is comparable in properties with that of LCD with the added advantages that it can be made into different colors without the addition of external dye or window, does not depend upon the viewing angle and can be easily prepared in the form of large area windows. Widespread applications of ECDs depend on reducing costs, increasing device lifetime and overcoming the problem of ECD degradation. Table A.1 gives a comparison between inorganic and polymeric electrochromic materials.

TABLE A.1   Comparison between inorganic and polymeric electrochromic materials.

Property	Inorganic materials	Polymers
Method of preparation	Need sophisticated techniques such as vacuum evaporation, spray pyrolysis, sputtering, etc.	The material can be easily prepared by simple chemical, electrochemical polymerization and the films can be obtained by simple techniques such as dip-coating, spin coating, etc.
Processibility of the materials	The materials are poor in processibility	The materials can be processed very easily
Cost for making the final product (device)	High as compared to the polymer based devices	Low cost as compared to the inorganic materials
Colors obtainable	Limited number of colors are available from a given material	Colors depends on the doping percentage, choice of the monomer, operating potential, etc. Hence, large number of colors are available with the polymeric materials
Contrast	Contrast is moderate	Very high contrast can be optained
Switching time (ms)	10–750	10–120
Lifetime	103–105	104–106 cycles

PPy film is blue-violet in doped (oxidized) stet. Electrochemical reduction yields the yellow-green undoped form. The schematic of the doping/dedoping process can be given as

$$Ppy_{ox} + ne^- \Leftrightarrow Ppy_{red}$$

The electrochromism of PPy is unlikely to be exploited, mainly due to the degradation of the film on repetitive color switching.

The electrical and electrochromic properties of PANi depend not only on its oxidation state but also on its protonation state, and hence the pH value of the electrolyte used. As shown in Fig. A.1 PANi exhibits electrochromic behavior. Electrochromic behavior of PANi is shown in Fig. A.9. As demonstrated, small change in the pH of the solution and/or potential could create color and conductivity changes. PANi is green in the oxidized state and is transparent yellow in the reduced states. It has violet color in very acidic and yellow brownish in very basic media.

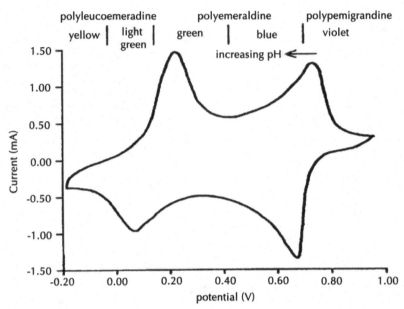

FIGURE A.9 Multicolor performance of conducting PANi at different oxidation/reduction states.

Scientists constructed a portable electrochromic display window using a PANi coated SnO_2 glass plate and $AlCl_3$ melt. The potential excursions from 0 to 1.0 V produced sharp color changes on the electrode from colorless to green, suggesting that surface activity is helpful for electrochromic display. It exhibited a fast response time (10ms) and a long cycle life (1000). A composite of two conducting polymers, PANi and PPy, could result in much improved physical properties with respect to their applications in electrochromic devices. Although PANi composite materials have unique photo/chromic properties they have their specific problems with respect to their reproducibility and reusability. Problems exist due to the dynamic nature of these polymers thereby preventing their successful application as novel display devices. Using the structure shown in Fig. A.10 a conducting PANi was polymerized on the top of a Pt electrode previously coated with PPy. By measuring the conductivity of the polymer we are able to predict the color of the polymer and voce versa.

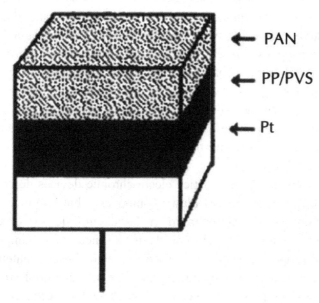

← PAN

← PP/PVS

← Pt

FIGURE A.10 A polymeric composite electrode. PAN (Polyaniline), PP (polypyrrole) and Pt (Platinum).

It was thought that by combining the PPy and PANi with Prussian blue (PB) not only the adhesion of PB could be improved but also this combination would exhibit a wide range of electrochromic colors. Scientists studied the electrochromic response of PPy/PB and PANi/PB composite films in different electrolytes by depositing the PB films on top of conducting PPy and PANi films, all prepared by electrochemical methods. They concluded that the use of PB on PPy and PANi not only yields high contrast but also extends the electrochromic response to a wider region of the visible spectrum, thus working as a sensitizer for improving the electrochromic response of the conducting PPy and PANi. It was noted that during electrochromic cycling, PPy changed its color from dark greenish blue (in oxidized state) to pale brown/yellow and PB from blue to bleached transparent white state (when reduced). On the other hand, PPy/PB film changed its color from bluish green (in oxidized state) to pale brown (in the reduced state).

Scientists described the assembly method of a solid-state electrochromic device using organic cinductive polymer as the active material and a transition metal oxides the electrochromic layer. The electrochromic films (WO_3, V_2O_5) prepared by sol-gel spin-coating and the vacuum evaporation process. The PPy films were coated by electropolymerization in a pyrrole solution. The polymer product obtained by electro-oxidizing the pyrrole monomer depended on the fabrication conditions used during the electropolymerization, for example, the potential employed for the deposition, the temperature, the solvent and the counter ion present in the pyrrole-containing solution.

Most of the so-called "all plastic" electrochromic devices described in the literature comprise not only plastic components, but also inorganic compounds as one of the electrochromic materials. In order to assemble an all-plastic and flexible electrochromic device it is necessary to improve the mechanical properties of the ICPs used as electrochromic materials. This problem can be overcome by mixing the ICPs with thermoplastics or elastomers to produce polymeric blends, which show the electrochromic properties of the ICPs associated with the mechanical properties of common polymers. Another attempt has been made to use an electrochromic device for time–temperature integration. Such a device could be used for "smart" labeling of frozen good. An electrochromic window (EW) is an ECD, which allows electrochemically driven modulations of light transmission and reflections. Such electrochromic optical switching device is usually called as "smart window". It can be used for a variety of applications where the optical modulation effect can be used significantly.

ICP thin films were used as driving electrodes for polymer-Dispersed Liquid-Crystals (PDLC) display devices. Liquid-crystalline–based display devices, which are commonly made of a liquid-crystal compound sandwiched between two substrates coated with a conducting layer of indium tin oxide (ITO), whose substitution with ICP electrodes could improve the optical and mechanical properties of the display devices. On the way to all-organic displays, PDLC sandwiched between two plastic substrates coated with ICP layers are promising devices for paper-like displays for electronic books, which require flexibility, lightness, and low-power consumption. The electro-optical characteristics (transmission properties, drive voltages and switching times) of the PDLC devices depend on the nature of the ICP substrate used.

Scientists fabricated All-polymer thin film transistors by inkjet printing technique. They used these transistors as active-matrix backplane for information displays. This field has been dominated by amorphous Si TFTs and large liquid crystal displays with an amorphous Si TFT active matrix backplane have been manufactured at a reasonable cost. An organic TFT is expected to reduce the cost even more, and to be applied to flexible displays based on a plastic substrate. The TFT characteristics required for active-matrix displays are (1) sufficient drain current, (2) low off current, (3) low gate leakage current through an insulator, (4) small gate overlap capacitance and (5) uniform characteristics.

BIOSENSORS

The estimation of metabolites such as glucose, urea, cholesterol and lactate in whole blood is of central importance in clinical diagnostics. A biosensor is a device having a biological sensing element either intimately connected to or integrated within a transducer. The aim is to produce a digital electronic signal, which is proportional to the concentration of a specific chemical or set of chemicals. A general schematic of a biosensor is shown in Fig. A.11.

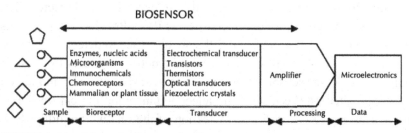

FIGURE A.11 Schematic diagram of a biosensor.

A transducer converts the biochemical signal to an electronic signal. The transducer of an electrical device responds in a way that a signal can be electronically amplified, stored and displayed. Biocomponents function as biochemical transducers. This biocomponent can be enzymes, tissues, bacteria, yeast, antiboies/antigens liposomes and arganelles. Most of the biological molecules such as enzymes, receptors, antibodies, cells etc. have to be fixed in a suitable matrix. The activity of immobilized molecules depends upon surface area, porosity, hydrophilic character of immobilizing matrix, reaction conditions and the methodology chosen for immobilization. ICPs have attracted much interest as a suitable matrix of enzymes, ICPs are used to enhance speed, sensitivity and versatility of biosensors in diagnostics to measure vital analytes.

Another advantage offered by ICPs is that the electrochemical synthesis allows the direct deposition of the polymer on the electrode surface, while simultaneously trapping the protein molecules. It is thus possible to control the spatial distribution of the immobilized enzymes, the film thickness and modulate the enzyme activity by changing the state of the polymer. These ICPs have been used in the fabrication of biosensors in various fields such as: Health care, immuno sensors, DNA sensors, environmental monitoring, and food analysis.

In the field of environmental works scientists. developed a sensor array based measurement system for continuous monitoring of water and waste water. They tested an on-line measurement system that incorporated an array of ICPs for continuous analysis both in laboratory, and in field. Other types of sensors: The emission of gaseous pollutants has become a serious environmental concern. Sensors are needed to detect and measure the concentration of such gaseous pollutants. ICPs showed promising applications for sensing gases having acid-base or oxidizing characteristics. ICP composites with other polymers such as PVC, PMMA, etc. polymers with active functional groups and SPEs are also used to detect such gases.

Scientists investigated on optical inteferometric structures that can be applied in toxic gas sensors. The sensor head consists of PANi and nafion layers deposited on the face of the telecommunication optical fiber. Humidity sensors are useful for the detection of the relative humidity (RH) in various environments. Polymer composites and modified polymers with hydrophilic properties have been used in humidity sensor devices. Researchers prepared nanocomposite pallets of iron oxide and PPy for humidity and gas sensing by a simultaneous gelation and polymerization process. This resulted in the formation of a mixed iron oxide phase for lower PPy concentration, stabilizing to a single cubic iron oxide phase at higher PPy concentrations, sensitivity to humidity increased with increasing PPy concentration.

Researchers constructed micro-humidity sensors from polyester-insulated platinum wire substrate to have a thickness of no more than 150 μm. These electrodes were dip-coated in ICP. Two processable PANi blends were developed with polyvinyl alcohol (PVA) and a butyl acrylate/vinyl acetate copolymer. The sensors showed high sensitivity, low resistance, and good reversibility without hystersis. Scientists have deposited SPANi using layer by layer and spin coating techniques for humidity sensing application. The layer-by-layer assembled sensors showed better sensing performance in terms of response time, sensitivity and repeatability as compared to the spin-coated sensors. Researchers used inkjet-printed films of the conductive polymer PPy for vapor sensing at room temperature.

Other important type of sensors are ion sensors generally, ion sensors have been developed taking the polymer as the conductive system/component, or as a matrix for the conducting system. When such systems come in contact with analysts to be sensed, some ionic exchange/interaction occurs, which in turn is transmitted as an electronic signal for display. Ion sensors find wide applications in medical, environmental and industrial analysis. They are also used in measuring the hardness of water. Ion-selective chemical transduction is based on ion selectivity conveyed by ionophore ion-exchange agents, charged carriers and neutral carriers doped in polymeric membranes.

A new Ca^{2+}-selective PANi based membrane has been developed for all-solid-state sensor applications. PANi is used as the membrane matrix, which transforms the ionic response to an electronic signal. Researchers used PPy as a component in all-solid-state ion sensors. PPy(DBSA) modified electrodes showed good reproducible cationic response to Ca^{2+} with the sensitivity of 27.2 ± 0.2 mV decade^{-1} which remained practically constant over 10 d. The standard potential, however, was found to drift 70 mV over the same time period.

Scientists prepared amperometric ammonium ion sensor based on PANi-poly(styrene sulfonate-co-maleic acid) composite conducting polymeric electrode. Two kinds of PANi- PSSMA composite films were prepared and their ammonium ion–sensing were studied. PANi- PSSMA(I) was prepared by incorporating PSSMA anions into a PANi/Au/Al$_2$O$_3$ plate was used to prepare another composite electrode of PANi-PSSMA(II)/Au/Al2o3.Compared with PANi-PSSMA(I), PANi-PSSMA(II) exhibited sensitivity to ammonium ion due to its fibrous morphology and high porosity.

In Summary, ICPs have attracted much attention in recent years because of the large number of possible applications in various electronic devices such as in sensors, light emitting diodes (LEDs), organic thin film transistors (OTFTs), electrochromic display devices (ECDs). It is even said that ICPs can be considered as alternatives for carrying out the same functions that are presently being performed by conventional polymers. Their conductivity, besides retaining the main properties of an organic polymer such as strength and low density and packed structure can make them a good choice for future works. OTFTs have a potential as the active matrix driving element for many electronic devices such as flexible paper-like displays, biological plastic chips or low-cost identification tags, whose driving elements may not be achieved from conventional inorganic materials and processing technologies. Conjugated polymer based FETs are much cheaper than Si based device. In the recent years, ICPs have gained a lot of attention for ECDs. This is due to the fact that all electroactive and ICPs are potentially electrochromic materials, and are more processable than inorganic electrochromic materials and offer the advantage of a high degree of color tailorability. Conductive polymers enhance speed, sensitivity and versatility of sensors and they have an increasing use in producing more reliable sensors.

KEYWORDS

- **Electrochromism**
- **Electronic nose**
- **Emeraldine base**
- **Friedel-Craft**
- **PANI**

REFERENCES

1. Schultze, J. W; Karabulut, H, Electrochim. Acta, **2005**, 50, 1739.
2. Skoheim, T. A; Elesenbaumer, R. L; Renolds, J. R, In *Hand Book of Conducting Polymers*, Second Ed.; Marcel Dekker: New York, 1998.
3. Gustafsson, G; Cao, Y; Treacy, G. M ; Klavetter, F; Colaneri, N ; Heeger, A. G. J. Nature **1992**, *357*, 477.
4. Jaeger, W. H; Inganas, O; Lundstrom, I, *Science* **2000**, *288*, 2335.
5. Wang, W; Sotzing, G. A.; Weiss, R. A., *Chem. Mater.*, **2003**, *15*, 375.
6. MacDiarmid, A. G.; Mu, S. L.; Somarisi, M. L. D.; Wu, W., *Mol. Crist. Liq. Crist.*, **1985**, *121*, 187.
7. Nakajima, T.; Kawagoe, K., *Synth. Met.*, **1989**, 28, C 629.
8. Shoji, E., *Chem. Sensor,* **2005**, *21(4)*, 120.
9. Zeng, K.; Tachikawa, H.; Zhu, Z., *Anal. Chem.* **2000**, *72*, 2211.
10. English, J. T.; Deore, B. A.; Freund, M. S.; *Sensor Actuat B: Chem.*, **2006**, *B 115(2)*, 666.
11. Michelotti , F. ; Morelli , M. ; Cataldo ,F. ; Petrocco, G. ; Bertolotti, M., *SPIE Proceedings Series,* **1994**, *2042*, 186.
12. Paul, E. W.; Ricco, A. J.; Wrighton, M. S., *J. Phys. Chem.*, **1985**, *89*, 1441.
13. Chao, S.; Wrighton, M. S., *J. Am. Chem. Soc.* **1987**; *109*, 6627.
14. Kobayashi, T; Yoneyama, H.; Tamaru, H., J. Electroanal. Chem., **1986**, *209*, 227.
15. Helay, F. M.; Darwich, W. M.; Elghaffar, M. A., Poly. Degrad. Stab, **1999**, *64*, 251.
16. Krinichnyi, V. I.; Yeremenko , O. N; Rukhman, G. G.; Letuchii , Y. A.; Geskin , V. M., *Polym. Sci. USSR*, **1989**, *31*, 1819.
17. Gospodinova , N.; Mokreva, P.; Tsanov, T.; Terlemezyan , L.; Polymer, **1997**, *38*, 743.
18. Banerjee, P., Eur. Polym. J., **1998**, *34*, 1557.
19. Sobczak, J. W.; Kosinski, A; Biliniski, A.; Pielaszek, J.; Palczewska, W., *Adv. Mater. Opt. Electron.*, **1998**, *8*, 213.
20. Athawale, A. A.; Kulkarni, M. V., *Sensor Actuator, B*, **2000**, *67*, 173.
21. Lin C. W.; Hwang, B. J.; Lee, C. R., *Mater. Chem. Phys.* **1998**, *55*, 139.
22. Wood, A. S., *Modern plastic Int.*, **1991**, Aug. 3.
23. Epstein, A. J.; Yue, J., US Patent No. 5137991, 1992.
24. Bissessur, R.; White, W., *Mat. Chem. Phys.*, **2006**, *99*, 214.
25. Kitani, A.; Satoguchi, K.; Tang, H-Q.; Ito, S.; Sasaki, K., *Synth. Met.*, **1995**, *69*, 131.
26. Lin, D. S..; Yang, S. M., *Synth. Met.*, **2001**, *119*, 14.
27. Cattarin, S.; Doubova, L.; Mengoli, G.; Zotti, G., *Electrochemica Acta,* **1988**, *33*, 1077.
28. Ranger, M.; Leclerc, M., *Synth. Met.*, **1997**, *84*, 85.
29. Sazou, D., *Synth. Met.,* **2001**, *118*, 133.
30. Planes, G. A.; Morales, G. M.; Miras, M. C.; Barbero, C., *Synth. Met.*, **1998**, *97*, 223.
31. Pringsheim, E.; Terpetsching, E.; Wolfbeis, O. S., *Anal. Chim. Acta* **1997**, *357*, 247.
32. Shoji, E.; Freund, M. S., *Langmuir*, **2001**, *17*, 7183.
33. Chiang, J-C.; MacDiarmid, A. G., *Synth. Met.*, **1986**, *13*, 193.
34. Wudl, F.; Angus, R. O., Jr.; Allemand, P. M.; Vachon, D. J.; Norwak, M.; Liu, Z. X.; Heeger, A. J., *J. Am. Chem. Soc.*, **1987**, *109*, 3677.

35. Tang, H. T.; Kitani, A.; Yamashita, T.; Ito, S. *Synth. Met.,* **1998**, *96*, 43.
36. Nakanishi, K.; Solomon, P. H. "Infrared Absorption Spectroscopy", 2nd ed.; Nankodo, Tokyo, 1977; p. 50.
37. Silverstein, R. M.; Bassler. G. C.; Morrill, T. C. In *Spectrometric Identification of Organic compounds*, 4th ed.; John Wiley & Sons, 1981; p. 131.

CHAPTER 30

EXPERIMENTAL AND THEORETICAL STUDY OF THE EFFECTIVENESS OF CENTRIFUGAL SEPARATOR

R. R. USMANOVA and G. E. ZAIKOV

CONTENTS

30.1 Introduction 504

30.2 Statement of the Research Problem 504

30.3 Analysis of Mathematical Models 505

30.4 Experimental Study of the Separator 509

30.5 Comparison of the Results of Theoretical Calculations and
     Experimental Results 511

30.6 Conclusion 512

Keywords 512

References 513

30.1 INTRODUCTION

One of the most troubling contemporary issues was local air pollution by industrial emissions. The share of industrial emissions in the overall air pollution is increasing, both in connection with the expansion of the chemical industry, and by the transfer of electricity or transport less harmful fuels. Most of the pollutants have on gaseous and dust that can travel long distances and accumulate. At high concentrations on the surface of the Earth, they can influence the condition of plant and animal life, both locally and globally. Therefore, the development of many industries in conjunction with the objectives of nature protection and rational use of its resources, are often closely related to the separation of dusty flows. Relevance increases with the spatial concentration of industrial enterprises and increasing their capacity. The problem of reducing dust emissions to ensure the allowable concentration of dust in the air of industrial zones and residential areas can be solved if each case reasonably choose an economical and effective enough dust separator.

30.2 STATEMENT OF THE RESEARCH PROBLEM

To calculate the efficiency of the separator used techniques based on theoretical and experimental studies. The most complete and accurate results provide experimental studies separators, which are held mostly in the physical models. These costly experiments can provide comprehensive information about the processes occurring in the cage, but they apply only to the specific study design of the separator, and other structures should be examined again in full.

Much more general findings and recommendations can be made using mathematical models of hydrodynamical processes in the separator. Creating a mathematical model of the motion of a particle of dust in the swirling flow will evaluate the impact of various factors on the collection efficiency of dust in the separators, as well as to create a methodology to assess the effectiveness of the dust collector.

To date, the theoretical calculation is carried out only total efficiency separator, without calculating the removal efficiency of each fraction as well as the calculation of the minimum diameter of particles deposited in the housing cage completely. Thus we have shown the scope of this work:

Creation and study of mathematical models of centrifugal motion of the particles.

Development of methodology for evaluating the effectiveness of the separator.

Verification of the effectiveness of this method of calculation.

30.3 ANALYSIS OF MATHEMATICAL MODELS

Creating a mathematical model of aerosol particles in the swirling flow will assess the effectiveness of dust control and identify the influencing factors (Fig. 1).

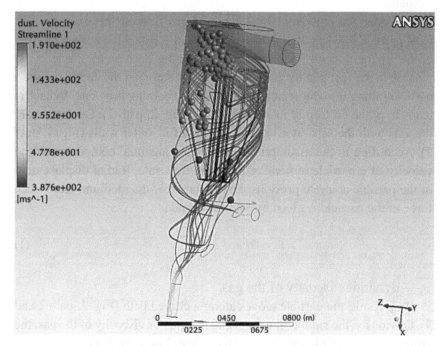

FIGURE 1 Mathematical model of particle motion in a swirling flow.

The use of computer technology and software to compute the hydrody-namic characteristics of eddy currents during the development and design of industrial devices. This avoids the need for costly field tests of gas pu-rification apparatus.

Software suite is a modern ANSYS-14/CFX-modeling tool, based on the numerical solution of the equations of hydrodynamics [1, 2]. Hydro-dynamic calculation makes it possible to determine the flow resistance of the device and to predict the efficiency of the separation process in the design stage.

Numerical analysis of the gas inside the dynamic scrubber reduces to solving the Navier-Stokes equations [3]. For the solution of equations of Navier-Stokes equations with a standard (k-ε)-turbulence model. To find the scalar parameters k and ε are two additional model equations containing empirical constants [4, 5]. The computational grid was built in the grid generator ANSYS ICEM CFD. The grid consists of 1247 542 elements.

Consider the conventional scheme of gas flow in the separator [6]. Dust particle flows at high velocity tangentially enters the cylindrical part of the separator, centrifugal separation effect has been on the circle.

Due to the fact that the particle density greater than the density of air, a particle falling into the separator with some incoming flow rate, by inertia tends to move uniformly in a straight line (Fig. 2, path 1). Deviates from the axis with the separator in the direction of its outer walls (Fig. 2, path 2). According to the assumptions made, the tangential component of the velocity of a particle moving at the same flow rate. Radial displacement of the particle at a rate prevents the resistance of the medium. Resistance force of the particle is given by Stokes formula.

$$F_{st} = 3\pi\mu_g d_p \left[\vec{v}_g - \vec{v}_p \right] \qquad (1)$$

$\vec{\mu_g}$ – dynamic viscosity of the gas.

As a result, the particle moves along a curved path (Fig. 2, path 2 and 3). Obviously, the smaller the mass and the greater viscosity of the gas, the closer it will be to the trajectory of the circle (Fig. 2, path 3).

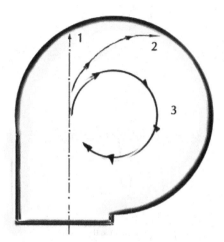

FIGURE 2 The trajectory of the particle.

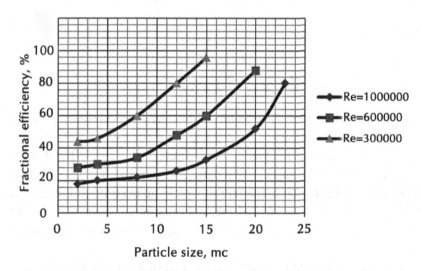

FIGURE 3 Fractional efficiency gas cleaning depending on the Reynolds number Re for gas flow.

Curved trajectory of a particle corresponds force centripetal acceleration

The method of calculation is similar to the method of calculating the cyclone [7], in which the total and fractional dust collection efficiency can be determined analytically:

$$\eta = 50 \cdot \left[1 + \Phi(x')\right], \tag{2}$$

where

$$x' = \cfrac{\lg\left[\cfrac{d'_{50}}{d_{50} \cdot k \cdot 10^3 \cdot \sqrt{D \cdot \cfrac{\mu_g}{\rho_p} \cdot \vartheta_g}}\right]}{\sqrt{\sigma^2 + \lg^2\left(\cfrac{d_{50}}{d_{16}}\right)}},$$

$$\eta_f = \frac{1}{2} \cdot \left[1 + \varphi(x)\right], \tag{3}$$

where

$$x = \cfrac{\lg\left[\cfrac{d_\delta}{d_\delta \cdot k \cdot 10^3 \cdot \sqrt{D \cdot \cfrac{\mu_g}{\rho_p} \cdot \vartheta_g}}\right]}{\sigma},$$

where – d'_{50} the median distribution of dust particles entering the device, m; d_{50} – diameter particles captured with an efficiency of 50%, m; ϑ_g – a common velocity of the gas in the apparatus, m / s; μ_g – the dynamic viscosity of gas, Pa ˙ m/s² ; ρ_p – particle density, kg/m³ ; d_{16} – particle diameter at the entrance to the unit in which the mass of all the particles having a size less than 16% of the total dust mass, m; σ – value that characterizes the dispersion of the particles; κ – coefficient for this unit received $\kappa = 34,76$.

Consideration of this formula allows identifying the factors that determine the effectiveness of catching dust separators. This formula allows us to determine the effectiveness of dust control on fractions. Figure 3 shows the fractional efficiency of the separator with a tangential inlet air when the gas flow rate at the inlet of the device, which affects the change in the value of the Reynolds number and increased turbulence. Increase the gas flow rate at the entrance to the separator improves dust collection, but to a certain velocity value for that species of dust. A further increase in speed is not only does not increase the collection efficiency, but significantly reduces it. The reason is that with the increasing speed of the separator increases turbulence. Turbulence separation prevents dust and even promotes the transition of dust in suspension and removal of her separator.

The increase in weight (density) of the dust particles facilitates their capture and, therefore, improves the efficiency of separation. Thus, the coagulation of dust in the highly desirable.

The viscosity of the gas increases with increasing temperature. This reduces the separation efficiency of dust.

30.4   EXPERIMENTAL STUDY OF THE SEPARATOR

For the validation of calculations using the developed methodology for assessing the effectiveness of the separators was used dynamic separator with tangential air supply. The tests were conducted on the separator installation consisting of the following equipment: 1 – starting motor; 2 – blow fan; 3 – impeller; 4 – guide vanes; 5 – dust collector; 6 – air duct; 7 – dynamic scrubber; 8 – cyclone; 9 – dust hopper; 10 – slime pond.

Stand scheme to test separator is shown in Fig. 4.

To determine the fractional separation efficiency based on experimental tests of dust samples were taken for laboratory analysis. Dustiness and gas mixture was determined by the direct method. On the dust collector before and after the sampling unit to the gas-air mixture. After the establishment of the appropriate mode of the apparatus the gas sample are selected by gas probe. Full capture dusts sampled gas mixture, produced by external filtering the mixture with a soakage dust-aspirator A-30 through specific analytical filters AFA-5, which is inserted into the filter cartridges. The selection process was recorded with a stopwatch, and speed – rotameter dust-aspirator A-30. Sampling before and after the device was carried out simultaneously. Before starting measurements analytical filters were placed in a desiccator and stand for 12 hours, then weighed on an analytical balance with an accuracy of up to 0.0001 g after sampling analytical filters are also kept in a desiccator 12 h and weighed. Then calculates the change in mass filter based on the weight change of control filters.

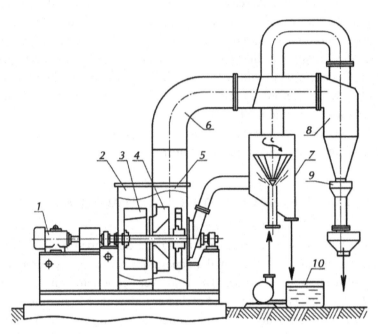

FIGURE 4   Installation for air purification from dust.

For each mode of the device was carried out eight measurements with different time of sampling.

Efficiency air cleaning of dust can be determined if the value of the concentration of dust in the air to clear z_1 and after cleaning z_2 (mg/m³). Efficiency is expressed as a percentage:

$$\eta \approx \frac{z_1 - z_2}{z_1} \cdot 100, \%$$

(4)

Processing of the results was carried out by a known method. For the study we used a dust white black.

Dispersity of dust were measured by sedimentation, which is based on the deposition of particles in a fluid.

The experiment revealed that the tested dust contains no particles greater than 70 microns. Results are presented in Fig. 5 and suggest that the tested dust contains 70% of particles with a diameter less than 25 microns, about half of them with a diameter up to 10 mm.

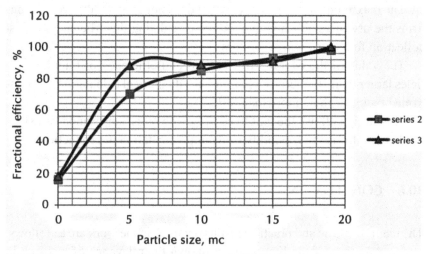

FIGURE 5 Comparison of dust separation efficiency values calculated theoretically-2 with the effective separation of dust, the experimental-3.

30.5 COMPARISON OF THE RESULTS OF THEORETICAL CALCULATIONS AND EXPERIMENTAL RESULTS

The result of this practical and theoretical work was to determine the fractional composition of the trapped particles. It is possible to determine the effectiveness of the fractional test separator, and to compare its capacity for separating particles of different diameters.

Efficiency, calculated from the experimental results was compared with the results obtained in the theoretical calculation. Efficacy parameters were the same as in the proposed method of calculation.

Thus: The overall efficiency is calculated by the Eq. (4) – 93.26%. Fractional efficiency, calculated by Eq. (3), for particle size:

<5 mc: 16.55%; 5–10 mc: 57.20%; 10–20 mc: 100%; 20–50 mc: 100%; 50–100 mc: 100%

We compare the results obtained from the theoretical calculation and experiment.

Value of the difference between the amount of dust collection efficiency calculated according to the Eq. (3) and obtained experimentally is shown in the graph (Fig. 5).

The maximum value of the deviation does not exceed 20%, which confirms the use of established methods of calculating the efficiency of dust collection for centrifugal separators.

The value of the overall effectiveness and efficiency of trapping particles larger than 10 microns, resulting in a theoretical calculation and confirmed by experiment more than 80%.

This design allows the unit to comply with emissions standards [8].

This confirms the feasibility of its implementation in production.

30.6   CONCLUSION

The main scientific and practical results obtained in the work are as follows:

1. A review of the literature has shown the relevance of research in the field of air purification, and the need to develop a methodology for calculating the efficiency of separating apparatus.

2. A mathematical model of the motion of a particle of dust in the swirling turbulent flow, making it possible to calculate the trajectory of a particle.

3. An estimation method for fractional separation efficiency for centrifugal machines.

4. An experiment to determine the effectiveness of dust control dynamic separator. The analysis shows a qualitative agreement of the theoretical and experimental results.

5. Showing the technical parameters of the dynamic separator for separating dust in different diameters. Technical solutions implemented in the production company "Soda."

KEYWORDS

- **Industrial emissions**
- **Navier-Stokes**
- **Soda**
- **Swirling flow**

REFERENCES

1. Kaplun, A. B.; Morozov, E.; Olfereva, M. A. *ANSYS in the Hands of the Engineer: A Practical Guide*. VNIIMP, Moscow, 2003; 272.
2. Basow K. A. *ANSYS and LMS Virtual Lab. Geometric Modeling*. DMK Press: Moscow, 2005, 640.
3. Patent 2339435 RF, B01 D47/06, Opubl.27.11.2008. Bull. 33.
4. Bulgakov, V. K.; Potapov, I. I. *Finite-Element Schemes of High Order for the Navier-Stokes Equations*. SUPG-modified method. Publishing House of St. Petersburg State, Tech. University, St. Petersburg, 2003; 129–132;
5. Goncharov, A. L.; Fryazinov, I. V. *Construction of Monotone Difference Schemes for the Navier-Stokes Equations on a Nine-Templates*. Institute of Applied Mathematics. Keldysh Institute, Moscow, **1986**, *93*, 14–16.
6. Pat. RF 2339435, 2008.
7. Uzhov, V. N.; Valdberg, A. *Chemistry*, **1975**, 216.
8. Barahtenko, G. M.; Idelchik, I. E. *Industr. Sanitation Gas*, **1974**.

INDEX

A

Ab initio calculations, 206
  caffeine and calcium ions, transport properties of, 209–210, 216–217
    in aqueous solutions, 216–217
  of energies for conversions of nitrogen dioxide dimmers, 59–60
Absolute temperature, 4
AC-electrospinning, 304–305
ACEPHATE, 123
Acetanilides, 135
Acid black ATT, 119
Acid blue ZK, 118
Acid chromium black, 119
Acid chromium green, 119
Acid chromium yelow, 120
Acidic phosphatase, 167
Acid leaching of chalcopyrite, 143
Acid phophatase, 165
Acid sulfonic blue, 119
Acireductone dioxygenases (ARDs) Ni(Fe) ARD, 186
Acireductone oxidations, 186
Acrylo urethane resins, 139
Actin fibers, 21
Active measuring system, 314–315
Acylalkylaminoxyl radicals, 43, 52
Adaptogenes, 452
Adhesion-producing materials, 150
Adiabatic systems, 5
Adriblastin, 161
Advanced Oxidation Process (AOP), 127
Air contaminants, 111
Albumin, 24
Aldehydes, 130
Aleksandrova-Gurevich's equation, 321
Algotoxine, 124
Alkylate DNA, 232

Alkyl macroradicals, 47
Allometric equations in electrospinning, 305–306
Allometric scaling law, 306
All-polymer FET, 490
All-Union Scientific Research Chemical-Pharmaceutical Institute, 234
Allyl C–N bonds, 49
Alternaria, 252
Alveolar macrophage (AM), 168
  anion peroxide radical, formation of, 169
  ozone exposure and, 168–169
  primary function of, 168
Amaryllidaceae, 440
Amberlyst-15, 132
Amberlyst A26
  (BER), 141
  (SPIR), 141
a-aminoaldehydes, 134
2-Amino-4-ethylamino-6-chloro-1,3,5-thriazine, 122
a-aminoketones, 134
Aminoxyl macroradicals, 51
Aminoxyl radical formation, 50
  in polyperfluoroalkanes, 45
Ammonium ion, 114
Ana-telophases, 233
Andreev, V. S., 233
Aniline hydroxylase, 163
Anti-gerontological indicator of foodstuff, 14
Antioxidants (AO), study of, 420, 452
  experiment setup
    content of investigated preparations, 423–424
    expenditure time, determination of, 422

results and discussion, 423–425
    samples, 421–422
ANZEMET anti-emetik, 132
Aromatic compounds, nitration of, 136
Aromatic polyamide (AP), 52
    scheme of Rim formation in, 55–57
Aromatic polyamidoimides, 52
Arrhenius equation, 320
Arsenic acid, 144
Ascorbic acid, physico-chemical properties
    of aqueous solutions of, 452–459
    cathetometric method of substances
        screening, 453
    meniscus height, 453–456
    SSD effect, 453
    values of absorption, 456–458
Astetionella algae, 125
ASULAM, 122
Atomic force microscopy (AFM), 286
Atomic oxygen. See Ozone
Autohemotherapy, 157–158, 161, 171
Auxinnaphthaleneacetic acid, 440
Avance-500 Bruker, 222

B

Barbators, 127
Benzopyrene hydroxylase, 163
Benzyladenine (BA), 440
Benzyl arginine b-naphthylamide amido-
    hydrolase, 167
3-benzyloxy-1-nonene, 134
2-benzyloxyoctanoate, 134
Benzyl penicillin, 394
Berkovich indentor, 79
Betaglucoronidase, 165
Bioactive substances, 452
Biocybernetical system (BCS), geometric
    parameters of, 462–469
    signal/noise function, 466–469
    sun emission reflection, 464–466
Bioinorganic interface, 20
BIOK-022, 28
Biological nanoparticles, 20
Biopolymers, thermodynamic compatibil-
    ity o, 354

Biosensors, 497–500
Blemaren, 420–421, 423
Blood cell vitality and ozone, 151
B3LYP restricted method, 58
Bovine serum albumin (BSA) system. See
    also Gelatin-BSA mixtures, study of
    thermoaggregation of, 354
    thermodynamic compatibility of, 355
    thermomodification of, 355
Buffered deoionized water, 122
Bu4NBr, 197
Bu4NI, 197
Butadiene-styrene rubber (SKS-30), 78
3-butylonitrile, 132

C

Caffeine (3,7-dihydro-1,3,7-trimethyl-1H-
    purine-2,6-dione), 206
    calcium ions and, study of
        ab initio studies, 209–210
        binary diffusion coefficients, 208
        cross diffusion coefficients, 208
        density measurements, 209
        diffusion measurements, 207–209
        efflux time, 209
        refractive-index profiles, 209
        ternary diffusion coefficients, 208
        viscosity measurements, 209
    experimental results and discussion
        ab initio calculations, 216–217
        density values, 213–215
        diffusion coefficients, 210–213
        viscosity values, 215–216
Carbonaceous material, 144
Carbonyl oxides, 133
Castor-oil ozonides, 171
Catalase, 165
Catalytic ozonation, 113
Cathepsins, 21
    cathepsin A and D, 167
Caveolae, 23
Caveolae-independent endocytosis
    (CvME), 22–24
    ligands internalized, 24
Caveolae-mediated endocytosis, 22

C2ClF3 – polymer film, 140
Cell drinking method, 22
Cellulose fibers, 137
Cellulosics, 20
Centrifugal separator, study of
  analysis of mathematical models, 505–509
  assessment of effectiveness, 509–510
  background, 504–505
  experimental results, 511–512
Cephalosporins, 394
Ceramic coatings on substrates, 149
Ceric ammonium nitrate (CAN), 47
  photolysis of, 47
  polyvinylpyrrolidone (PVP) with, 47
C0-ferrite stainless steel, 142
Chalcopyrite, acid leaching of, 143
Characteristic functions of the state of system, 3–4
Chemical bonds and substance stability principle, 13
Chemical mutagenesis, 232
  of asynchronous cell cultures, 233
  chain processes in, 233
  cytogenetic studies of, 233
C2H4-ethyl acrylate-maleic anhydride copolymer film, 140
Chloroacetanylide, 123
6-Chloro-2-hexanone, 132
Cholesterol, 24
Chromosomal rearrangements, 233, 239, 242
Chromosome damages, study of, 232
  experimental setup
    analysis of rearrangements, 235–236
    doubling of chromatid bridges and, 235
    environmental contamination and, 233
    from ionizing radiation damages, 232–233
  materials and methodology
    chromosomes in metaphase plates, 237

    mitotic activity of seedlings, 237–238
    seeds of Crepis capillaris, 237
  result and discusssion, 238–247
    chromatid type rearrangements, 242
    chromosome rearrangements, 239
    level of metaphases, 238
    level of mitotic activity-control, 238, 242–245
    phosphemid-induced, 234, 240–247
    thymine + ethylene imine, influence of, 246
    uracil + ethylene imine, influence of, 246
1,4-cis-polyisoprene (PI), 49
Classical (equilibrium) thermodynamics, 2–3, 6
  irreversibility, 6–7
  physical sense of, 5
Clathrin-mediated endocytosis (CME), 22–23
Cleavage O–H communication, 227
Climatic change reaction of plant organism, 262
Closed systems, 5
CO_2, 110
Colchicum laetum Steven, 440
Colchicum speciosum Steven, 440
Colchidaceae, 440
Colchis flora, 377
Colchis forest area, 378
C8–30 olefins, 135
Colibacterin probiotics preparations, 389
Collagen, 20
Colony forming units, 388
Comet, 253
Complement receptor (CR), 21
Composite materials, 66
Concanavalin A (Con A), 169
Condensed ozone, 145
Conducting polymers, 474
Contaminated smelling gases, 123–124
Contaminated soils, treatment of, 129
Contrinol, 420–421, 423
Conventional oxidizers, 111
CO oxidation by ozone, 112

Corrosion resistance and ozone, 143
Coulombic force, 285
Coulomb's law for electric field, 302
Creatine phosphokinase (CPK), 166
Crepis capillaris, 234–236
Criegee rearrangement, 189
Crop growing conditions, 252
Cu-containing Quercetin 2,3-Dioxygen-
    ases, 189
Cumene (isopropyl benzene), 422
Cu-30 Ni, 143
Cu-30 Ni – alloys, 143
Cyanoacetylaldehyde (3-oxopropyloni-
    trile), 132
Cycloalkenes, 133
(cyclohexa-1,4-dienyl)-L-alanine, 134
Cyston, 420–421, 423

D

Dark reactions, 11
DC-electrospinning, 304
Decolorization, 110
Decolorization of natural water, 124
Decubitus, 171
Dendritic cell, 21
Density values, in caffeine and calcium
    ions system, 209, 213–215
    in aqueous solutions, 213–215
Deodorization, 110
Deodorization method of ill-smelling air,
    123
Desinfection of drinking water, 127
Detoxicants, 452
D-glucose ozonolysis, 134
Dialkylaminoxyl radicals, 43
Dialkylaminoxyl radicals, 50
Diazinone, 123
DICAMBA, 123
Dicarboxylic acids, 136–137
Dietary nutrition, 14
Diets, thermodynamically focused, 14
Diffusion coefficients, in caffeine and
    calcium ions system, 207–213
    in aqueous solutions, 210–213
    binary diffusion coefficients, 208

cross diffusion coefficients, 208
    ternary diffusion coefficients, 208,
    211–212
b-diketonates complexes, 188
Dimer forms of NO2, 54–55
Dimethyl acetal (3,3-dimethoxypropyloni-
    trile), 132
Dinitro compounds, 49
1,2-dioxalanes, 133
Diphenylphosphine, 141
1,3-diphosphoglycerate, 160
2,3-diphosphoglycerate (2,3DPG), 160
Direct black Z, 120
Direct blue KM, 118
Direct Congo red, 118
Disinfection of swimming pool water, 128
Disorder (chaos) in system, 6
2,6-di-tert-butylphenol inhibitor, 153
DL-a-tocopherol content, 152–153
DNA macromolecules, functions of, 12
Dodecanedicarboxylic acid, 136
Dredging wastes, 128
Drinking water, 124–128
    coloring of, 124
    conditions for Fe and Mn removal, 128
    desinfection of, 127
    organic admixture, 124
    ozonator stations for treatment of,
    125–126
    ozone treatment of, 124
    ozonolysis, 126–127
    UV-radiation treatment of, 127
Drop dynamics, 300
Drosophila, 232, 260
Drug delivery methods, 18–19
Dry/wet solution spinning, 285
Dubinin, N. P., 233–235
Durymanova, S. E., 233
Dust, 110

E

Easy Scan DFM, 78
Ecology
    drinking water, 124–128
    soils, 128–130

waste gases, 110–113
waste water, 113–124
(E)-1,4-dicyano-2-butene, 132
Ediphenphos (EDDP), 123
E-infinite theory, 301
Elastomeric polymer materials, 77
Electrochromic device, 492
Electrochromic display devices (ECDs), 485, 492
Electrochromism, 491
Electrophilic ozone molecule, 159
Electrospinning, 283
    AC-electrospinning, 304–305
    allometric equations, 305–306
    applied voltage, 285
    beaded fibers, 287
    driving force, 284
    elongation of jet during, 284
    essential mechanism of, 285
    fiber diameter, determination of, 286
    industrial applications of, 288–289
    magnetic, 307
    modeling and simulation of, 289
        basics, 289–296
        capillary pressure, equation of, 292
        continuity, equation of, 292
        detection of X-ray generated by, 296
        dimensionless non-newtonian fluid mechanics, 295–296
        electric forces on surfaces of liquids, 295
        hydrodynamics, 292–293
        jet kinematic measurements, 293–294
        surface tension, equation of, 293
        time dependent flows, 292
        viscoelastic finite element computations, 292
        viscoelastic flow analysis, 291
        viscosity, equation of, 293
    multiple jet modeling, 306–307
    nanofibers, modeling of, 297–308
        assumptions in modeling jets and drops, 302
        branching, 297

droplet shape evolution, 300
        instabilities, 297
        lattice Boltzmann method (LBM), 304
        leaky dielectric model, 298–299
        linear momentum balance, 302
        mathematical model, 301
        nonlinear viscoelasticity models, 300–301
        slender-body model, 301–303
        for viscoelastic fluids, 304
        whipping model, 299–300
    nanoporous materials model, 307–308
    nozzle vs nozzle-less, 288
    parameters, 285–286
    of polymer solutions, 284
    porosity aspects, 287
    weak point of, 287
Electrospun nanofiber webs, 286
Electrospun nanoporous microspheres, 308
Electrospun PANi/PEO nanofibers, 490
Electrospun polymer nanofibers, 283
Embriogeniya, 446
Encapsulation, 354
Endocytosis, 22
Enthalpy, 4
Entropy, 4–5, 314
Enzyme immobilization, 354
Epoxy resins, 139
EPR spectrum of iminoxyl radicals, 45
EPR tomography, 51
Esterases, 21
3-ethoxycarbonylglutaric dialdehyde, 132
7-ethoxycumarine-o-diethylase, 163
Ethyl benzene oxidation, 186
Ethylene imine, 232
Ethylene-propylene copolymer rubber, 66
Ethylene-propylene copolymer rubber (EP), 330. See also PHB-EPC blends
Et4NBr, 197
Et3PhNCl, 197
Euler's equation, 292
European Pharmacopoeia (EP), 340
Evolution of biological systems, 12
External ozone gas therapy, 158

F

Fatty acids content, determination of, 152
Fc receptor (FcR), 21
Fe-containing Quercetin 2,3-Dioxygen-
ases, 189
FEC-56PM, 29
Fex(acac)yCTABm, 197
Fex(acac)yCTABm(CHCl3)p, 198
Fiber-reinforced plastics, 142
Field effect transistors (FETs) operate, 486
Fireproof coverings, 40
Fistulae, 171
Flora Orchids Sochi Black Sea region, 374
Flotation, 110
Fluoroalkyl polymers, 46
Folic acid, 24
Fractal network, 76–77
Fulerene (C-60) ozonolysis, 134
Fusarium, 253
   wheat samples resistant to, 253
Fusarium nivale Ces., 252, 255, 262
   development of fungus, 255
   fungal colonies, 255
   phytopathogenic fungi, 252
   snow mold conditions, 256

G

G. platiphyllus Traub et Moldenke, 440
G. risechense Stern, 440
Galanthus alpinus Sosn., 440
Galanthus woronowii Losinsk., 440
Galanthus woronowi Losinsk, 440
Gaschromatography, 152
Gaussian 98 program, 58
Gelatin-BSA mixtures, study of
   materials and methods, 356–359
      BSA Fraction V, 356
      CD measurements of BSA solution,
      357–358
      DISMIC-25cs (cellulose acetate)
      filters, 357
      fluorescence emission spectra, 358
      gelatin sample, 356
      intensity-size distribution functions,
      determination of, 357

Malvern Zetasizer Nano instrument,
357
   Rosenberg's method, 358–359
   ternary water-gelatin-BSA systems,
   357
   objectives, 355–356
   results and discussion
      addition of NaCl, effect of, 363–364
      destabilization of secondary struc-
      ture, 368
      electrostatic interactions, 363
      fluorescent emission, 368
      intensity-size distribution functions,
      359–360
      negative band, 366–367
      PEG on the solubility, effect of, 365
      structural changes, 365–366, 369
Generalova, M. V., 233
Gentamicin, 394
Gerontological healthiness, 14
Gibberelic acid, 446
Gibbs' specific (per unit of volume or
   mass) functions of formation, 7
   of chemical substances, 11
   environment conditions and, 13
   of food biomass, 13
   of foodstuff, 14
   life expectancy of biostructures, 11
   of supramolecular structures, 11
Gibbs' thermodynamic potential, 3–4
Glutathione redox system, 160
Graft polymerization
   of acrylic acid, 141
   of vinyl monomers, 140
Granular activated carbon (GAC) columns,
116
Grinih, L. I., 233

H

Halogenated organic compounds (HOC),
128
Harmful substances emissions, 110
Helmholtz' free energy, 3–4
Helminthosporium, 252–253

Hematological ozone therapy (HOT), 157–158
Hemoglobin antioxidative resistance, 151
Hepatic ascorbic acid (HAA), 166
Hexafluoropropylene, 46
Hexamethylbenzene, 62
High-Tc superconductors, 145
Histamine lactodehydrogenase, 166
Holl-Petsch formula, 77
Hormones, 452
H2SO4/H2O2/H2O cleaning, 148
Hyacinthaceae, 440
Hydrometallurgy, 40
b-hydroxycarbonyl species, 133
Hydroxylethyl starch (HES), 340
Hydroxyl groups, 141
HyperChem v7.5 software package, 210
Hypoxia, 152

I

ICP based FETs, 487
Iminoxyl radicals, 44, 52
    EPR spectrum of, 45
Immunoglobins (Ig) G, M, 21
Immunomodulators, 452
In-compounds, 144
Indium rich phosphate, 149
Indium tin oxide (ITO), 496
Industrial phosphonic acid, 145
Infinitesimal changes of functions of state, 3
Inorganic electrochromic materials, 493
Inorganic–organic hybrid FET devices, 487
Interbiopolymer complexes, 354
Intermacromolecular interactions, 354
Intrinsically Conducting Polymer (ICP), 483–484, 492
Ion–radical mechanism, 61
IR (FTIR) spectroscopy, 141
IR-spectroscopy of phosphorus-containing polypeptides, 35–36
    characteristic vibrations, 35
Isolated systems, 5
Isopropyl-14C-atrazine, 116

Isopropylthiolane, 122
Isotonic NaCl solution, 153

K

Kagramanyan, R. G., 233
Knauer, 341
Kobylyansky, V. D., 256

L

Laminates, 142
Lattice Boltzmann method (LBM), 304
Law of increasing entropy, 5
Law of temporary hierarchies, 7, 13
Layer-by-Layer (LbL) technique, 400
Lead zirconate, 149
Leaky dielectric theory, 298–299
Legionella pneumoniae, 172
Leica Confocal Laser Scanning Microscope TCS SP, 404
Leucojum aestivum L., 440
Life expectancy of biostructures, 11
Light-emitting diodes (LEDs), 485
Light scattering theory, 340
Liliaceae plants, 440
    micropropagation, study of
        material and method, 440–441
        results and discussion, 441–447
Lincomycin, 394
    antimicrobic activity of, 394
    inhibitory concentration of, 394
Lipid-based nanoparticles, 19
Liposomal benzoyl peroxide, 395
Liposomal benzyl penicillin, 394
Liposomal benzyl penicillin, effect on S. aureus
    background, 394–395
    materials and methods, 395–396
    results and discussion, 396
        minimum inhibitory concentrations (MIC), 396
Liquid-crystalline–based display devices, 496
Liquid phase deposition, 149
Live/DeadBaclight™, 388–389

Local thermodynamics of supramolecular processes, 11
Low temperature silicon surface cleaning, 147
Luschenko, G.V., 264
Lyophilized biological preparations, assessment of viability rates of
    materials and methods, 388–389
    results and discussion, 389–390
Lyophilized cells, 388
    E. coli M17 cells, 389
Lysosomal enzymes, 23
Lysozyme, structure and functions of
    hen egg white, 429
    5-methylresorcinol (MR), effect of, 428–429
        adsorption of methylresorcinol, 431–432
        sinusoidal oscillation for surface dilatational measurement, 430–431
        surface dilatational properties, 433–436
        surface rheological parameters, 430–431
        surface tension measurements, 429–430
Lysozymes, 165

M

M. dolichanthum Woronow et Tron, 440
M. neglectum Guss., 440
M. pallens M. Bieb., 440
Macrokinetics of live matter, 13–14
Macrophages, 21
Macropinocytosis, 22, 24–25
Macropinosomes, 25
Magnetic electrospinning process, 307
MALLS-RI-SEC approach, 340
Malvern Zetasizer Nano ZS, 354
Mass conservation, 302
MECOPROP, 122
Me4NBr, 197
Me3(n-C16H33)NBr (CTAB), 197
Mercury from waste gases, removal of, 113

Mercury(I)chloride, 144
Metal-based nanoparticles, 19
Metal hydrides, 130
Methicillin, 394
Methionine salvage pathway (MSP), 186
Methoxycarbonyl radicals, 43–44
1-methylcyclopentane, 132
1-methylcyclopentyl hypochloride, 133
4-methyl-2,6-di-tert.butylphenol, 222
Methyl ester 3-(3¢,5¢-di-tert.butyl-4¢-hydroxyphenyl)-propionic acid, 224
Methylmethacrylate polymers, 140
1-methyl-1-phenylhydrazine, 132
5-methylresorcinol (MR), effect on lysozyme. See Lysozyme, structure and functions of
1-metoxy-4,8-dimethylnone-1,7 diene, 133
Microdochium nivale (Fr.), 255, 262
Micronization of a material, 20
MII xL1 y(L1 ox)z(L2)n(H2O)m complexes, 187
Miniaturization of sensing elements, 400
Minor autohemotherapy, 157
Mir 11, 253–254
Mitotic cycle, 232, 235
Mn-containing compounds, 144
Mobile hydrogen atom abstraction, 44
Modern (non-equilibrium) thermodynamics, 6
Moisture-proof packaging material, 140
Molecular beam epitaxy (MBE), 145
Molybdenite flotation, 144
Monocytes, 21
MPP-sulfone, 123
MPP-sulfoxide, 123
MPP-sulfoxide-oxone, 123
Multi-angle laser light scattering detector (MALLS), 340
    albumin and immunoglobulin products, molecular weight distribution of, 343–350
    application, 340
    polysaccharides, molecular weight distribution of, 342–343
    signal, 340
    study of

experimental setup, 341–342
results and discussion, 341–350
Multicomponent medical preparations, 452
Multi-jet electrospinning systems,
306–307
Multi-step ozone flotation, 144
Muscari coeruleum Losinsk., 440

N

2-(N-acetylamid)-3-(3¢,5¢-di-tert.butyl-
4¢-hydroxyphenyl)-propionic acid
acetylamid group of, 222
antioxidant properties of, 222
discussion, 224–228
esterification reaction with iso-
propanol, 225
experimental setup, 222–224
inhibition of oxidation, 225
kinetics of oxygen uptake, 225
reactions peroxy radical, 226
value of homolytic cleavage, 227
ethyl ester, 223
methyl ester, 223
n-butyl ester, 224
n-nonyl ester, 224
propyl ester, 223
N-acetylglucosamine, 159
b-N-acetylglucose aminidase, 167
Nanocarriers in drug delivery, 20–22
non-phagocytic pathway, 22
phagocytic pathway, 21–22
Nanocomposites nanofiller aggregation
process, 76
Nanocrystallite, 19
Nano-DST, 78
Nanofibers, 283
Nanomedicine, 18
Nanoparticles
biological, 20
lipid-based, 19
metal-based, 19
polymer-based, 20
Nanotechnology, 18
Nanotherapeutic agents, 18
Nanotherapeutics

properties, 18
Navashin, Mikhail S., 235
Navier-Stokes equation, 305
Neumann, John, 314
Neupeptides, 452
Neutrophils, 21
Neutropic preparations, 452
Newtonian liquids hydrodynamics, 90
N.I. Vavilov Institute for Plant Growing,
262
NiIIARD reaction, 186
NiII(FeII)ARD dioxygenases, mechanism
of
background, 187–188
experiment setup, 188
AFM SOLVER P47/SMENA/ with
Silicon Cantilevers NSG11S, 188
silicone surface used, 188
taping mode, 188
results and discussion, 188–199
effect of FeII-acetylacetonate complexes, 190
electron-donating extra-ligand,
188–189
formation of a complex between
Fe(acac)3 and 18C6, 195
H-bonds, role of, 191–199
iron complex with CTAB:Fex(acac)
yCTABm(H2O)q, 199
OSCs (Outer Sphere Complexes),
196
protein structure, 194
transformation routes of a ligand,
189
Ni2(OAc)3(acac) MP·2H2O, 191, 193,
196
Nitrogen dioxide
dimmers, 58–60
reaction with organic compounds,
48–49
Nitrogen oxides, 42, 110
end peroxide radical and, 45
formation of radicals, 43–44
free radical and ion-radical reactions
mechanism of initiation, 52–58

obtaining spin-labelled rubbers, 49–51
interaction in photolysis and radiolysis of polymers, 43–46
nitrogen dioxide dimmers, 58–60
NN, 52
    radical cations, formation of, 60–62
NO3 conversion, 47
NO2 dimeric forms, 42
during photolysis of polymethylmethacrylate (PMMA), 43
to prepare spin-labelled macromolecules, 44–45
pyridine and, 60–61
radical reactions initiated by, 46–48
and reaction of nonactive arenes, 135
recombination with free radicals, 43
    conversion of PD into NN, 57–60
    EPR spectrum of radicals, 53–54
    oximes, formation of, 53
Nitrogen tetroxide, 52
Nitro nitrites, 49
Nitrosyl nitrate, 52
N.M. Emanuel Institute of Biochemical Physics, 261
N-methylpyrrolidone, 61
N-nitropyridinium nitrate, 60
Nondoped silicate glass film, 148
Nonelectrolytic nickel plating, 139
Non-equilibrium thermodynamics, 6
Nonhemolyzed plasma, preparation of, 150
Nonlinear non-equilibrium thermodynamics, 8
Non-phagocytic pathway, 22
Non-professional phagocytes, 21
Non-woven fiber fabric, 285
Nozzle-less (free liquid surface) technology, 288
N-phenylpyrazolyl, 134
Nton Paar DMA5000M densimeter, 209
Nylon 6, 140

O

Ocean power stations, 121
Olefins, 49

ozonolysis of, 130
Olive oil ozonides, 171
Oncological diseases, recommendations on dietary nutrition, 14
On-site recyclation of petroleum-contaminated soils, 129
Ontogenesis, 7
Open (biological) systems, 2
Open systems, 5
Ophris oestrifera, 374, 376
Opsonization of nanoparticlesis, 21
Optical information transmission, 462
Orchidaceae, family of, 374
Orchids
    as an object of ecoeducation, 379–380
    distribution, 374
    features, 377–379
        cycle of development, 377
        flowering, 377
        healing qualities of, 378
        medicinal properties of, 381–384
        green leaves of, 380
        wild, 379
Orchis mascula, 374–375
Orchis provincialis, 374, 376
Orchis purpurea, 374–375
Organic contaminants, removal of, 129
Organic electrochemical transistors (OECTs), 485
Organic field-effect transistors (OFETs), 485–486
Organic thin film transistors (OTFTs), 485
Oxadixyl, 122
Oxazole ring, 134
Oxidation catalysts, 112
Oxide optical crystals, 149
Oxides superconductors, preparation of, 146
Oximes, 132
Oxygen-ozone mixture, 149, 151
Ozonation
    of bromine water, 128
    cleaning of natural water, 113
    of concentrated aqueous solutions of phenol, 117
    of herbicides, 123

for mercury oxidation in chlorine producing, 118
molecular mass variations with, 141
of organophosphorous pesticides, 123
of PVC latex, 140
two-stage processes of, 127
of waste water from sulfate-cellulose production, 117
Ozonators, 113, 127–128, 172
portable, 172
stations for drinking water treatment, 125–126
Ozone, 110
for acid oxidizing leaching of chalcopyrite, 143
with activated carbon, effect of, 145
as an oxidizer, 143
applications, 110
blood cell vitality and, 151
cell membrane resistance and, 150
cells, effect on, 154
condensed, 145
CO oxidation by, 112
in correcting respiratory systems, 151
covering of semiconductor devices with silica films, 148
for decomposing petrochemical waste, 129
deodorization effect of, 124
destruction of double C=C bonds, 136
disinfectant and sterilizing properties, 169–170
electrophilic reaction of, 159
etching, effect of, 139
in fabrication of high-Tc superconductors, 145
and functional activity of neutrophilic cells, 153
in growth of superconductive oxides under vacuum conditions, 145
human organism, effect on, 150
for improvement Ag extraction, 144
in impurity removal on silicon surface, 148
induced oxidative conversion of methane, 135

induced reaction of polychloro benzenes, 135
industrial application
    inorganic compounds, manufacture of, 142–150
    organic compounds, manufacture of, 130–142
leukocytes antibacterial function and, 151
light-absorption characteristics of, 149
mediated reaction of aromatic acetals and acylal, 136
mediated reaction of nonactive arenes, 135
modification of surface properties, role of, 139
multi-step flotation with, 144
oxidative, disinfective, decolorizative and deodorizative properties of, 125
oxidative degradation of waste polyethylene/polypropylene (I), 142
oxidization pf toxic products, 151
passivating ability of, 149
in pharmaceuticals, 137
on POL of rabbit brain and liver, 152
preparation of nonhemolyzed plasma, 150
prior adsorption, effect of, 145
reaction with NO, 112
reaction with olefins, 130
for reducing the temperature and energy, 147
for removal of color and organic matter, 144
removal of organic pollutants, role in, 146
for saturation of blood with oxygen, 153
in selective decolorization of fabrics, 137
Si thermal oxidation with, 147
solutions, use of, 170–172
    in dermatology, 170
stainless steel parts, treatment of, 142
sterilization property, 172
therapeutic application, 154–165

activity of lactic dehydrogenase
(LDH), 162
antalgic effect, 157
arterial-metabolic circulatory disturbances, 156
bacterial effect of ozonized blood
serum on gram-negative bacteria,
163
bacteriological effect, 166
on central nervous system (CNS),
163
for changes in IgG levels, 161
concentration levels of prostoglandines PGFa and PGEa, 163
cytochrome P-450, 164
erythrocyte membranes stability,
162
on erythrocyte (RBC's) behavior,
161
erythrocyte shortness, 161
healing of skin lesions, 157
for hepatitis, 157
immune system, effect on, 165
immunologic defense systems, 160
intraarterial application of, 157
intravascular ozone/oxygen injection, 160
for local gas treatment, 157
on metabolite enzymes, 163
multiplex effect on endocrine system, 164
for myocardial damage, 164
prostoglandine synthesis, 158
pulmonary lysosomes and mitochondria, influence on, 167
subcutaneous injection of ozone
gas, 157
therapeutic doses, 155
thyroid hormone levels, 164
to treat gas gangrene, 158
for varicosis, 157
for venous circulatory disturbances,
157
virally produced diseases, 159
toxic action of, 165–170
cell destruction, 165

enzyme levels, changes in, 165
intoxication due to blocking of
alveolar capillary function, 167
on leucocyte cells, 169
polyunsaturated fatty acids (PUFA),
165
risk of asthmatic attack, 167
T-lymphocyte and B-lymphocyte
levels, changes, 163
tracheobronchial clearance, effect
of, 168
transitory effect of, 167
treatment of surfaces of propene polymer moldings, 140
use in an acid medium, 122
Ozone/air mixture, 140
Ozone enema, 157
Ozone-induced graft polymerization, 139
Ozone-oxygen mixture, 142, 146, 156, 158
Ozone peroxides, 159
Ozone therapy, 154–156
Ozonides, 171
Ozonized water components, 148
Ozonolysis
of alkenes, 131
of chemical and biochemical compounds, 110
of cis-endo-1,4-diones, 135
of CN-ions, 114
conversion of olefins into esters, 134
of copolymers containing 4-flurophenyl
butadiene units, 141
of cyclic alkenes in protic solvents, 132
cycloalkenes, 133
cyclohexene, 133
D-glucose, 134
and diffusion pattern of liquid monomers, 138
as ecological process, 134
of enol ethers, 133
3-ethoxycarbonylglutaric dialdehyde
by, 132
fulerene (C-60), 134
HIV inactivation, 151
of olefins, 130
of polyalkenes, 131

of poly(N,N-dimethylacrylamide-stat-2,3-dimethylbutadiene)s, 140
for preparing arsenic acid, 144
products, 133
of pyrols, oxazoles, imidazoles, and isooxazoles, 134
of regioregular methyl methacrylate-phenylacetylene copolymers, 136
of 1-substituted imidazoles, 135
with tetracyanoethylene, effect, 130
of vinyl ethers, 133
of vinyl halides, 134
of volatile organic compounds, 113
of waste water from coke processing, 116

P

Para-aminobenzoic acid (PABA), 260
in acetic acid, 262
advantages of dissolving, 268
in agriculture, 261
in ear treatment crops
concentrations, 264
discussion, 268–269
materials and methods, 263–264
results of experiments, 264–267
on inflorescences, 262, 268
with nutrients, 261
of pea and bunias seeds, 261
phenotypic activation of, 260
plants' adaptation mechanisms under, 262
salinity resistance, 261
on sowing qualities of seeds, 268
spring wheat plants and, 266–267
treatment of binary seed mixture, 261
treatment of vegetable seeds, 261
Paracetamol, physico-chemical properties of aqueous solutions of, 452–459
cathetometric method of substances screening, 453
meniscus height, 453–456
polymodal dependence effect, 454
SSD effect, 453
values of absorption, 456–458

Particulate-filled butadiene-styrene rubber, theoretical analysis of
experiment, 78–79
butadiene-styrene rubber (SKS-30), 78
nanofillers used, 78
scanning methodology, 78
tests on elastomeric nanocomposites, 79
results and discussion, 79–105
cluster (particles aggregate) mean density, 89
cross-sectional area of macromolecule, 80–81
deformation, 100
density fluctuation, 100–101
dependences of reinforcement degree En/Em on nanofiller particles size, 102–104
determination of interfacial layer thickness, 79–80
diffusion intensification, 90
dimensions of nanofiller particles (aggregates of particles), 93
elasticity modulus, 97
filler structure (distribution) determination, 90–91
fractal dimension of nanofiller surface, 81–82
fractal (Hausdorff) dimension, 91
growth in sites, 87
Grüneisen parameter, 98–99
intermolecular bonds anharmonicity level, 98–99
nanocomposites reinforcement degree, 102–104
nanofiller indentation in polymer matrix, 80
nanofiller particles cluster (aggregate) radius, 86–89
Poisson's ratio magnitude for non-filled rubber, 98–99
reinforcement degree of nanocomposite, 82–84
self-similarity iterations, 94–96

statistical walkers diffusion constant, 89
structure homogeneity condition, 100–102
three- and two-dimensional Euclidean spaces, 91–92
Particulate-filled elastomeric nanocomposites (rubbers) nanofiller particles, 76
Particulate-filled polymer composites, 76
Particulate-filled polymer nanocomposites, 76
Passive measuring system, 314
PEBBLE (Photonic Explorers for Bioanalyse with Biologically Localized Embedding) system, 400
Penicillin, 394
Peroxides, in cell processes, 159–160
Perrovski type oxides, 112
Phagocytic pathway, 21–22
Phagocytosis, 21
Phagolysosomes, 21
PHB-EPC blends, study of
    degree of crystallinity, 333
    DSC thermograms, 336–337
    experimental setup, 330–331
    IR spectra, 334
    repeated melting endotherms, 331–332
    thermal degradation, 335–337
    thermal stability, 337
    thermophysical characteristics, 331
    thermophysical properties of, 333
    value of W, 334–335
Phenols mixtures, 116
Phenomenological thermodynamics, 14
Phenthione (MPP), 123
Philogenesis, 7
Phosphemidum, 234
Phosphorus-containing polypeptides, synthesis of
    on base of peptides, 28–29
    at 21°C for 24 h, 32–33
    concentration of gelatin for hydrolysis, 31
    cross-linked phosphorus-containing polypeptides, formation of, 32
    Cu2+ cation of solution, 30

differential thermal analysis, 38–39
differential-thermal analysis (DTA) of product and initial peptides, 30–31
from fish collagen, 29, 34
full exchange capacity (FEC), 36–37
infrared spectroscopy of, 35–36
    characteristic vibrations, 35
intermediate production, 31–32
intial peptides, 31
investigation of, 31–35
methylol-derivations of peptide, 32
molar relation of peptide:formaldehyde:phosphinic acid, 33
pH changes, 38
reactionary centers, 38
sorption and ion-exchange activity of, 36–38
static ion-exchange capacity (SEC), 29–30
swelling of fish collagen during phosphonomethylation reaction, 34
synthesis conditions, 32–35
temperature influences, 33
template, 29, 37
water adsorption, 30
Photoconductive PbS films, 150
Physical theory of evolution of live beings, 9–10
Physicochemical system, 4
Pinocytosis, 22
Pitch-based fi bers, 138
Platanthera bifolia, 374
p-nitroanisole-N-dimethylase, 164
POL fluorescence products (FP), 152–153
Polyacrylamide, 20
Poly(acrylic acid), 20
Poly(acrylonitrile), 12
Polyacrylonitrile (PAN) nanofibers, 286
Polyaniline (PANI), 474
    3-chloropropylthiol ring substitution of, 474
    FTIR spectrum of, 481–483
    potential applications of, 474
    synthesis of propylthiol ring substitution of, study

characterization, 476–477
discussion, 477–480
instrument setup, 475
material and reagents, 475
reaction of 3-chloropropylthiol and
ferric chloride with, 475–476
Polyaniline (PANi), 484, 494–495
Poly (butyl) methacrylate, 141
Poly (butyl) methacrylate-copolymers, 141
Polycaproamide, 52
Polycaprolactone (PCL), 299
Polycarbonyl compounds, 133
Polycarboxylstyrene, 144
Polychloro benzenes, 135
Polyelectrolyte capsules, 400
Polyelectrolyte microsensors, study of
experimental setup
confocal images, 404
materials, 401–402
microcapsules, fabrication of,
402–403
SNARF-1 dextran and SNARF-1
dextran/urease solutions, 402
spectrofluorimetric studies, 404, 406
spectroscopic studies, 403
results and discussion
average intensity of a luminescence,
413–414
co-precipitating particles, concentration of, 405
dye and enzyme concentrations, 405
emission spectrum of SNARF-1,
405
fluorescence intensity and pH, ratio
between, 407–408
fluorescence spectra of capsules,
408–411
urea concentration of solution,
411–413
urea degradation, 404
Poly(ethylene-co-vinyl acetate), 20
Polyethylene dioxythiophene (PEDOT),
484
Polyethylene fi bers, 141
Polyethylene glycol monomethyl ether
methacrylate, 139

Polyethylene oxide (PEO), 299
Poly(3-hydroxyalkanoates), 66
Poly(3-hydroxybutyrate), 66
degradation of, 66
EP blend, structural features of
degree of crystallinity, 68–69
experiment, 66–67
first melting thermograms, 67
IR spectra, 69
methylene sequences, 69
phase inversion, effect of, 71
repeated melting endotherms, 68
temperature scale characteristic of
EPC, 72
thermophysical characteristics, 68
value of W plotted, 70
Poly(3-hydroxybu-tyrate), 330. See also
PHB-EPC blends
Poly(2-hydroxy ethyl methacrylate), 20
Polymer based FETs, 486
Polymer-based nanoparticles, 20
Polymer-Dispersed Liquid-Crystals
(PDLC) display devices, 496
Polymer FET type (metallic polymer)–insulator–(metallic polymer) (PIPFET),
488
Polymeric electrochromic materials, 493
Polymerization, 12
Polymer nanocomposites multicomponentness (multiphaseness), 76
Polymers
elastic and relaxation properties of,
315–316
relationship between stress and strain,
316
relaxation parameters for, 322
kinetic curves of variation, 322–326
rubbery (high elasticity) deformation
of, 316–317
activation energy of relaxation
process, 321
amplitude, 317
Boltzmann-Volterra model, 320
complex dynamic module of, 318
due to pressure and frequency of
loading, 318–319

elastic component of, 317
energy dissipation, 320
hysteresis loop and hysteresis loss
coefficient, 319
linear viscoelasticity, 319
relationship between stress and
strain, 316–317
relaxation time, 317, 320–321
Poly(methyl methacrylate), 20
Poly(N-vinyl pyrrolidone), 20
Polyperfluoroalkanes, 44
Poly (p-phenylene), 484
Polypropylene
fiber, 138
properties of, 139
Polypyrrole (PPy), 484
PPy film, 493–496
Polytetrafluoroethylene (PTFE), 44
Polyurethane (PU), 286
Poly(vinyl alcohol), 20
Polyvinylpyrrolidone (PVP), 47, 52
Potassium permanganate, 144
Potato tuber sprouts, 261
Pour point depressants, 142
Professional phagocyte, 21
Promena, 420–421, 423
Protopopova, E. M., 233
Psonins, 21
Pulmonary fibroses, 170
Pure ozone vapor, preparation of, 145
Purification. See also Waste water recy-
cling/purification
contaminated soils, treatment of, 129
decontamination of dredging wastes,
128–129
of drinking water with ionization radiation, 125–127
of exhaust gases from burners, 112
of gases containing condensable organic pollutants, 113
on-site recyclation of petroleum-contaminated soils, 129
organic contaminants, removal of, 129
removal of nitrogen oxides (NOx)
from, 112

removal of sulfur containing compounds, 112
synthesis of a material from zeolite,
113
UV-radiation method, 114–115
p-xylene, 62
Pyridine, 60

Q

Quasi-closed thermodynamic systems, 2,
13
(subsystems) in open biosystems, 7
Quasi-equilibrium, 7
Quinolinealdehyde, 132
Quinolones, 394

R

Radicals, formation from nitrogen trioxide,
46–48
Radioprotectors, 452
Rapoport, J. A., 232, 260
Rare earth barium copper oxide high-tem-
perature superconductor ceramics, 146
Reactive dyes, 139
Receptor-mediated endocytosis (RME), 23
Recycling of water. See Waste water recy-
cling/purification
Redox decomposition of peroxide, 138
Redox leaching of precious metals, 144
Reductive-oxidative decomposition, 138
Rehydrated cells, 389–390
Reinforcement effect, 77
Relativity, principle of, 5
Relaxation time, 317
Respiratory chain, 167
RHF algorithm, 209
Rotaviruses HRV, 154
Rubbers
butadiene-styrene rubber (SKS-30), 78
reinforcement of, 77
spin-labelled, 49–51
Russian Academy of Medical Science, 277

S

Salep, 374

Saratovskaya 57, 253–254
Scanning Electron Microscopy (SEM), 286
Scendezmus algae, 125
Scientifically-educational center (SEC), 274–275
    biodiversity preservation, 278
    limits of, 275–276
    participation of students, 277
    scientific researches, 277
    themes of degree works, 277
    themes of work, 276
Scilla bifolia L., 440
Second law of thermodynamics, 2, 5, 13
    applicability of, 7
    life in environment and, 8
Semiconductor materials, 19
Shevchenko, V. V., 233
Sidorov, B. N., 233, 242, 245
Silicon carbide ceramics, 144
Silicon surface, impurity removal on, 148
SiO2-films deposition, 149
Si TFTs, 497
Si thermal oxidation, 147
Sn-compounds, 144
SO2, 110
Sochi Black Sea Coast, 274–275, 277, 447
Sochi institute of RPFU, 274
Sodium hypophosphite, 28
Sofia Medical Academy, 154
Soils, 128–130
Sokolov, N. N., 233, 242, 245
Solar cell modules, 150
Solid lipid nanoparticles (SLN), 19
Sorbic acid ozonides, 171
Spasmex, 420–421, 423
SPECORD-M82, 29
Spin-labelled macromolecules, 42
    application of nitrogen oxide, 44
    of fluoroalkyl polymers, 46
Spin-labelled rubbers, 49–51
SPIP (Scanning Probe Image Processor, Denmark), 78
Spiranthes spiralis, 380–381
Spring crops, cultivation of, 262
Stabilization of nanoparticles, 20
Stable nitrogen-containing radicals, 43

Stable nitroxyl radicals, 42
Stable suprastructures, 11
Stainless steel parts, ozone treatment of, 142
    304L stainless steel, 143
Staphylococcus aureus (S. aureus), 394. See also Liposomal benzyl penicillin, effect on S. aureus
    planktonic cells of, 394
Static ion-exchange capacity (SEC) of polymer, 29–30
    experimental values, 37
Statistical walkers diffusion constant, 89
Sterilization, 110
Sterilization, using ozone, 172
Steroeselective synthesis
    of (2s, 3s) norstatine derivatives, 136
    of vinyl ethers, 136
Steveniella satyrioides, 378, 380
Stimulants, 452
S-triazoles, 123
Sub-micron colloidal carriers, 19
Substance stability principle, 13
Succinic acid preparation, 136
Sulfonamides, 394
Superconductive ceramic oxides, 146
Superconductive oxides, 145
Super-mutagens, 232
Superoxide dismutase, 165
Supersmall doses effect (SSD), 452
SV40 virus, 23
Swimming pool water, disinfection of, 128
Symazin, 122
Synchrotonic radiation optics, cleaning of, 147–148

T

Ta2O5 films, 147
Taylor cone, 285
Taylor dispersion technique, 207–208
TBA-reacting products (TBA-rp), 152
Telechelic oligomers, 141
Telehelic methylmethacrylate, 141
Tertiary amine aminooxides, 138
Tertiary nitroso compounds, 53

Tetracycline, 394
Tetraethylorthosilicate, 148
Tetrafluoroethylene, 45
Tetraquinone oxa-cages, 135
Thermal degradation
  of peroxides, 138–139
  of polymers, 140
Thermodynamically equilibrium process, 3
Thermodynamically focused diets, 14
Thermodynamically irreversible process, 3
Thermodynamically non-equilibrium
  process, 3
Thermodynamically reversible process, 3
Thermodynamic compatibility of biopoly-
  mers, 354
Thermodynamic conservatism of DNA, 12
Thermodynamics, 2–3
  of biological systems, 7–14
  classes of systems, 5
  empirical provisions of, 4
  of live matter, 13–14
  model of evolution of live beings, 11
  theory of biological evolution, 12
Thin-film transistors (FETs), 485
Thymidine diphospho-6-deoxy-a-D-ribo-
  3-hexulose, 134
Thyrotropin, 164
Tilletia, 252
Titanate thin films, 149
Tobolsk complex scientific station, 264
Tocopherol, 153
Tonurol, 420–421, 423
Total room temperature wet cleaning
  method, 147
Transition metal complexes, catalytic
  activity of, 186
Trialkylphosphate, 123
Tricarboxylic acid cycle, 167
Trichothecium, 252
1,2,4-trioxalanes, 130
Triticale, 261
Triticum aestivum (L.), 263
Triticum aestivum L., 267
Trofimova, U.B., 255
Trypsin, 31
Tsomartova, F.T., 264

Tyumen region, 252, 261, 264
Tyumenskaya 80, 253–254

U

UL-14C – atrazine, 116
Ulcer cruries, 171
Ultrapure water, 148
Upper-Convected Maxwell (UCM) model,
  307
Urease based enzymatic methods, 401
Urethra, 420
Urinary bladder, 420
Urinary system, 420
Urolyzin, 420–421, 423
Urotractin, 420, 423
UV/O3 cleaning method, 147
UV/ozone exposure, effects of, 141
UV/ozone modified wool fiber surfaces,
  137
UV/ozone treatment, 146
UV-radiation method of purification,
  114–115
    of water containing Cr(III), 114
    of water containing propyleneglycol
    nitrate or nitrotoluenes, 114–115

V

Vacuolar membrane, 21
Vacuolar proton pump ATPase, 21
Very large scale integration (VLSI), pro-
  cess of, 147
Vinyl ethers, stereoselective synthesis of,
  136
Viscosity values, in caffeine and calcium
  ions system, 209, 215–216
  in aqueous solutions, 215–216
Vitamin A, 163
Vitamin C, 163
Vitamin E, 163, 165, 172

W

Waste gases, 110–113
  air contaminants, 111
  CO2, 110
  dust, 110

removal of nitrogen oxides (NOx)
from, 112
SO2, 110
Waste water, 113–124
from dyes manufacturing, 118
from paper pulp production, 115
ozone decomposition, 115–116
from petrochemical industry, 116
level of petrochemicals, 117
Waste water recycling/purification,
113–114. See also Purification
of concentrated aqueous solutions of
phenol, 117
containing ClCH2COOH and PhOH,
114
containing propyleneglycol nitrate or
nitrotoluenes, 114–115
Cr(III), 114
of cyanide-based waste water, 113–114
decolourization and destruction of
waste water, 115
deodorization of petrochemicals contaminated water, 117
from the electrostatic and galvanic
coatings, 114
means of liquid chromatography, 120
from paper manufacture, 115
removal of benzofurenes, 122

in a soil decontamination plant, 122
technology for, 120–122
using polyurethane foams, 115
Water-soluble agrochemicals, 122
Weber—Fechner's law, 9
Wet-scrubber method, 113, 130
Wheat-germ oil, 172
Wheat samples
resistant to Fusarium, 253
traits indicators of, 254
Winter rye, resistant varieties of, 256

X

Xoloxan, 161
X-ray photoelectron (ESCA), 141

Y

YBa2Cu3O7-delta fi lms, 145
Yttrium barium copper oxide (Y Ba2C-
uO7) thin films, 146

Z

Zazhoginsk's deposit, 78
Zinc oxide light-shielding film, 149
Zn/AcOH, 132
Zn cyanides, 117

Printed in the United States
by Baker & Taylor Publisher Services